Conference on Drug Design and Discovery Technologies

Conference on Drug Design and Discovery Technologies

Edited by

Manikanta Murahari
MS Ramaiah University of Applied Sciences, India
Email: manikanta.murahari@gmail.com

Lakshmi Sundar
MS Ramaiah University of Applied Sciences, India
Email: lakshmi.py.ph@msruas.ac.in

Soma Chaki
MS Ramaiah University of Applied Sciences, India
Email: soma.bt.ls@msruas.ac.in

Vasanthanathan Poongavanam
Uppsala University, Sweden
Email: vasanthanathan.poongavanam@kemi.uu.se

Pritesh Bhat
Schrodinger, India
Email: pritesh.bhat@schrodinger.com

and

Usha Y. Nayak
Manipal Academy of Higher Education, India
Email: usha.nayak@manipal.edu

ROYAL SOCIETY
OF **CHEMISTRY**

This publication is based on peer-reviewed manuscripts submitted for the Conference on Drug Design and Discovery Technologies (CDDT) held at Ramaiah University of Applied Sciences, India, 21–22 November 2019.

Special Publication No. 355

Print ISBN: 978-1-78801-862-3
PDF ISBN: 978-1-83916-078-3

A catalogue record for this book is available from the British Library

The Royal Society of Chemistry is a charity, registered in England and Wales, Number 207890, and a company incorporated in England by Royal Charter (Registered No. RC000524), registered office: Burlington House, Piccadilly, London W1J 0BA, UK, Telephone: +44 (0) 20 7437 8656.

Visit our website at www.rsc.org/books

Contents

Part 1: Structure Based Drug Design Approach

Part 2: Structure Based Drug Design and Biological Evaluation

Part 7: Updates in Analytical Techniques

Part 8: Full-length Articles

Part 1: Structure Based Drug Design Approach

MOLECULAR DYNAMICS SIMULATIONS OF *BACILLUS SUBTILIS* EngA – FOR EXPLORING NUCLEOTIDE DEPENDENT CONFORMATIONS

N. Upendra and S. Krishnaveni*

Department of Studies in Physics, Manasagangothri, University of Mysore, Mysuru-570006, India

*Corresponding Author: Email: sk@physics.uni-mysore.ac.in

1 INTRODUCTION

Essential Neisserial GTPase A (EngA) is a member of the TRAFAC class of GTPase superfamily. GTPases are molecular switches that switch between the GTP bound (On) state and the GDP bound (Off) state. This switching regulates many biological functions such as protein synthesis, ribosome biogenesis, cell multiplication etc. GTPases possess four highly conserved G-motifs (G1((GxxxxGK[ST])), G2(T), G3(DxxG), G4([NT][KQ]xD)) and one weakly conserved G5 (SAK) motif, which provides the facility to interact with guanine nucleotides. The P-loop of GTPases with a G1-motif interacts with the beta and gamma phosphates of GTP/GDP, the G2 and G3 motifs that interacts with gamma phosphate undergo large conformational changes termed Switch-I (Sw-I) and Switch-II (Sw-II). The G4 and G5 motifs stabilize the interaction with guanine bases.

EngA consists of two contiguous GTP binding domains / G-domains (GD1 and GD2) at the N-terminus. The RNA binding domain is at the C-terminus (CD). EngA is involved in the late steps of ribosome maturation. Therefore, mutation or depletion of EngA results in incomplete ribosome biogenesis. Ribosome is a complex macromolecule which is the site for protein synthesis; it consists of two ribosomal subunits – 30S small subunit and 50S large subunit for prokaryotic 70S ribosome. These units are made up of ribosomal RNA's (rRNA's) and ribosomal proteins (rproteins), assembling these components (rRNA's and rproteins) into a functional ribosome is refereed as ribosome maturation or ribosome assembly or ribosome biogenesis.[1]

The crystal structures of EngA from 1) *Bacillus subtilis* (also known as YphC) bound to GDP at both GD1 and GD2 [GDP:GDP], 2) *Thermotoga maritima* (*Tm*Der) bound to [GTP:GDP][2,3] and the cryo-electron microscopy structure of EngA from *Escherichia coli (Ec)* bound to ribosome 50S subunit are available.[4] All these structures exhibit different conformations.[2-4] Some studies have shown that nucleotide bound combinations dictate the binding of EngA to the ribosomal units. The [GTP:GTP] nucleotide-bound state binds with 50S subunit, where as [GDP:GTP] nucleotide-bound state binds with both 50S and

30S to form the 70S ribosome. Finally the [GDP:GDP] nucleotide-bound EngA, detaches from the ribosome.[5,6]

Despite the available structural and biochemical information on EngA-ribosome association, the role of EngA and its implications in ribosome assembly at the molecular level is still elusive. It is suggested that the EngA-ribosome association is governed by the different conformational states of EngA, and these conformational states and function are dictated by the combination of the bound guanine nucleotides. Understanding the role of EngA will help to design inhibitors to arrest the ribosome maturation and thereby control the multiplication of harmful bacteria. Therefore, to explore the conformations of EngA in different nucleotide bound combinations, we have carried out molecular dynamics (MD) simulations of EngA in three different nucleotide ([GDP:GDP], [GTP:GDP] and [GDP:GTP] bound states. The simulations show that the guanine nucleotide dictates local fluctuations, which may in long term, disrupt the interactions between the domain-domain interface and trigger large globular domain motions.

2 MATERIALS AND METHODS

MD simulations for YphC in different nucleotide bound combinations – [GDP:GDP], [GTP:GDP] and [GDP:GTP] was carried using GROMACS-2016.5 package.[7] The crystal structure of YphC (PDB-ID 4DCU) was retrieved from RCSB-PDB. The missing atoms and amino acid residues were modeled using the Modeller-9.0 program.[8] EngA-nucleotide complexes were built using UCSF-Chimera software.[9] Topologies for proteins and guanine nucleotides were generated using Charmm36 force-field. EngA-nucleotide complex was placed in the Triclinic box, using periodic boundary conditions with a distance of 1.2 nm from the box edges, then solvent filled using the tip3p water model. To make the simulating system neutral, Na^+ and Cl^- ions were replaced with some water molecules. Next to remove steric clashes, energy minimization for 50000 steps was carried out using the 'steepest descent algorithm' with an energy tolerance of 1000 kJ/mol. One hundred ps NVT and 100 ps NPT equilibration simulations were performed using a V-rescale thermostat and Parrinello-Rahman barostat to bring the system temperature to 300 K and pressure of 1 bar. Finally, 10 ns production runs with an NPT ensemble were carried out with 2 fs as each time step. The coordinates were saved at every 10 ps. Long range electrostatic interactions were handled using a Particle Mesh Ewald scheme with a cutoff of 1.2 nm. All the analyses were performed using the in-built Gromacs tools. Visualizations and image rendering were processed with VMD[10] and CHIMERA software. Graphs were plotted from XMgrace.

3 RESULTS AND DISCUSSION

3.1 Structural aspects

YphC with 436 amino acid residues consists of two N-terminal GTP binding domains, GD1 (residues: 1-166), and GD2 (residues: 175-352) connected by a linker (residues: 167-174) and a C-terminal KH domain (residues: 353-436) (Figure 1a). Comparison of the structures of YphC, *Tm*Der and *Ec*EngA reveal different conformations (Figure 1b). Further YphC is in the [GDP:GDP] bound combination and *Tm*Der in [GTP:GDP] bound combination. This suggests that the guanine nucleotides play a role in the conformational states of EngA.

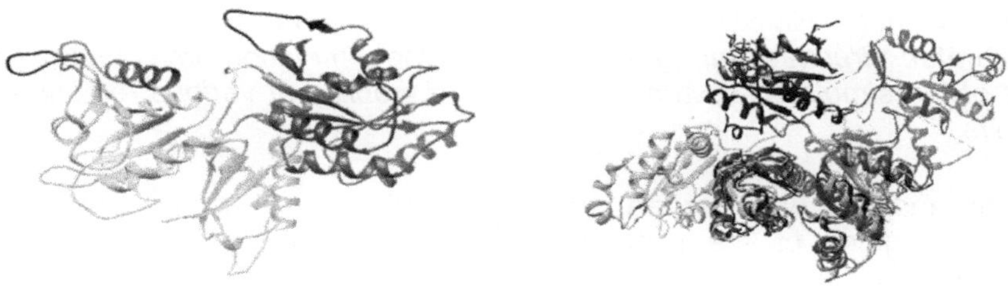

Figure *1* *(left) Crystal structure of YphC. (right) Structural comparisons of EngA from Bs (PDB:4DCU), Tm (PDB:1MKY) and Ec (PDB:3J8G) EngA.*

3.2 Simulation results

The MD simulations of YphC in [GDP:GDP], [GTP:GDP] and [GDP:GTP] nucleotide bound combinations were explored to understand the nucleotide dependent conformations of EngA. Initially to validate our simulations, RMSD of c-alpha atoms was calculated for the simulated systems, with respect to the equilibrated-structure during the course of the simulation. RMSD plots show that all the simulated systems converge towards equilibration (Figure 2a). RMSF of c-alpha atoms was plotted to check the mobility or fluctuations at the internal regions of the protein, with respect to the average position of the coordinates (Figure 2b). RMSF plots reveal large fluctuations at the switch regions for all the simulated systems, which are basic features exhibited by GTPases. Further, it is observed that in the [GTP:GDP] bound system larger fluctuations occur around residue number 125 compare to [GDP:GDP] and [GDP:GTP]. Noticeably, the region around residue 125 is at the CD interface. This large mobility observed at the CD interface suggests that larger fluctuations at the interface region destabilize the interactions between GD1 and CD domain, resulting in GD1 domain undergoing a large domain motion (Figure 1b). In the [GDP:GTP] bound system, a larger mobility at the linker region compared to [GTP:GDP] and [GDP:GDP] is observed. The presence of

GTP at the GD2 domain probably triggers the linker region, which drives GD1 to undergo a large globular domain motion.

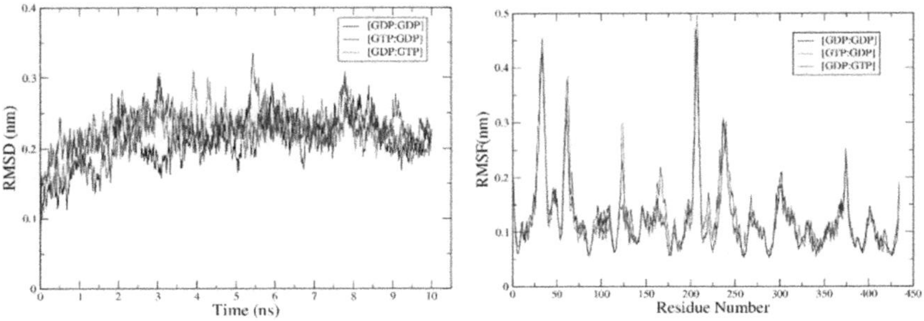

Figure 2 *(left) RMSD plots of simulated systems with respect to the equilibrated structure during the course of the simulation time. (right) RMSF plots of the simulated systems showing the mobility at various regions of the protein.*

Cluster analysis was carried out to evaluate the structural variations at different regions of the simulated systems. Clustering for the simulated trajectory of 1000 frames was done using the GROMOS method with a RMSD cutoff of 0.2 nm. Three, six and four clusters were obtained for [GDP:GTP], [GTP:GDP] and [GDP:GTP] nucleotide bound systems respectively (Figure 3). The larger shift observed in the helical region of GD1-linker interface for the [GTP:GDP] bound system may further trigger the linker loop to fluctuate and undergo large domain motion.

4 CONCLUSION

EngA is a GTPase involved in the later steps of ribosome maturation. Despite the available structural and biochemical data, on EngA-ribosome association, the mechanism of EngA-ribosome interaction at the molecular level is still elusive. In the present study the conformations of EngA in different nucleotide bound combinations using MD simulations were attempted. The results show that the combinations of the bound guanine nucleotides dictate local fluctuations, which probably disrupt interface interactions between the domains to undergo large domain motion. Further simulation studies using enhanced sampling techniques and/or longer simulations are in progress to trace the mechanism of large domain motion and conformational transitions.

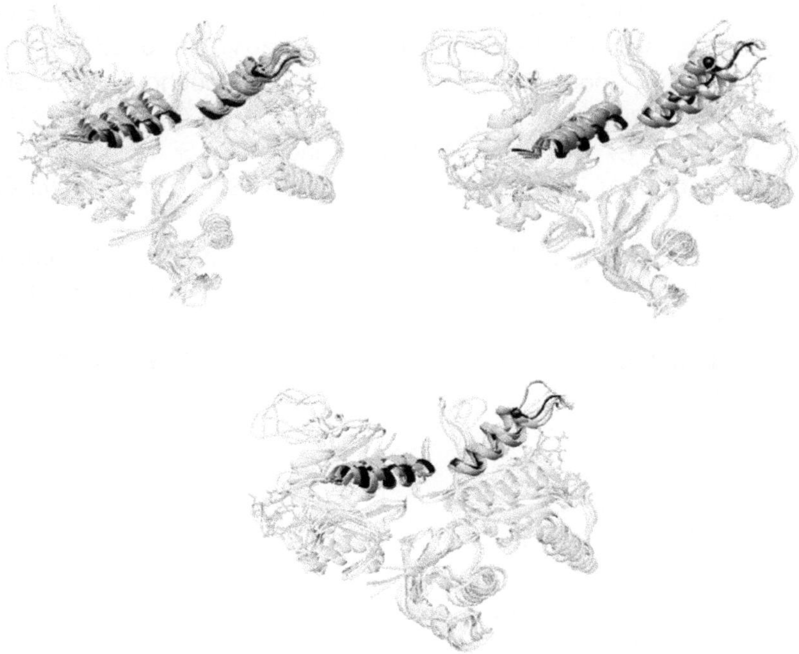

Figure 3 *Simulated clusters extracted using GROMOS method with a RMSD cutoff of 0.20 nm for (left) [GDP:GDP], (right) [GTP:GDP] and (bottom) [GDP:GTP] guanine nucleotide bound systems.*

ACKNOWLEDGEMENT

The author NU thank UGC for providing BSR-RFSMS fellowship, the authors SK and NU thank University of Mysore for sanctioning minor research project.

References

1. R. A. Britton, *Annu. Rev. Microbiol.*, 2009, **63**, 155.
2. A. E. Foucher, J. B. Reiser, C. Ebel, D. Housset and J. M. Jault, *PLoS One*, 2012, **7**, 46795.
3. V. L. Robinson, J. Hwang, E. Fox, M. Inouye and A. M. Stock, *Structure*, 2002, **10**, 1649.
4. X. Zhang, K. Yan, Y. Zhang, N. Li, C. Ma, Z. Li, Y. Zhang, B. Feng, J. Liu, Y. Sun and Y. Xu, *Nucleic Acids Res.*, 2014, **42**, 13430.
5. S. K. Tomar, N. Dhimole, M. Chatterjee and B. Prakash, *Nucleic Acids Res.*, 2009, **37**, 2359.

6. S. Majumdar, A. Acharya, S. K. Tomar and B. Prakash, *Biochim. Biophys. Acta - Proteins and Proteomics*, 2017, **1865**, 289.
7. D. Van Der Spoel, E. Lindahl, B. Hess, G. Groenhof, A. E. Mark and H. J. Berendsen, *J. Comput. Chem.*, 2005, **26**, 1701.
8. N. Eswar, B. Webb, M. A. Marti-Renom, M. S. Madhusudhan, D. Eramian, M. Y. Shen, M, U. Pieper and A. Sali, *Curr. Protoc. Bioinf.,* 2006, **15**, 5.
9. E. F. Pettersen, T. D. Goddard, C. C. Huang, G. S. Couch, D. M. Greenblatt, E. C. Meng and T. E. Ferrin, *J. Comput. Chem.,* 2004, **25**, 1605.
10. W. Humphrey, A. Dalke and K. J. Schulten, *J. Mol. Graphics*, 1996, **14**, 33.

ERRORS MADE IN *AB INITIO*+ HOMOLOGY-BASED PROTEIN STRUCTURE PREDICTION TOOLS ARISE DUE TO LACK OF CONSIDERATION OF IMPACT FORCES

Aishwarya Chander and Priyanka Selvaraj

Department of Bioinformatics, College of Biomedical Sciences, research and technology, Sri Ramachandra Institute of Higher education and research, Porur, Chennai, Tamil Nadu, India

1 INTRODUCTION

There is a requirement for accurate tertiary structureprediction to understand a protein's conformation. Forces like electrostatic and Van der Waals interaction, hydrogen bonds and hydrophobic interaction are often considered when it comes to tertiary structure modelling by computer algorithms[1]. Even though the resulting structure is well modelled and shows some resemblance to native structures, it always permits slight inaccuracies at very low resolutions. Until now, no algorithm has efficiently predicted protein structures with 100% accuracy.

With the advent of modern medical technologies such as precision medicine[2], we require highly resolved protein structures predicted to almost 100% accuracy. The specificity of structures accounts for the various protein functions, protein interaction networks, their mutations, and how the mutation cause diseases. Variations in the sequence will, in turn, affect protein functionality and their pathways[3]. Thus, the structure of the encoded protein, as well as its mutations, can provide a full understanding of the pathogenesis, targeted therapies and additionally, potential strategies for drug development based on molecular structure. The success of structure-based targeted drug therapy relies heavily on the availability of accurate 3D structures. For this, detailed knowledge of protein structure at the atomic level of resolution is required[4].

Over the last century, it has been proven that physical forces are exerted upon all objects continually.We will refer to these forces as *impact* forces due to the direct impact we believe they have on proteins. Published data involving high altitude, zero-gravity, low gravity protein modelling[5] has shown that there is a consistent variation between these proteins and proteins modelled in standard protocol conditions. These impact forces vary along with geographical location, temperature, atmospheric pressure and magnetic fields, and it is essential to standardize a value based on their location and geo-political conditions.

Our project seeks to prove that there is a consistentand direct repercussion of these impact forces on protein structures as observed in natural biological synthesis.

2 METHOD AND RESULTS

Using various *ab Initio* and homology combination-based modelling software program, SWISS-MODEL predicted the tertiary structure of proteins from their primary amino acid sequences[6]. For this experiment, 25 amino acids from the species *Homo sapiens* were selected at random, keeping in mind that a majority of them were well experimented upon and characterized using high profile imaging techniques, such as X-Ray Crystallography, electron microscopy or NMR Spectroscopy, from Protein Data Bank. All 25 amino acids had highly cited tertiary sequence structures and were ranked from *'Best to worst'* in terms of*'Resolution'* under the category of *'Filter'* in Protein Data Bank[7].

Upon obtaining the 3D coordinates for the tertiary structures from the structuring tool, they were modelled in a molecule mapper, Edu PyMol v2.0x. Molecule mappers utilize tertiary structure coordinate information to provide an exact model of the input protein information.

Subsequently, the nascent structural information, as obtained from protein data archives, was also modelled using the same tool. Then, a comparison of the two was made, and the results obtained regarding structural overlap was tabulated. Root-Mean Square deviation (RMSD) values were taken as a standard calculation parameter to understand the variance between the nascent and acquired structures. Although RMSD at lower value ranges is considered negligible; the value must theoretically be 0 for correctly modelled structures.

Utilizing this information, and going through published data, the following conclusions were drawn.

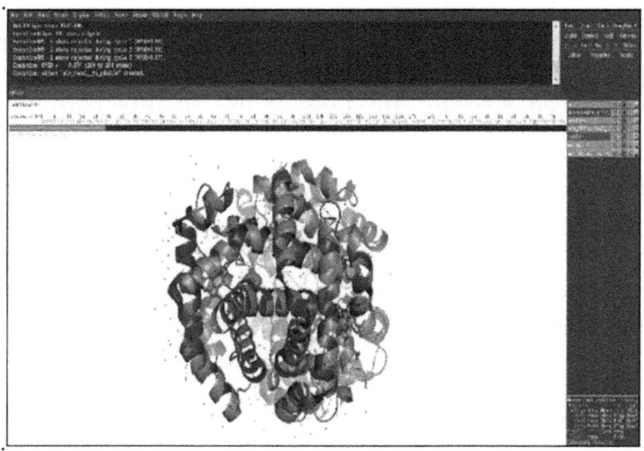

Figure 1 *Screenshot showing software modelled and nascent Structure overlap of PDB ID 1i3D. Modelling software is a majorly homology-based algorithmThe incomplete overlap is observed, RMSD Variation is at 0.074.*

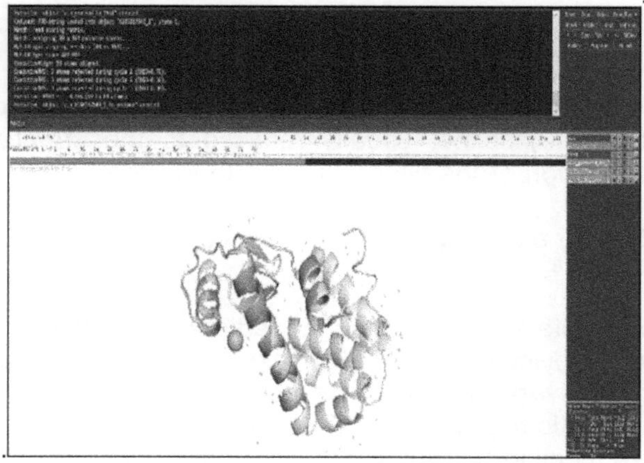

Figure 2 *Screenshot showing software modelled and nascent Structure overlap of PDB ID 253L Modelling software is a majorly combination-based algorithm. The incomplete overlap is observed, RMSD Variation is at 0.246.*

3 RESULTS

Individual RMSD deviations were tabulated, and the mean RMSD deviation was calculated to show a rough indication of structure prediction accuracy of SWISS-MODEL[6]. This value was 0.426. As all selected structures were of similar nature, this calculation is justified.

The RMSD range lies from 0.060-0.790. A graph is plotted to show the fluctuation in RMSD values as observed (Figure 3).

We have found a nearly consistent gap in all the structured comparative models, indicating a similar error in every model generated by the software.

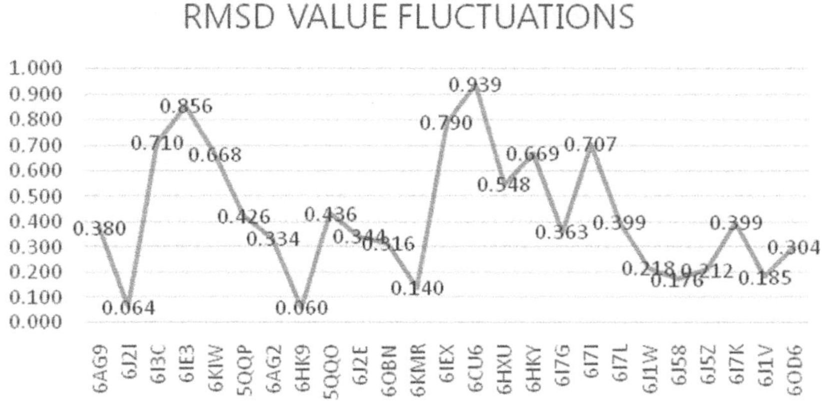

Figure 3 *RMSD value fluctuations as recorded after a comparative analysis of modelled structures Mean RMSD value is 0.426.*

3 CONCLUSION

Experimental data suggest a slight scope for improvement in modelling algorithms. This can be done by the inclusion of a filter for impact forces in the modelling software's pipeline; usually referred to the *structure refinement* segment[1]. The function of the filter would be to include impact factor data as supplemented by a predefined database curated by us. This is backed up by existing literature evidence and has been overlooked by current algorithms[1,2,8].

By the inclusion of filters that enhance the tertiary structure of proteins, we are one step closer to discovering an ideal algorithm that models with a 100% accuracy every at every account. This not only improves the chances of a successful therapeutic regime but is a ground-breaking success in the field of bioinformatics—including but not limited to evolutionary studies, convergence and divergence analysis, molecular phylogenetics and space biology[2,5].

References

1. B. Jayaram, P. Dhingra, A. Mishra, R. Kaushik, G. Mukherjee, A. Singh and S. Shekhar, *BMC Bioinf.*, 2014, **16**(16), S7.
2. A. Venkatesan, J. Gopal, M. Candavelou, S. Gollapalli and K. Karthikeyan, *Healthcare Inf. Res.*, 2013, **19**(2), 137.
3. R. Burkhard, R. Predrag and B. Yana, *FEBS Lett.*, 2016, 590.
4. B. Berger, N. M. Daniels and Y. W. Yu, *Commun. ACM*, 2016, **59**, 8.
5. R. C. Bi, L. L. Gui, Q. Han, F. L. Shen, K. Shi, Y. P. Wang, S. Z. Chen, Y. L. Hu, X. T. Niu, J. Dong, Y. C. Zhou and N. Q. Lin, *Microgravity Sci. Technol.*, 1994, **7**(2), 203.
6. T. Schwede, J. Kopp, N. Guex and M. C. Peitsch, *Nucleic Acids Res.*, 2003, **31**(13), 3381.
7. A. Kouranov, L. Xie, J. de la Cruz, L. Chen, J. Westbrook, P. E. Bourne and H. M. Berman, *Nucleic Acids Res.*, 2006, **34**, D302.
8. P. Aloy, M. Pichaud and R. B Russell, *Struct. Biol.,* 2005, **15**(1), 15.

PARADIGM SHIFT IN RESEARCH FOCUS: COMPUTATIONAL SOFTWARES IN CANCER BIOLOGY

B. N. Rajarathnam,[a] N. Rakesh[a] and M. Anita[b]

[a] Department of Oral Medicine and Radiology, Faculty of Dentistry, M S Ramaiah University of Applied Sciences, Bangalore, India; [b] Department of Pharmacology, Faculty of Pharmacy, M S Ramaiah University of Applied Sciences, Bangalore, India

1. INTRODUCTION

In the recent time field of analysis, systems biology has gained its momentum. Simulation can play a major role in selection of studies to be performed; clinical trial simulation covers multi-disciplinary individual's i.e. preclinical pharmacologist, statistician, computer programmer, etc. Although widely used, the physiologically-based pharmacokinetic approach requires intense resources to generate the data on the various parameters described in the models.[1] For example, an available software package called PK-Sim developed by Bayer Technology Services[2] can be employed for such purpose. Oral Submucous fibrosis (OSF) was recognized as a definitive lesion by Schwartz who described it as a fibrosing condition in 1952 and named it as "atrophicaidiopathica mucosae oris (tropica)".[4] It was described as precancerous condition by Paymaster in 1956.[5] This apparent incongruity was corrected by the WHO (2007), and the lesion along with other similar ones are now labelled as "potentially malignant oral disorders".[6] There is increased accumulation, decreased degradation of collagen fibres and altered collagenase activity is found in oral submucous fibrosis affected mucosa.[7] One enzyme that increasingly is associated with extra cellular matrix (ECM) is the protein– cross-linking enzyme tissue transglutaminase (TG2). Tissue transglutaminase (TG2) catalyzes the cross-linkage between glutamine (Gln) and lysine (Lys) side chains. Although TG2 is essential for the stabilization of the extracellular matrix.[8] G.S. Thangjam et al (2009), conducted a study to know the possible role of transglutaminase (TG2) in OSMF pathogenesis. Biopsy tissues were obtained from OSMF affected patients and subjected to RT-PCR. They found overregulation of TGM-2 suggesting the importance of TGM-2 in the pathogenesis of OSMF.[9] With the knowledge of molecular pathogenesis, role of transglutaminase enzyme in oral submucous fibrosis and use of pharmacokinetic simulation software – PK Sim, a research is formulated to find out the bioavailability and efficacy of cysteamine hydrochloride (transglutaminase inhibitor) in management of the progression of the disease.

2. METHODS AND RESULTS

2.1 Induction of oral sub mucous fibrosis and evaluation of transglutaminase

A study was carried out for induction of oral submucous fibrosis in male Wistar rats. Ethical clearance was obtained from the Institutional Animal Ethics Committee (IAEC). Registration number 220/PO/abc/2000/CPSEA and reference number MSRFPH/P/62/2015. Healthy Wistar rats (n=10, control-5; experimental animal-5) weighing approximately about 200mg were obtained from private animal breeders. They were kept in clean, hygienic cages and maintained under standard laboratory conditions. The rats were kept in groups of three per cage at controlled temperature (25°C ± 2°C) with 12 h light/dark cycle and humidity. They received standard diet and water *ad libitum*. Arecoline extract was procured from Sigma Aldrich Pharmaceuticals. Dosage of 10mg/kg of freshly prepared arecoline was injected into submucosa of right buccal mucosa in experimental animals over a period of 3 months on every alternate days. Whole blood of 1ml was collected from retro-orbital plexus in an EDTA coated vials at the starting of experiment from healthy wistar rats and also at the end of experiment of induction of oral sub mucous fibrosis. Experimental animals was sacrificed to check the induction of submucous fibrosis. Right buccal mucosa was dissected and subjected to histopathological examination. Tissue and blood samples were subjected to ELISA for estimation of levels of transglutaminase.

2.2 Simulation of efficacy of cysteamine hydrochloride

PkSim version 7.0 software release was used to compare the transglutaminase competitive amine inhibitors namely cysteamine, spermidine and putrescine. From the PK analysis, AUC and Cmax is compared for all the three drugs. The subsequent value of the information obtained from drug database.

2.3 Evalution of simulated results with experimental method

After 90 days of oral sub mucous fibrosis induction period, the wistar rats are subjected to treatment with transglutaminase inhibitor-cysteamine hydrochloride. From the simulated PK analysis it is evident that cysteamine AUC and Cmax is higher compared with other inhibitors. Same dosage protocol is followed in treatment process as with simulation process. For the oral sub mucous fibrosis induced wistar rats, 48mg/kg of cysteamine hydrocholoride is administered through drinking water. The concentration of transglutaminse is estimated from the serum samples at 7th, 14th and 21st day from the commencement of treatment procedure i.e., 97th, 104th, 111th day in total from the commencement of experiment.

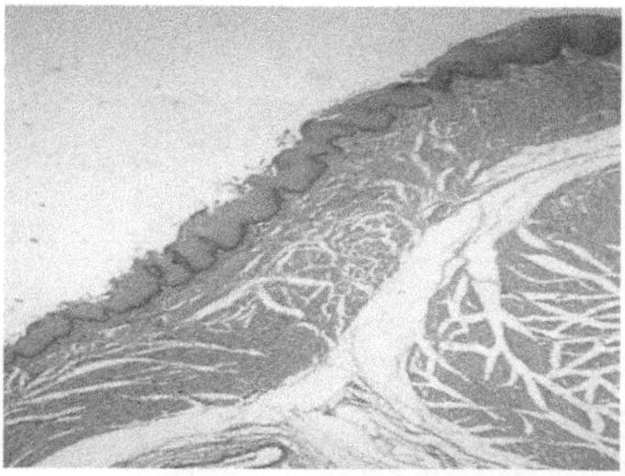

Figure 1 *Histopathology section of buccal mucosa of healthy male wistar rat. Epithelium - Ortho-keratinized stratified squamous epithelium with blunt rete ridges; Connective tissue -Normal component of collagen fibres, few inflammatory cells and normal vascularity.*

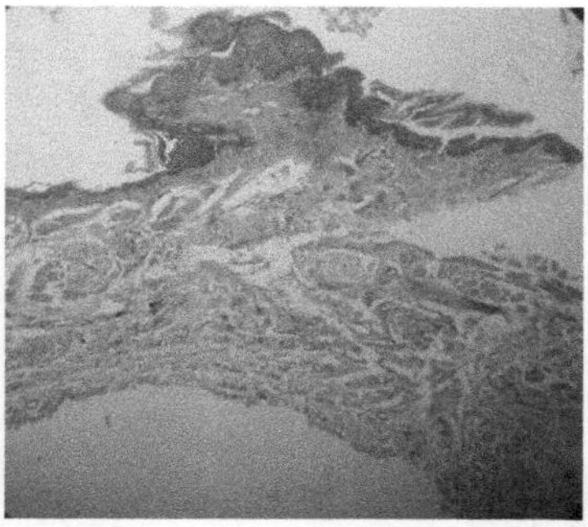

Figure 2 *Histopathology section of buccal mucosa of oral sub mucous fibrosis induced wistar rat.Epithelium-Areas of atrophy and hyperplasia are seen with sub epithelial hyalinization.Connective tissue-Increased density of collagen fibers and inflammatory cells.*

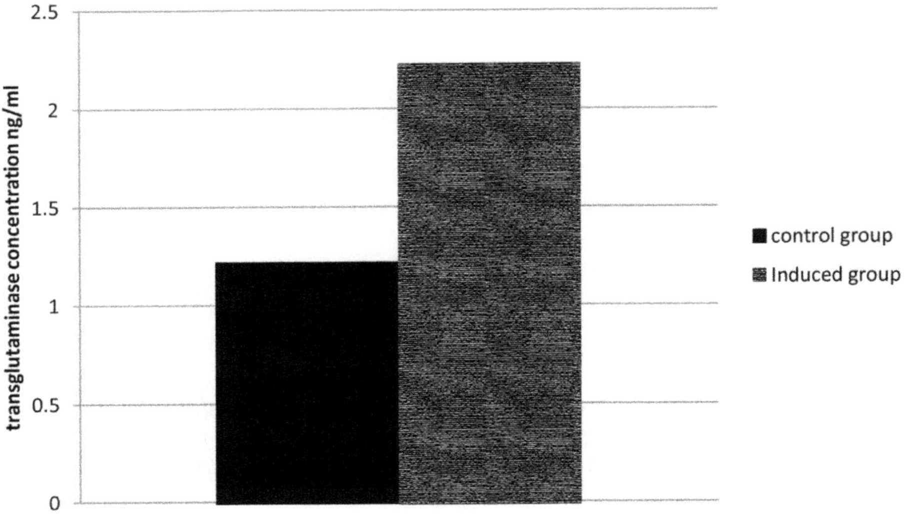

Figure 3 *Transglutaminase concentration levels in tissue samples of healthy control animals and oral submucous fibrosis induced animal.*

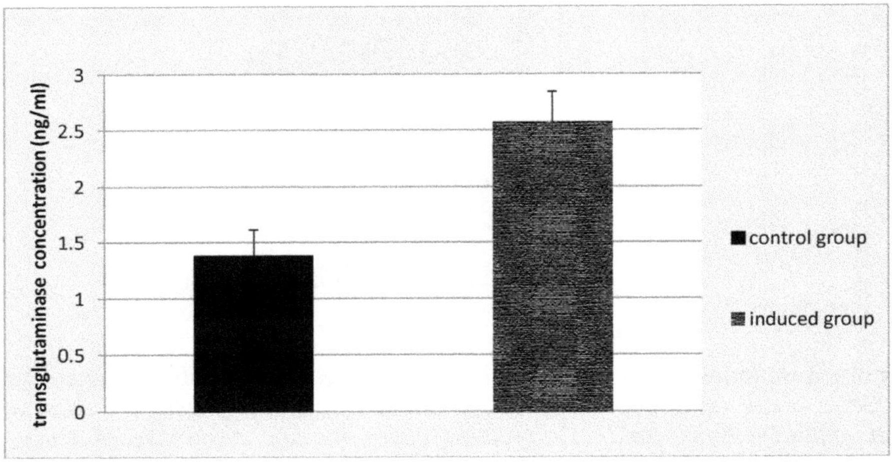

Figure 4 *Transglutaminase concentration levels in serum samples of healthy animals and oral sub mucous fibrosis induced animals.*

Table 1 *PK Analysis.*

Drugs	Cmax (μmol/l)	Cmax_norm (mg/l)	t_max (h)	C_tEnd (μmol/l)	AUC_tEnd (μmol*min/l)	AUC_tEnd_norm (μg*min/l)
Cysteamine hydrochloride	690.36	1.11×10^6	46.75	690.36	1.93×10^6	3.11×10^{12}
Putrescine	38.81	19009.42	48	38.81	91427.76	4.48×10^{10}
Spermidine	146.58	256523.49	48	146.58	376601.03	6.59×10^{11}

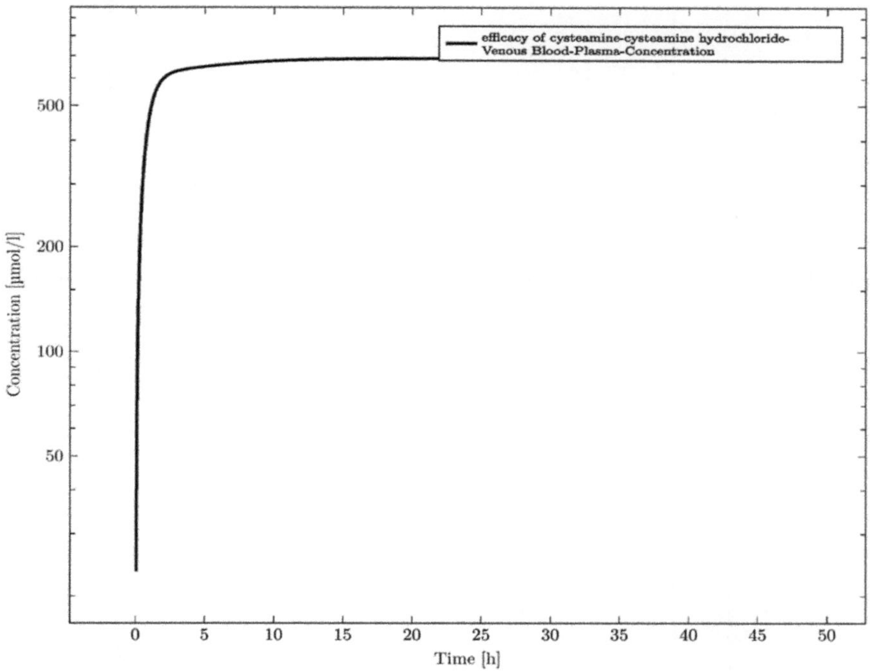

Figure 5 *The graph shows the intravenous cysteamine hydrochloride plasma concentration.*

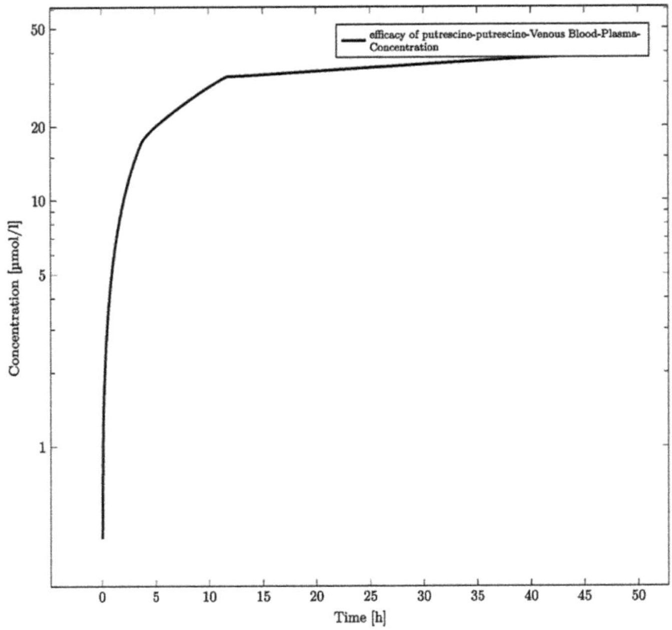

Figure 6 *The graph shows the intravenous putrescine plasma concentration.*

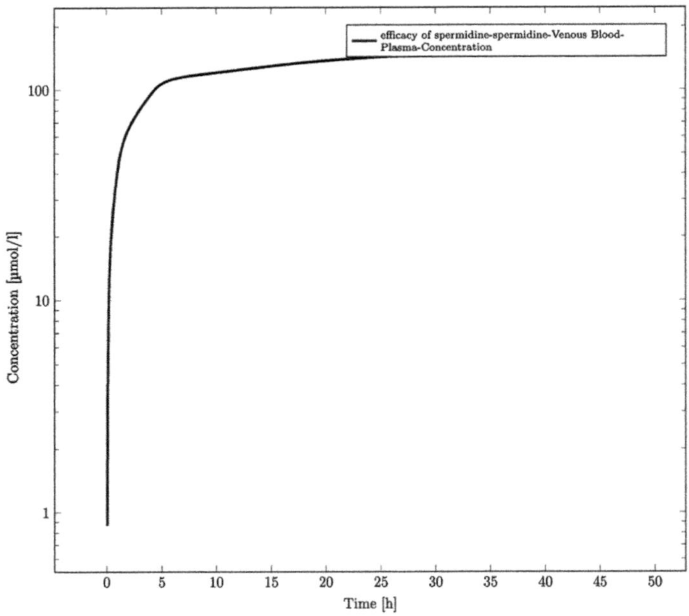

Figure 7 *The graph shows the intravenous spermidine plasma concentration.*

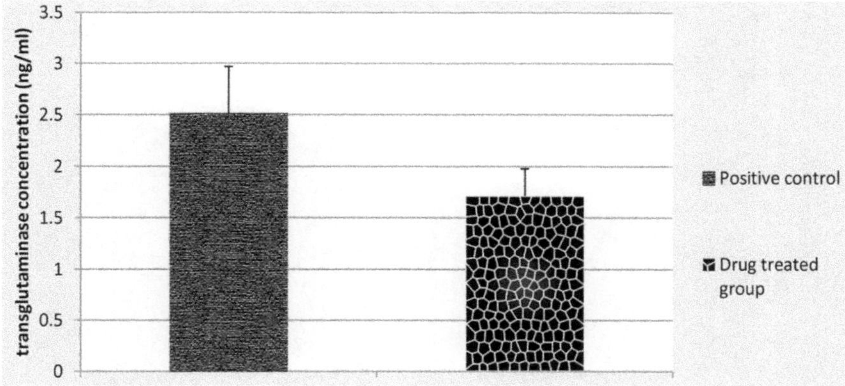

Figure 8 *Concentration of transglutaminase in serum samples on 97th day.*

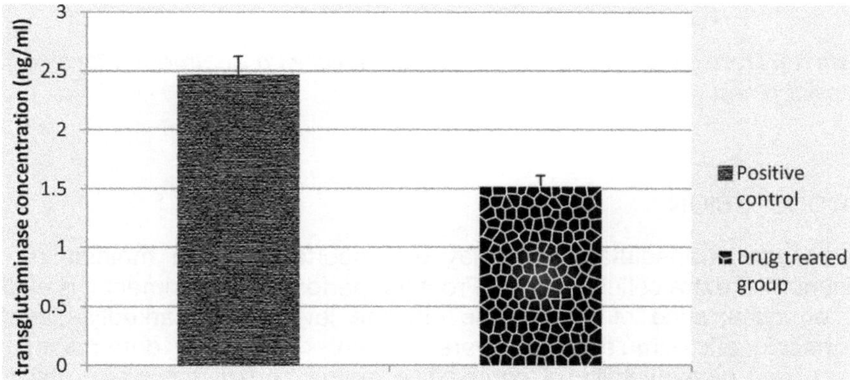

Figure 9 *Concentration of transglutaminase in serum samples on 104th day.*

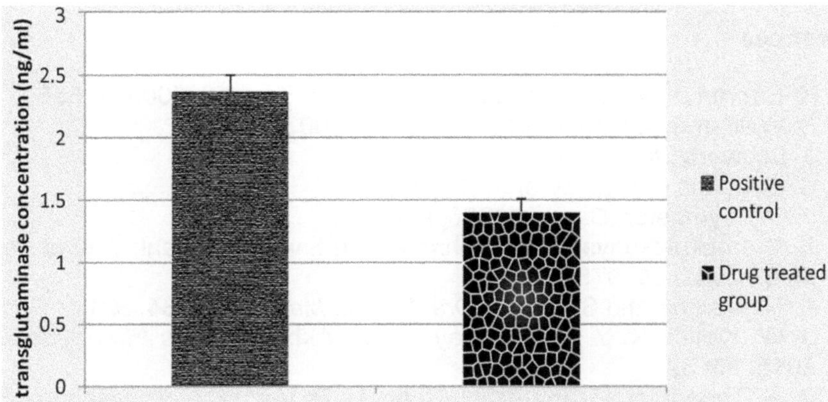

Figure 10 *Concentration of transglutaminase in serum samples on 111th day.*

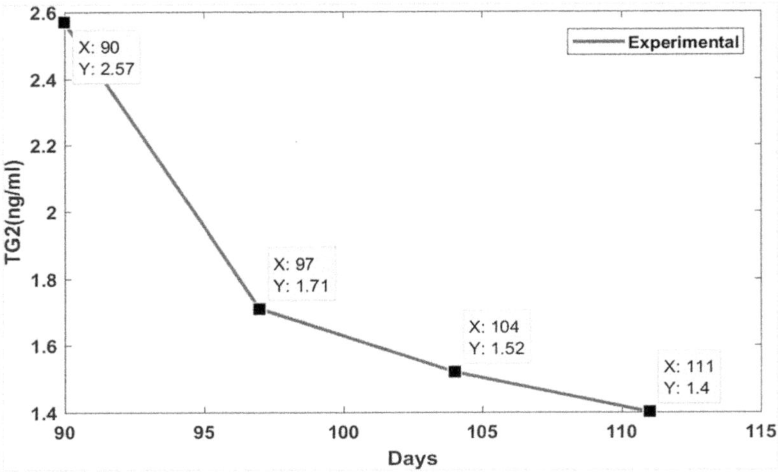

Figure 11 *Transglutaminase concentration levels with treatment of cysteamine hydrochlorident of cysteamine hydrochloride.*

3. CONCLUSION

The enzyme, transglutaminase play an important role in maintaining the equilibrium of extra cellular matrix. From the performed experiment it is evident that, on occurrence of fibrosis the enzyme levels are markedly elevated. Pharmacokinetic simulation software certainly helped in determining the efficacy and bioavailability of competitive amine transglutaminase inhibitors. For the first time in dental literature the simulation software is employed for study of drug characteristics and also evaluation of tranglutaminase inhibitor in management of oral sub mucous fibrosis is explored.

References

1. K. Subramanian, *Expert Opin. Drug Metab. Toxicol.*, 2005, **1**, 555.
2. S. Willmann, J. Lippert, *et al.*, *Biosilico,* 2003, **1**, 121.
3. J. Schwartz, *Ind. J. Med.,* 1952, **16**, 185.
4. S. G. Joshi, *Ind. J. Otolaryn,* 1953, **4**, 1.
5. J. C. Paymaster, *Cancer,* 1956, **9**, 431.
6. S. Warnakulasuriya, N. W. Johnson and I. van der Waal, *J. Oral Pathol. Med.,* 2007, **36**, 575.
7. P. Rajalalitha and S. Vali, *J. Oral Pathol. Med.*, 2005, **34**, 321.
8. J. W. Keillor, K. Y. P. Apperley and A. Akbar, *Trends Pharmacol. Sci.,* 2015, **36**, 32.
9. G. S. Thangjam, P. Agarwal, I. Khan, U. P. Verma, A. K. Balapure, S. G. Rao and P. Kondaiah, *J. Dent. Res.,* 2009, **88**, 170.

VIRTUAL SCREENING OF BIO-ACTIVE COMPOUNDS AS AN INHIBITOR OF C-FLIP PROTEIN FOR CANCER TREATMENT

Prashantha Nagaraja,[a] Arighni Guhathakurta,[b] Shweta Sukla Baidya[b] and Pujitha Krishnasamudram [b]

[a] Department of Biotechnology, School of Applied Sciences, REVA University, Bangalore-64, India; [b] Research Student, Department of biotechnology, School of Applied Sciences, REVA University, Bangalore-64, India

1 INTRODUCTION

Cellular FLICE (FADD-like IL-1β-converting enzyme)-inhibitory protein (c-FLIP) is an anti-apoptotic regulatory protein play an important role in cell maintenance and survival to regulate the activation of caspases-8 and -10 proteins in apoptotic signaling pathways [1-3]. cFLIP mediates death receptors to apoptosis via FAS, TNFR, DR4/DR5, and TNF-related apoptosis-inducing ligand (TRAIL) with the process and activation of procaspase-8 and -10 are over expressed in down process and is the major potential drug targets for cancer therapy [4,5]. Moreover, abnormal c-FLIP expression has been found in various diseases such as cancer, multiple sclerosis, Alzheimer's disease, diabetes mellitus, and rheumatoid arthritis [6-8]. Therefore, c-FLIP-has been identified as potential target for major cancer signaling pathways to inhibit the functions of c-FLIP by targeted therapy [9]. Using computational methods, identified active lead molecules from chemical library databases for the design of potential c-FLIP inhibitors [10].

2. METHODS AND RESULTS

2.1. Protein structure prediction and validation

The c-FLIP target protein sequence were selected from UniProt knowledge database (UniProt ID: O15519) using FASTA sequence format and the resultant sequences were used to search similar templates using BLASTp by structure based method using reference Protein Data Bank (PDB). With default BLASTp algorithm parameters. The physico-chemical properties were predicted to understand the stability of the protein using Protparam and the homology model structures were built using Swiss Model. The resultant model structures were analysed by the stereochemical activity of the amino acids arranged in the protein structure to predict the complexity of the protein using SAVES and the active site amino acids were identified using CastP calculation server. The best homology model protein structure was further used for molecular docking using selected active potential ligand structures.

Table 1 *Sequence similarity search and Quality prediction*

Template	% of identity	Z-score	Q-mean	Quality Score	Ramachandran plot	Seq Coverage
3H11	86%	0.87	0.12	88.793	87.4%	212-244

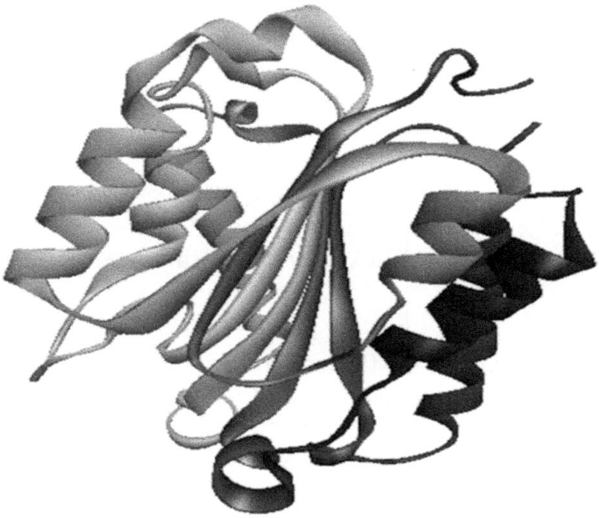

Figure 1 *C-FLIP Protein structure prediction using homology model method.*

2.2 Lead Identification and Docking Studies

PubChem compound database was screened to identify the potential ligands for Structure Based Drug Discovery (SBDD) approach. We fetched active chemicals from the chemical library database and have selected 48 compounds and further filtered for prediction of drug like properties such as Lipinski Rule of 5 properties to understand the physico-chemical properties of the compounds within biological system using Molinspiration online server. Based on the observation from Lipinski Rule of five, 5-FU and Rituximab has less LogP shows less liphophilicity of absorption in membrane to carry the chemicals to the blood stream and Kurarinone with highest logP. Molecular weights of the compounds were observed that shows Doxorubicin, Taxol and Geldanamycin have >500 kda which predicts the less aqueous dissolution. Based on Lipinski Rule of 5, we selected 10 compounds with strongly accepting drug-like properties and further taken for molecular docking. Auto Dock Tool (ADT) v 1.5.6 is used for molecular docking of c-FLIP protein with identified drug-like compounds using default parameters to study the interaction maps. Docking interaction of each compound is simulated using lamarckian genetic algorithm and the best docking scores and interaction energy is recorded for each ligand. The reference compounds such as Paclitaxel and Docetaxel demonstrated 5 and 6 hydrogen bonds with -12.8651 and -11.2532 kcal/mol of energy respectively. Compounds have exhibited 3-9 hydrogen bonds and the binding energy of -16.5412 to -5.06 kcal/mol. Results shows that all compounds can be screened for c-FLIP protein inhibition studies.

3. CONCLUSION

Based on the preliminary research on c-FLIP inhibitor study, we have predicted the protein-ligand interaction of selected chemical structures. The drug-like compounds demonstrated good inhibitory constant and binding energy values compared to reference drug compounds. Selected compounds can be tested further for *in vitro* and *in vivo* screening to characterize for better understanding of required features and correlation with experimental results. Present study gave us inputs for the future design of safe and potential c-FLIP inhibitors.

Table 2 *Molecular docking of selected active molecules with target c-FLIP protein.*

Ligands	LogP	MW	HBA	HBD	Vol	H Bonds	IC	Binding Energy
5-FU	-0.587	130.078	4	2	96.913	4	195.15	-5.06
LY303511	3.087	306.365	4	1	281.547	3	98.6374	-2.90514
MS275	2.03	376.416	7	4	339.353	5	69.3749	-16.5412
Camptothecin	2.028	348.358	6	1	297.41	6	71.0242	-7.28057
Doxorubicin	0.567	543.525	12	7	459.18	7	80.7132	-15.6801
Taxol	4.945	853.918	15	4	756.598	8	215.697	-5.51687
9-NC	2.242	393.355	9	1	320.745	6	79.5304	-15.2905
Flavopiridol	3.455	401.846	6	3	340.148	7	106.326	-17.5494
Kurarinone	6.185	428.52	6	3	411.069	8	91.4253	-7.45988
Geldanamycin	0.429	560.644	11	4	522.114	8	110.389	-12.3561
Rituximab	-0.575	389.213	5	1	276.791	2	90.213	-5.02202
PIK-75	3.384	452.29	9	0	322.472	9	120.155	-11.7128
SAHA	2.467	264.325	5	3	255.644	6	120.687	-15.7876
Paclitaxel	2.566	853.9	4	14	853.333	5	170.965	-12.8651
Docetaxel	1.636	807.9	5	14	807.346	6	160.620	-11.2532

LogP: Octonol and water dissolution (0<5); MW: molecular weight (180<500kda); HBA: hydrogen bond acceptor (1<5); HBD: Hydrogen bond donor (1<10); Vol: Volume (150<500); H-bonds: hydrogen bonds (2<8); IC: inhibitory constant

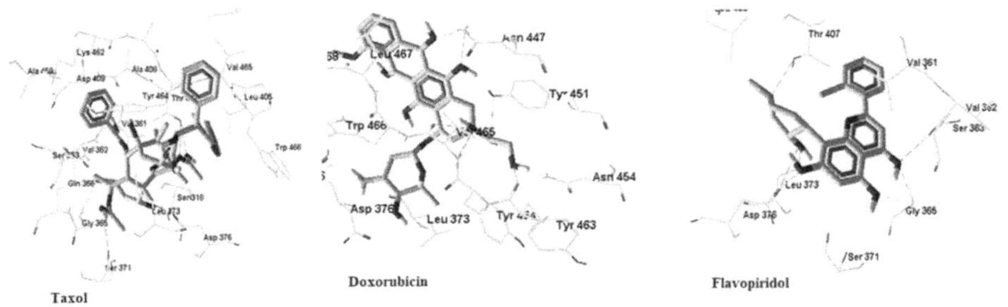

Figure 2 *Molecular docking of selected best ligand structures with target receptor.*

References

1. L. Czerski and G. Nunez, Apoptosome formation and Caspase activation: is it different in the heart?. *J. Mol. Cell. Cardiol.*. 2004, **37**(3), 643–652.
2. D. Green, *Means to an End: Apoptosis and other Cell Death Mechanisms*, Cold Spring Harbor Laboratory Press, Cold Spring Harbor, NY, 2011.
3. A. Strasser, L. O'Connor and V. M. Dixit, Apoptosis signaling, *Annu. Rev. Biochem.*, 2000, **69**, 217–245.
4. Y. Fuchs and H. Steller, Programmed cell death in animal development and disease, *Cell*, 2011, **147**(4), 742–758.
5. T. G. Cotter, Apoptosis and cancer: the genesis of a research field, *Nat. Rev. Cancer*, 2009, **9**, 501–507.
6. J. Y. Jung and W. J. Kim, Involvement of mitochondrial- and Fas-mediated dual mechanism in $CoCl_2$-induced apoptosis of rat PC12 cells, *Neurosci. Lett.*, 2004, **371**(2–3), 85–90.
7. F. Pirnia, E. Schneider, D. C. Betticher, M. M. Borner, Mitomycin C induces apoptosis and caspase-8 and -9 processing through a caspase-3 and Fas-independent pathway, *Cell Death Differ.*, 2002, (9), 905–914.
8. R. F. Kelley, K. Totpal, S. H. Lindstrom, M. Mathieu, K. Billeci, L. Deforge, R. Pai, S. G. Hymowitz and A. Ashkenazi, Receptor-selective mutants of Apo2L/TRAIL reveal a greater contribution of DR5 than DR4 to apoptosis signaling, *J. Biol. Chem.*, 2004, **280**, 2205–2212.
9. E. A. Perez, Microtubule inhibitors: Differentiating tubulin-inhibiting agents based on mechanisms of action, clinical activity, and resistance, *Mol. Cancer Ther.*, 2009, **8**, 2086–2095.
10. R. Clarke, F. Leonessa and B. Trock, Multidrug resistance/P-glycoprotein and breast cancer: Review and meta-analysis, *Semin. Oncol.*, 2005, **32**, S9–S15.

FRAGMENT-BASED DESIGN OF NOVEL INHIBITORS OF HPV 16 E6 ONCOPROTEIN: MOLECULAR DOCKING, MOLECULAR DYNAMICS SIMULATION AND *IN SILICO* ADME ANALYSIS

A. Kumar, E. Rathi and S. G. Kini

Department of Pharmaceutical Chemistry, Manipal College of Pharmaceutical Sciences, MAHE, Manipal, Karnataka, India

1 INTRODUCTION

Cervical cancer (CC) is the second most common cancer type in females in India. It has been four decades since the discovery of human papilloma virus (HPV) as the causative agent of CC. The race to develop drugs against HPV and HIV (human immunodeficiency virus) began almost at the same time. But even today not a single drug has been discovered against HPV while there are more than twenty anti-retroviral drugs. Although we have vaccines against HPV but the high prevalence rate in many of the women after the start of sexual intercourse, HPV type specificity of these vaccines, long term efficacy and huge logistics involved in running mass vaccination programmes poses a major challenge in winning the battle against CC. Moreover, vaccines are also ineffective for already infected people. Therefore, it is indispensable to develop new therapies.

In this quest of developing novel small organic molecules against CC, viral E6 protein has gathered the attention of many researchers. It's a well-established fact now that the early viral proteins of high-risk HPV mainly E6 and E7 work together to make normal cells immortal. E7 has been found to push the differentiated cells back in the division phase by targeting pRb protein.[1] Normally such abnormal cells are tackled by p53 (an apoptotic protein). But with E6 protein operating in tandem with E7, p53 proteins are degraded by the complex of E6-E6AP proteins.[2] This mechanism of tumorigenicity can be extrapolated as a case of "gene addiction".[3]

Recently, we have also tried to target E6 protein to develop its novel inhibitors. With the help of e-pharmacophore modelling and high throughput virtual screening of diverse libraries, we have identified some chromone derivatives as its potential inhibitors.[4] Our group has also reported drug repurposing approach to develop its inhibitors.[5] In this paper we have reported a completely novel approach employing fragment-based drug design employing the amino acid residues present in the predicted active pocket by SiteMap tool of Maestro(Schrodinger).[6] To the best of our knowledge there is no such report in the literature for E6 protein.

2 METHODS AND RESULTS

2.1 Prediction of active site

For this study X-ray crystal structure of E6 protein was retrieved from protein data bank (PDB ID 4GIZ). This study was conducted on a hp computer with Linux Ubuntu 18.04.1 LTS platform, 8 GB RAM, Intel Core i3-4160 processor and Intel Haswell graphics card. In silico studies were carried out using Maestro (version 11.8), a small-molecule drug discovery suite of Schrödinger Inc.

The retrieved protein was prepared with the help of Protein Preparation Wizard of Maestro which involved steps like pre-processing (add hydrogen atoms, addition of missing residues or loops, assigning proper bond order, creating zero-order bonds with metals etc.), review and modify. The last step of this exercise was energy minimization of the refined protein (in this case chain C). Once the protein was ready, SiteMap tool was used to predict possible active sites. The predicted active sites were ranked based on site score and D-score (Table 1). SiteMap identifies potential binding sites by linking together 'site points' that are most likely to contribute to tight protein–ligand or protein–protein binding. These site points are useful for visualizing the extent of a site and can be employed to define the active site for virtual screening experiments using Glide. When the results are incorporated in SiteMap, six map types like, Hydrophobic map (yellow mesh), Hydrophilic map (green mesh), Hydrogen-bond donor map (blue mesh), Hydrogen-bond acceptor map (red mesh), Metal-binding map (pink mesh), Surface map (grey surface) are generated (Figure 1). The amino acid residues present in the predicted cavity were identified and based on them a receptor grid was generated. Then we manually measured all the amino acid residues in the vicinity of 4Å for four major site map features like hydrophobic map, hydrophilic map, hydrogen bond donor and hydrogen bond acceptor. Based on these amino acid residues four smaller receptor grids were generated for each site map feature.

Table 1 *Predicted active site by SiteMap tool of Maestro.*

Predicted site	Site 1
Site score	0.934
D score	0.969
Volume(Å^2)	265.285
Donor/acceptor	0.971
Residue number (Chain C)	31,32,45,50,51,53,61,62,67,70,71,73, 74,75,100,102,104,107,128,131,132,133

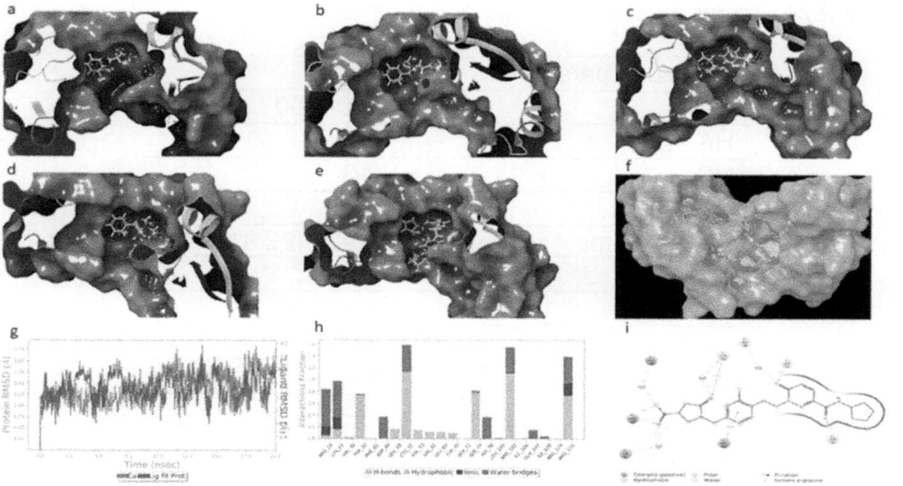

Figure 1 *Top ranked fragment in the predicted cavities of (a) hydrogen bond acceptor, (b) hydrogen bond donor, (c) hydrophilic (d) hydrophobic. (e) Top ranked hit generated from combining fragments (f) All the site map features generated by SiteMap (g) RMSD plot of protein-ligand (A1) (h) stacked bar chart of hit A1-E6 interaction (i) A schematic of detailed ligand atom interactions of A1 with the protein (E6) residues.*

2.2 Computational fragment-based design and molecular docking

Fragment-based drug design finds its origin around two decades ago when "SAR by NMR" was introduced by Fesik et al at Abbott laboratories.[7] It is based on a basic concept postulated by Jencks around three decades ago.[8] He argued that the Gibbs free energy changes upon ligand (L) binding to a protein (P). It may be broken into intrinsic binding energy of protein (ΔG_P^O) ligand (ΔG_L^O) and a connection Gibbs energy (ΔG_S^O) related to the loss of rotational/translational entropy upon binding. This equation can be represented as follows:

$$\Delta G_{PL}^O = \Delta G_P^O + \Delta G_L^O + \Delta G_S^O$$

For the present study a golden fragment library from Enamine was downloaded. It consists of 1795 fragments which is a set of most representative and diverse small molecules available from Enamine HTS Collection and building blocks. This fragment library was first prepared using LigPrep module which in turn generated 4478 structures. The molecular parameters of this fragment library have been summarised in Table 2.

Table 2 *Molecular parameters of golden fragment library.*

Molecular parameters	Range
MW	140- 300
HA	0-3
HD	0-3
Rot bonds	0-3
Heavy atoms	10-21 (92.8% <19)
ClogP	-0.5-3
TPSA	≤ 100 Å²
Rings	1-4
Aromatic rings	≤ 2
Fused rings	≤ 3
N+O+S	1-6
Chiral centres	≤ 2

All these fragments were put for screening using molecular docking in high throughput virtual screening (HTVS) mode. The fragments were docked in HTVS (high throughput virtual screening) mode against the each of smaller receptor grids of four feature maps i.e. hydrophobic, hydrophilic, hydrogen-bond donor, hydrogen-bond acceptor as explained in sub-section 2.1. From each receptor grid the fragments with more than -6 dock score were selected and grouped together (Figure 1). This step lead to generation of 518 fragments. Then by using combine fragments tool of the Ligand preparation and library design module of Maestro, all these 518 fragments were combined to give a set of novel hit compounds. Total 95 hits were generated, and all these hits were put for ligand preparation and docked in HTVS mode against the receptor grid generated using all the amino acid residues present in the predicted active site by the SiteMap. Top 30 ranked compounds based on dock score were further selected for SP docking and finally top 10 compounds were selected for XP docking. The detailed results of best three hits have been compiled in Table 3. Ligand interaction diagram of best three hits have been illustrated in Figure 3.

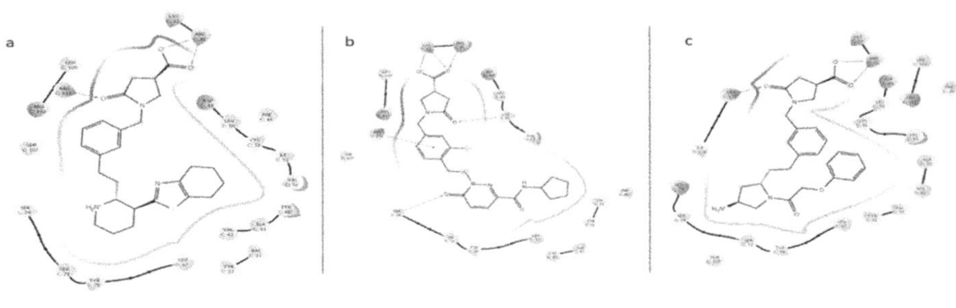

Figure 3 *2-D interaction diagram for top three hits (a) A1 (b) A2 (c) A3.*

2.3 Molecular dynamics simulations and ADMET analysis

Top ten hits based on dock score in XP mode were put for ADME analysis using Qikprop module of Maestro. There are several descriptors for which ADME analysis was done but here major parameters like molecular weight, lipophilicity (QPlogPo/w) cardiac toxicity (QPlogHERG) and violation of rule of five have been reported in Table 3. Then the best identified hit was put for molecular dynamics (MD) simulations study using Desmond module of Maestro. The simulation was carried out for 20 ns and a total of 1000 frames were generated after capturing a frame every 20 ps. MD simulations study showed that the complex between E6 protein and best identified hit (A1) was stable for the whole simulation period. Detailed results have been illustrated in Figure 1.

Table 3 *Dock score, h-bond interactions, free binding energy and ADMET analysis of best identified hits.*

Compound code	A1	A2	A3
Dock score	-8.33	-8.02	-7.90
H-bonding Interaction (amino acid residues)	CYS51, ARG131	ARG10, ARG131	ARG10, ARG131
MM-GBSA (Kcal/mol)	-65.658	-62.296	-58.782
Join Gscore	-11.16	-10.32	-10.27
Mol. Wt.	470.49	465.54	46762
QPlogHERG	-1.560	-2.389	-2.280
Rule of five	0	0	0
QPlogPo/w	2.752	-0.138	1.601
% Human oral absorption	65.026	28.146	54.716

3 CONCLUSION

In the present study, this is for the first time that the concept of fragment-based drug design was used to design novel E6 inhibitors against CC. Here in we are reporting some potential hits based on molecular docking and molecular dynamics simulations study. Compound A1 showed a dock score of -8.339 and retained the crucial H-bond interactions with TYR31, CYS51 amino acid residues as reported in earlier works. These compounds can be further taken up for synthesis and in vitro testing. This approach of fragment-based design also opens new avenues for designing drugs against other proteins for which no co-crystallized ligands have been reported.

References

1. S. Teissier, C. L. Pang and F. Thierry, *Oncogene*, 2010, **29**, 5061–5070.
2. M. Scheffner, J. M. Huibregtse, R. D. Vierstra and P. M. Howley, *Cell*, 1993, **75**, 495–505.

3. I. B. Weinstein, A. Joe and D. Felsher, *Cancer Res.*, 2008, **68**, 3077–3080.
4. A. Kumar, E. Rathi and S. G. Kini, *J. Mol. Struct.*, 2019, **1189**, 299–306.
5. A. Kumar, E. Rathi and S. G. Kini, *Struct. Chem.*, 2019, 1–13.
6. T. A. Halgren, *J. Chem. Inf. Model.*, 2009, **49**, 377–389.
7. G. D. Van Duyne, R. F. Standaert, P. A. Karplus, S. L. Schreiber and J. Clardy, *Science*, 1991, **252**, 839–842.
8. W. P. Jencks, *Proc. Natl. Acad. Sci. U. S. A.*, 1981, **78**, 4046–4050.

DESIGN AND SYNTHESIS OF NOVEL POLY ADP-RIBOSE POLYMERASE 1 (PARP1) INHIBITORS FOR THE TREATMENT OF SOLID TUMORS

J. Priyancy and D. P. Bhumika *

Department of Pharmaceutical Chemistry, Institute of Pharmacy, Nirma University, Ahmedabad 382481, Gujarat, India

Email: bhumika.patel@nirmauni.ac.in

1. INTRODUCTION

Poly ADP-Ribose Polymerase 1 (PARP1) is one of the promising target in cancer biology, specifically for BRCA mutated breast and ovarian cancer. It is a member of ADP-ribosyl transferases (ART) family of enzyme and does post-translational modification of the target protein. PARP1 uses NAD^+ as a substrate, cleave it catalytically and transfers the ADP-ribose moiety to acceptor proteins to create long chains of linear and/or branched poly (ADP-ribose) (PAR) as post-translational modification. This is called parylation of the target protein[1-5]. One of such target proteins is Histone. PARP1 gets up-regulated during DNA damage and it repairs the DNA through involvement of natural Homologous recombination (HR) pathway. Two genes, BRCA1 and BRCA2 are the key components of this pathway and so in cancer patients with BRCA-mutant genes, the HR pathway of DNA repair does not work. So, such patients can be treated with great potential and milder side effects using PARP1 inhibitors based on the synthetic lethality concept. At present, four PARP1 inhibitors; Olaparib, Niraparib, Rucaparib and Talazoparib are approved by USFDA and are in clinical use. All these marketed drugs equally inhibit PARP2 along with PARP1 with almost same potency. Also they all are NAD^+ analogues and so inhibit other metabolic pathways as well where NAD^+ is a natural substrate. These lead to toxicities like GI toxicity, myelodysplastic syndrome, acute myeloid leukemia etc. Overall, looking at the success rate of PARP1 inhibitors into various solid tumors, there is an urge for discovery of novel and selective PARP1 inhibitors[6-8].

2. METHODS AND RESULTS

2.1 Designing of Compounds using Various *In Silico* methods

With the aim of designing next generation PARP1 inhibitors, a ligand based drug design approach; 3D-QSAR study was performed on a series of 2, 3-Dihydrobenzofuran-7-carboxamide derivatives reported as PARP1 inhibitors using SybylX. A dataset of 31 molecules were divided into training set (23) and test set (8) in 75:25 ratio. Three different alignment methods were used and amongst them distill based alignment was found to be significant. Contour map analysis of best COMFA and COMSIA model suggested donor, acceptor, steric and hydrophobic favorable groups as important ligand features [9-11]. In parallel, 11 diversified molecules with similar PARP1 assay method were used to develop a ligand based pharmacophore model using DISCOTECH module, which was then refined by GASP in Sybyl X software[12,13]. The best model generated three types of pharmacophoric features namely, hydrogen bond donor (HBD), hydrogen bond acceptor (HBA) and one hydrophobic (HYD). The predictive power of pharmacophore model was then validated using GH score and ROC curve method. Finally, this model was screened against NCI library and total 6992 hits were obtained after applying various filters. Hits with higher Q_{fit} values compared to training set molecules were then subjected to scaffold hopping and 30 novel substituted benzoxazin-4-one derivatives were designed. All the compounds were subjected to molecular docking in the active site of PARP1 (PDB ID: 4PJT, BMN-673) and PARP2 (PDB ID: 4TVJ, Olaparib) using GOLD 5.2. The summary of the processes is shown in Figure 1.

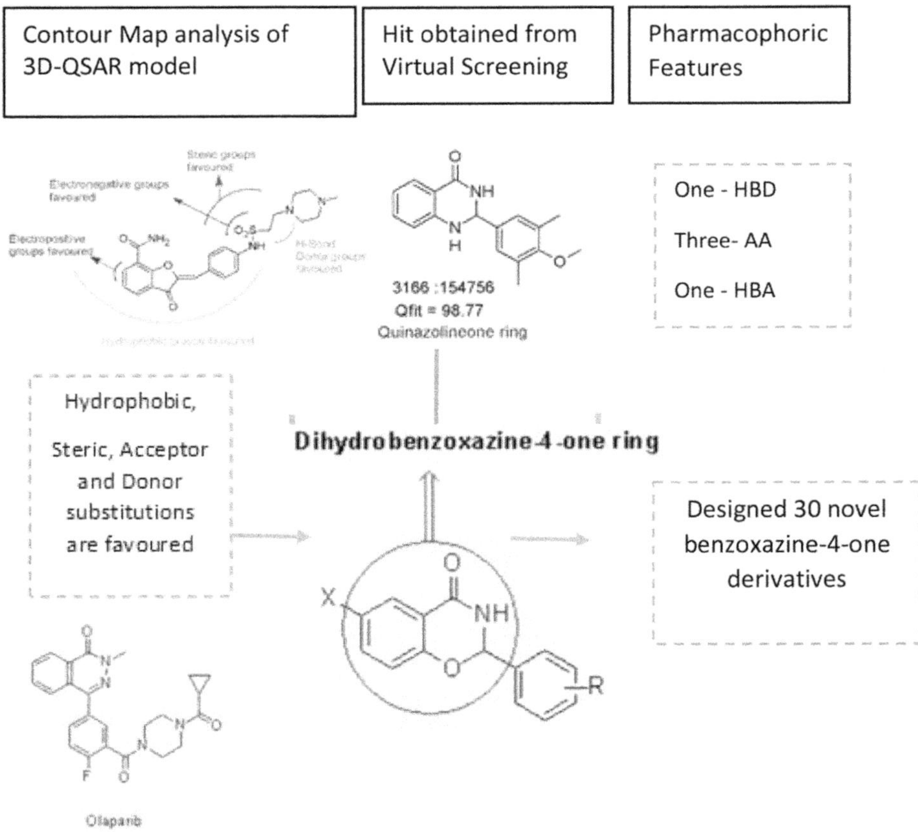

Figure 1 *Designing of Novel PARP1 Inhibitors.*

2.2. Synthetic Scheme of dihydrobenzo[e][1,3]oxazin-4-one derivatives [14,15]

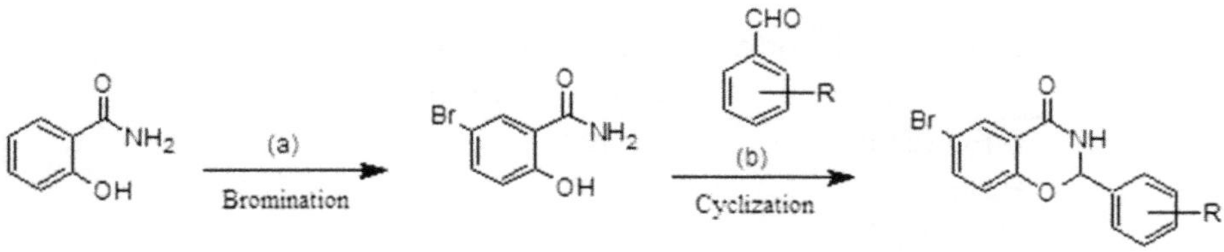

Scheme 1 *(a) NBS (1 eq.), ACN, RT, 0.5h (b) Polyphosphoric acid (PPA), 120°C, reflux, 22 h.*

Table 1 *Structure of dihydrobenzo[e][1,3]oxazin-4-one derivatives.*

Compound Code	R	M.W	Yield (%)
PJ-BD-01	-H	304.14	67
PJ-BD-02	3-NO_2	349.14	85
PJ-BD-03	2-NO_2	349.14	72
PJ-BD-04	4-NO_2	349.14	74
PJ-BD-05	4-CN	329.15	82
PJ-BD-06	3-F	322.13	94
PJ-BD-07	4-F	322.13	81
PJ-BD-08	4-Cl	338.58	90

2.3 In vitro MTT Assay[16]

All synthesized benzoxazinone derivatives were screened against MDA-MB-436 (BRCA mutant breast cancer cell line) at 100µM concentration. Out of 8 derivatives, compound PJ-BD-04 was found to be promising showing 51.241% inhibition as compared with standard 5-FU (100µM) displaying 59.593% as shown in Table 2. The obtained results needs experimental validation. So, further work is under progress.

Table 2 *In vitro* MTT Assay Results against MDA-MB-436 (BRCA mutant breast cancer cell line).

Compound Code	%Inhibition at 100μM Concentration
5-FU (Std)	59.593
PJ-BD-01	25.733
PJ-BD-02	50.383
PJ-BD-03	0.812
PJ-BD-04	51.241
PJ-BD-05	43.476
PJ-BD-06	47.313
PJ-BD-07	15.169
PJ-BD-08	27.945

3. CONCLUSION

PARP1 is an important molecular target for drug discovery. PARP1 inhibitors have shown their potential in treating BRCAm breast and ovarian cancer and also trials are going on for treatment of solid tumours like pancreatic, prostate, colorectal etc. Due to problems associated with existing PARP1 drugs, efforts were made to design and synthesize novel next generation PARP1 inhibitors using various *in-silico* methods like ligand and structure based approaches. Based on results obtained from computational studies, 30 novel hits were deigned and docked in crystal structure of PARP1 (PDB-4PJT) and PARP2 (PDB- 4TVJ). Highest GOLD score compounds, having dihydrobenzo-oxazinone as core moiety were chosen and then synthesized. Total eight derivatives were synthesized and checked for *in-vitro* biological screening against mutated breast cancer cell line (MDA-MB-436) using MTT assay. Amongst all, compound PJ-BD-04 was found to be potent showing 51.241% inhibition as compared with standard 5-FU displaying 59.593% inhibition at 100μM concentration. The obtained results needs experimental validation. So, further work is under progress.

4. ACKNOWLEDGEMENTS

Authors are thankful to Nirma University, Ahmedabad for providing financial assistance and necessary facilities to carry out the research work

References

1. P. A. Jeggo, *Curr. Biol.*, 1998, **8**, R49
2. M. Y. Kim, T. Zhang and W. L. Kraus, *Genes Dev.*, 2005, **19**, 1951
3. J. B. C. Papers, M. Doi, S. L. Oei and M. Ziegler, *J. Biol. Chem.*, 2000, **275**, 23234

4. H. Farmer, N. McCabe, C. J. Lord, A. N. J. Tutt, D. A. Johnson, T. B. Richardson, M. Santarosa, K. J. Dillon, I. Hickson, C. Knights, N. M. B. Martin, S. P. Jackson, G. C. M. Smith and A. Ashworth, *Nature*, 2005, **434**, 917.
5. F. Megnin-Chanet, M. A. Bollet and J. Hall, *Cell. Mol. Life Sci.*, 2010, **67**, 3649.
6. P. G. Jain and B. D. Patel, *Eur. J. Med. Chem.,* 2019, **165**, 198
7. D. Passeri, E. Camaioni, P. Liscio, P. Sabbatini, M. Ferri, A. Carotti, N. Giacch, R. Pellicciari, A. Gioiello and A. Macchiarulo, *ChemMedChem,* 2016, **11**, 1219.
8. M. S. Murahari and M. C. Yergeri, *Clin. Cancer Drugs*, 2015, **1**, 127.
9. B. D. Patel and M. D. Ghate, *Med. Chem. Res.*, 2015, **24**, 1060.
10. R. Ruslin, R. Amelia, Y. Yamin, S. Megantara, C. Wu and M. Arba, *J. Appl. Pharm. Sci.*, 2019, **9**, 89.
11. A. Ghaleb, A. Aouidate, A. Sbai, M. Bouachrine and T. Lakhlifi, *J. Struct. Chem.*, 2018, **59**, 1544.
12. H. Bhatt, P. Patel and C. Pannecouque. *Chem. Biol. Drug. Des.,* 2014, **83**, 154.
13. S. Shah, B. Patel and J. K. Savjani. *Mult. Scler. Relat. Dis.*, 2018, **21**, 35.
14. M. J. Mphahlele, M. M Maluleka, L. Rhyman, P. Ramasami and R. M. Mampa, *Molecules*, 2017, **22**, 83.
15. D. W. Hein, R. J. Alheim and J. J. Leavitt, *J. Am. Chem. Soc.,* 1957, **79**, 427.

IN SILICO EVALUATION OF THE ROLE OF PHYTOCHEMICALS IN ALZHEIMER'S DISEASE THROUGH WNT SIGNALING MODULATION

S. Manandhar,[a] S. P Kabekkodu[b] and K. S. R. Pai[a].

[a] Department of Pharmacology, Manipal College of Pharmaceutical Sciences, MAHE, Manipal, India

[b] Department of Cell and Molecular Biology, School of Life Sciences, MAHE, Manipal, India

1. INTRODUCTION

Alzheimer's disease (AD) is an irreversible, progressive and the most common reason for neurodegenerative disease. Wnt signaling plays important role during embryonic CNS development and is involved in neuronal induction, self-renewal, polarity, and synaptogenesis.[1] Alteration in the Wnt Signaling pathway has been linked with age-related diseases like osteoporosis, colon cancer, Parkinson and AD.[2] Several studies indicate that disruption of Wnt signaling particularly, canonical Wnt pathway contributes to AD pathogenesis.

This study aims to evaluate the role of phytochemical in modulating Wnt signaling pathway by targeting the negative regulators of Wnt signaling. WIF1 (Wnt inhibitory factor 1) is the protein that binds to Wnt proteins making them unavailable and hence inhibiting the Wnt signaling.[3] Low density lipoprotein receptor related protein 6 (Lrp6) is the co-receptor that along with Frizzled receptor is involved in the activation of Wnt signaling. Dickkopf 1(Dkk1) is the antagonist of Wnt signaling pathway that act by binding with LRP6 and leading to its degradation. [4]Glycogen synthase kinase-3β (GSK-3 β) forms multimeric degradation complex that results in the degradation of β-catenin and inhibition of Wnt signaling pathway.[5] All these four proteins are important targets for Wnt signaling inhibition that serves the purpose for modulating Wnt pathway.

2. METHOD AND RESULTS

2.1. Selection of the protein structures and Protein preparation

Phytochemicals database were selected for *in silico* evaluation of their potential to inhibit the known inhibitors of Canonical Wnt signaling pathway like Wif1, Dkk1, GSK-3β, LRP6 using Maestro v11.7. Acetylcholine esterase (AChE) enzyme was taken as control. The protein crystal structures 2YGO (WIF1), 1Q5K (GSK-3β), 4M0F (AChE), 3S2K (DKK1 and LRP6) were selected based on the literature and Ramachandran plot. The selected proteins were obtained from Protein Data Bank and prepared using Protein preparation wizard.

2.2. Docking of the phytochemical library from zinc database

The phytochemical library containing 48164 was obtained from the Biogenic subset section of ZINC database. The ligands were prepared using Ligprep tool at neutral pH for energy optimization. For protein GSK-3β, AChE grid was generated based on the bound inhibitor whereas the grid for other proteins was generated based on active druggable sites obtained from the site map. XP docking was performed for all the proteins. The docking study was followed by the MMGBSA calculation for Gibb's free energy of the ligand-protein complex.

The docking study of the several phytochemicals on five PDB targets confirmed that Mangiferin (Figure 1) showed good docking score and binding energy among all the molecules. Among all the PDB structures, Mangiferin showed best docking score which is shown in Table 1. Acetylcholine (4M0f) has maximum docking score of -14.205 followed by WIF1, Dkk1, Lrp6 and GSK3β has lowest docking score of -10.344. Reported *in vivo* studies shows the potential of Mangiferin for the neuroprotective activity through anti-cholinesterase and antioxidant property.[6,7] This study elucidated the possible role of Canonical Wnt Signaling Pathway as an additional mechanism for the activity of Mangiferin in Alzheimer's disease.

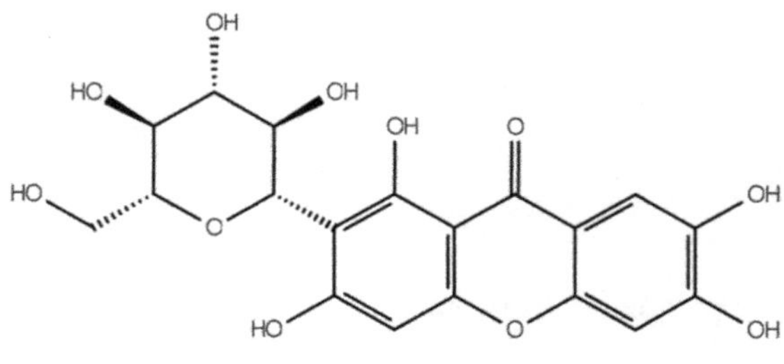

Figure 1 *Structure of Mangiferin.*

Table 1 *Docking score, binding energy, binding residues and different interactions shown by Mangiferin.*

Targets	*XP score*	*MMGBSA dG Bind*	*Interaction diagram*
Wif1(2YGO)	-13.546	-53.75	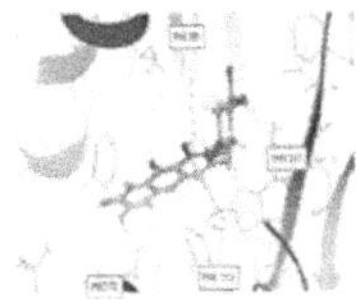
AchE(4M0F)	-14.205	-52.65	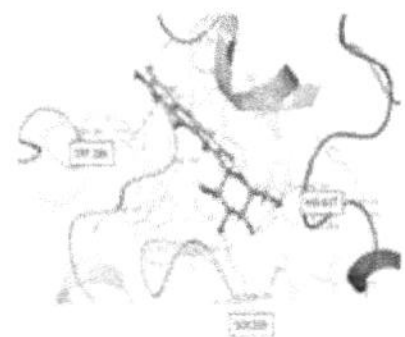
GSK3β(1Q5K)	-10.344	-49.49	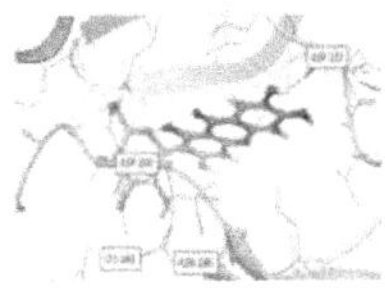
Dkk1(3S2K)	-11.155	-48.53	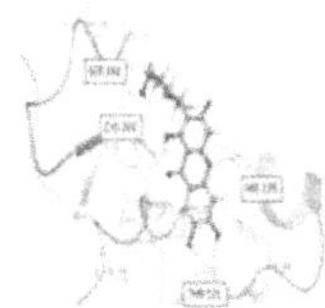
Lrp6(3S2K)	-10.694	-15.38	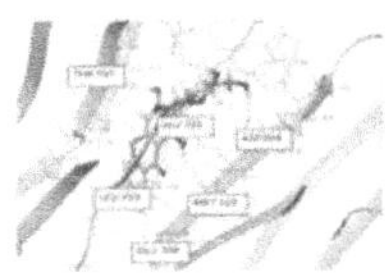

3. CONCLUSION

In silico docking study identifies Mangiferin as an agent having a potential to target Canonical signaling molecules. Also it has good docking score for AChE enzyme, showing its potential to inhibit Acetylcholine esterase enzyme, key target for Alzheimer's disease. This opens the possibility of Mangiferin to have activity for Alzheimer's disease mediated through Canonical Signaling pathway. This can be further proven by *in vitro* and *in vivo* studies.

References

1. A. Caricasole, A. Copani, A. Caruso, F. Caraci, L. Iacovelli, M. A. Sortino, G. C. Terstappen and F. Nicoletti, *Trends Pharmacol. Sci.*, 2003, **24**, 233–238.
2. C. Tapia-Rojas and N. Inestrosa, *Neural Regener. Res.*, 2018, **13**, 1705.
3. C. Surmann-Schmitt, N. Widmann, U. Dietz, B. Saeger, N. Eitzinger, Y. Nakamura, M. Rattel, R. Latham, C. Hartmann, H. von der Mark, G. Schett, K. von der Mark and M. Stock, *J. Cell Sci.*, 2009, **122**, 3627–3637.
4. M. V. Semënov, X. Zhang and X. He, *J. Biol. Chem.*, 2008, **283**, 21427–21432.
5. K. Leroy, Z. Yilmaz and J.-P. Brion, *Neuropathol. Appl. Neurobiol.*, DOI:10.1111/j.1365-2990.2006.00795.x.
6. S. M. Biradar, H. Joshi and T. K. Chheda, *Eur. J. Pharmacol.*, 2012, **683**, 140–147.
7. S.-T. Feng, Z.-Z. Wang, Y.-H. Yuan, H.-M. Sun, N.-H. Chen and Y. Zhang, *Pharmacol. Res.*, 2019, **146**, 104336.

IDENTIFICATION OF POTENTIAL VEGFR2 INHIBITORS EMPLOYING E-PHARMACOPHORE MODEL, VIRTUAL SCREENING, MOLECULAR DYNAMIC SIMULATION AND ADME ANALYSIS

E. Rathi, A. Kumar and S. G. Kini

Department of Pharmaceutical Chemistry, Manipal College of Pharmaceutical Sciences, MAHE, Manipal, Karnataka, India

1 INTRODUCTION

Angiogenesis is an epochal phenomenon which can be targeted to restrain the growth and metastasis of tumor cells. How to target the angiogenesis? Though Angiogenesis is regulated by multiple molecular mediators, VEGF (vascular endothelial growth factor) is a key mediator of angiogenesis. Hence, targeting of VEGF mediated signal transduction pathway is a pivotal way to supress angiogenesis in tumor cells. Next question is how to target this signal transduction pathway and what are the promising targets to inhibit this process? Before we will address this question, it is essential to understand that VEGF consist of VEGF-A, VEGF-B, VEGF-C, VEGF-D and VEGF-E. Out of all, VEGF-A is a key factor to regulate the angiogenesis phenomenon. However, VEGF-A mediates its biological function through VEGFR-2 (Vascular endothelial growth factor receptor). Hence, VEGFR-2 is the most promising target to supress angiogenesis in tumor cells [1-3].

Till date, Pazopanib, Sunitinib, Sorafenib, Axitinib, Regorafenib, Cabozantinib, Vandetanib and Lenvatinib are the small molecule FDA-approved drugs as VEGFR inhibitors. Still, intensive research is progressing in the field of VEGF-A mediated VEGFR2 signal transduction pathway. Why intensive research is going on in this field? Though FDA-approved drugs are potent towards VEGFR2 target, the major challenge of these drugs are resistance and serious side effects like hepatoxicity, haemorrhage, gastro-intestinal toxicity, proteinuria etc. Hence, it is crucial to design a drug with better potency and less toxicity [4].

In order to rationally design the potent molecule for VEGFR2 target, thorough knowledge of protein structure of VEGFR targets which classified under the tyrosine kinase receptor is essential. Findings of literature survey informed that tyrosine kinase protein inhibitors are classified based on the conformation of DFG motif. Inhibitors which will bind to the DFG-in conformations are known as Type I inhibitors whereas which will bind to the DFG-out conformation are known as Type II inhibitors. Type II inhibitors are more selective and have more chemical space to exploit. Hence, our main endeavour of this study is to design potent type II VEGFR2 inhibitors. How to define that the designed molecule is potent toward this target? The design molecules should retain the interactions with crucial amino acids like CYS919, ASP1046 and GLU885. Further, the molecules should have terminal aryl moiety to occupy the allosteric binding site to design as Type II inhibitors along with these crucial interactions [5-8].

Hence, in our study we have developed a five-feature receptor-ligand complex based e-pharmacophore model. For this study, 4ASD PDB file was retrieved from RCSB protein data bank which is bounded with potent FDA-approved drug, Sorafenib. Therefore, this model will help to design the potent compounds retaining all the crucial interactions.

2 METHODS AND RESULTS

2.1 Receptor-ligand complex based e-pharmacophore model and Virtual Screening

A commercial software Maestro 11.8 (Schrodinger Inc.), on a hpcomputer with a Linux Ubuntu 18.04.1 LTS as the operating system, Intel Haswell graphics card, 8GB RAM and Intel Core i3 processor was employed to carry out this study. For this study, X-ray crystal structure of VEGFR2 protein (PDB Code: 4ASD) was retrieved from RCSB protein data bank. Further, Protein preparation wizard tool was used to prepare the retrieved protein 4ASD. The minimized protein is bounded with the potent FDA-approved drug (sorafenib), therefore e-pharmacophore model was built by receptor- ligand complex employing develop pharmacophore hypothesis of PHASE module. In this study a five featured e-pharmacophore model (AADRR) (Figure 1) was built around the vicinity of ligand to virtually screen the compounds as VEGFR2 inhibitor.

Further, Pharmacophore-based virtual screening was performed against focussed library of Kinase inhibitors retrieved from the Asinex database. Retrieved compounds were prepared using ligprep panel of Maestro 11.8. The prepared ligands were screened against the e-pharmacophore model and the compounds mapping to the four chemical feature and had a fitness score >1.6 were selected for structure based virtual screening.

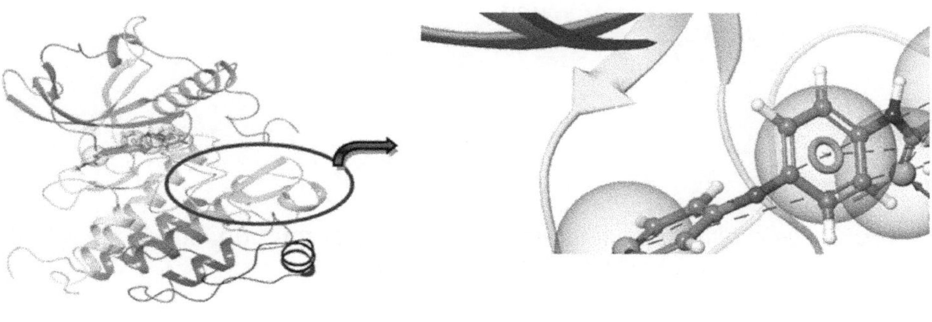

Figure 1 *Five featured e-pharmacophore model (AADRR), built around the vicinity of bounded ligand of 4ASD protein.*

2.2. Structure-based Virtual screening

Structure-based virtual screening was accessed to identify the binding affinity of selected lead compounds in terms of Dock score by docking them against the co-crystallized ligand of 4ASD. During this approach, 2000 molecules were subjected to the HTVS (High-throughput virtual screening) docking mode against the receptor-grid of 4ASD generated using Receptor Grid Generation panel. Further, 120 compounds having docking score greater than -10 were subjected to SP (Standard precision) docking mode of GLIDE panel of Maestro 11.8. Among 120, 50 molecules having docking score >-11 were further analysed by the more extensive XP (Extra precision) mode of GLIDE module to obtain the more reliable results by reducing the possibilities of false positives. Out of all, top four hits were chosen based on docking score and interactions. It was observed that all the hits are having good docking score and retaining interaction with CYS 919 and ASP1046 amino acids (Table 1 and Figure 2) except with the GLU885. Hence, to retain the interaction with GLU885, two molecules (D1 and D2) were designed from the top first hit (A1) by introducing the amino group. The introduction of amino group in D1 and D2 molecules enhanced the docking score but only D2 molecule retained the interaction with GLU885 amino acid residue (Figure 2). To further access the binding affinity of designed hit D1 and D2, binding free energy (Δ G) was calculated employing Prime MM-GBSA method (Table 1).

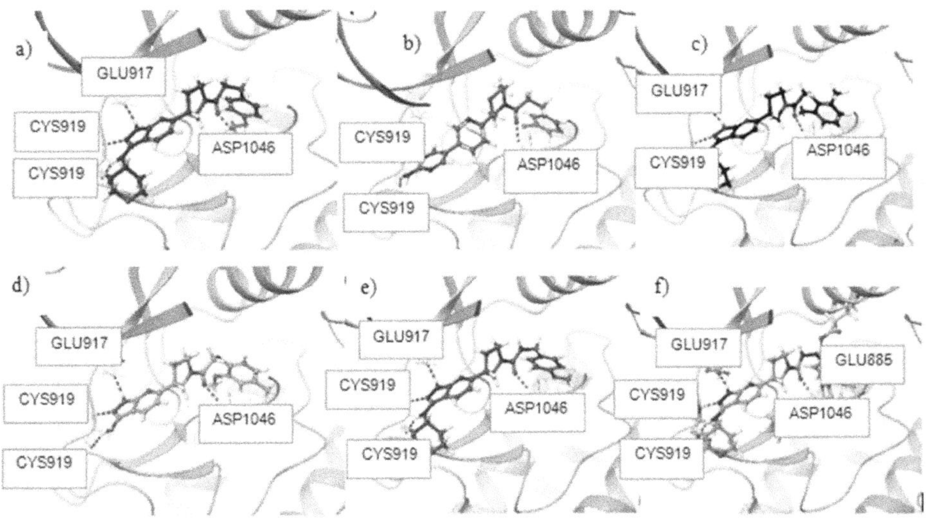

Figure 2 *Three-dimensional interaction diagram of four top hits A1(a), A2(b), A3(c), A4(d) and designed hits D1(e) and D2(f).*

Top four hits identified by molecular docking study were further subjected to the ADME analysis by QikProp Panel of Maestro 11.8. The various descriptors were analysed by QikProp tool and the most promising properties have been reported in Table 2. Further, the binding stability of ligand-protein complexes of top hit (A1) and the designed hits (D1 and D2) were assessed by Molecular Dynamic (MD) simulation study. The simulation study of these hits was performed employing Desmond panel of Schrodinger. The simulation was carried out for 20ns and a frame were captured after every 20 ps for 500 frames. MD simulation study inferred that the complex between VEGFR2 protein and hits (A1, D1 and D2) were stable throughout simulation period.

Table 2 *Dock score, h-bond interactions, free binding energy and ADMET analysis of best identified hit A1.*

Compound code	Dock score	H-bonding Interaction (amino acid residues)	MM-GBSA (Kcal/mol)	Mol. Wt.	QPlogHERG	Rule of five	QPlogPo/w	% Human oral absorption
A1	-13.765	CYS919, ASP1046, GLU917	-58.41	449.552	-4.649	0	3.546	100
A2	-13.108	CYS919, ASP1046	-49.41	423.901	-5.218	0	2.856	87.373
A3	-12.512	CYS919, ASP1046, GLU917	-52.25	449.552	-5.468	0	3.580	100
A4	-12.466	CYS919, ASP1046, GLU917	-47.42	381.433	-4.387	0	2.142	81.578
D1	-14.098	CYS919, ASP1046, GLU917	-59.46	450.539	-4.754	0	3.640	96.571
D2	-14.030	CYS919, ASP1046, GLU917, GLU885	-56.05	451.527	-4.552	0	2.856	88.267

3. CONCLUSION

Receptor-ligand complex based e-pharmacophore model were built for the first time in this study to design the potent compounds as VEGFR2 inhibitors. During this study, we have identified the potential hit A1 (Docking score = -13.765) based on the docking score and MM-GBSA. Further, two molecules (D1 and D2) were designed from the potent A1 hit which had shown more potency towards the VEGFR2 target in comparison of A1 hit. Further, analysis of binding affinity of designed molecules and A1 hit by MD simulation study represented that these compounds can be taken up further for the lead optimization.

References

1. N. Ferrara, *Oncologist*, 2004, **9**, 2–10.
2. H. F. Dvorak, *J. Clin. Oncol.*, 2002, **20**, 4368–4380.
3. M. Terme, S. Pernot, E. Marcheteau, F. Sandoval, N. Benhamouda, O. Colussi, O. Dubreuil, A. F. Carpentier, E. Tartour and J. Taieb, *Cancer Res.*, 2013, **73**, 539–549.
4. P. Wu, T. E. Nielsen and M. H. Clausen, *Trends Pharmacol Sci.*, 2015, **36**, 1–18.
5. A. Morabito, E. De Maio, D. Maio, N. Normanno, Oncologist, 2006, **11**, 753–764.
6. Y. H. Peng, H. Y. Shiao, C. H. Tu, P. M. Liu, J. T. Hsu, P. K. Amancha, J. S. Wu, M. S. Coumar, C. H. Chen, S.Y. Wang and W. H. Lin, *J. Med. Chem.*, 2013, **56**, 3889–3903.
7. M. Okaniwa, T. Imada, T. Ohashi, T. Miyazaki, T. Arita, M. Yabuki, A. Sumita, S. Tsutsumi, K. Higashikawa, T. Takagi and T. Kawamoto, *Bioorg. Med. Chem.*, 2012, **20**, 4680– 4692.
8. Y. Liu and N. S. Gray, *Nat. Chem. Biol.*, 2006, **2**, 358–364.

DEVELOPMENT OF NEW NEPRILYSIN INHIBITOR AS A MODULATOR OF CHRONIC KIDNEY AND HEART DISEASE USING *IN SILICO* DRUG REPURPOSING APPROACH

R. Sankhe[a], A. Kumar[b], E. Rathi[b], A. Kishore[a]

[a] Department of Pharmacology, Manipal College of Pharmaceutical Sciences, MAHE, Manipal, Karnataka, India.

[b] Department of Pharmaceutical Chemistry, Manipal College of Pharmaceutical Sciences, MAHE, Manipal, Karnataka, India.

1 INTRODUCTION

The pathophysiology of chronic heart and kidney disease is interlinked and associated with different signalling pathways. The global burden of chronic kidney and heart diseases are increasing rapidly due to different interlinked factors. An increased risk of kidney diseases was observed in heart failure patients. Similarly, chronic kidney disease-induced a disturbance in homeostasis may directly affect cardiovascular disease by increasing blood pressure. These similarities in the pathogenesis of chronic kidney and heart disease, have led to the hypothesis that treatments proven for heart failure can be used as a treatment for kidney diseases. Recently in 2015, US FDA approved Sacubitril/Valsartan, the combination of Neprilysin (NEP) inhibitor (Sacubitril) and angiotensin receptor blocker (ARB) (Valsartan) for the treatment of heart failure with reduced ejection fraction. Further, in 2017, the combination of NEP inhibitor with ARB was also found to be more effective in renal failure patient as compared to ARB alone[1].

In human, NEP is an endogenous peptide known for modulating degradation and production of various endogenous peptides such as natriuretic peptides, amyloid-β, endothelin-I and II, angiotensin (1-7), substance P, bradykinin etc... The highest concentration of NEP is present in the kidney and regulates the intrinsic homeostatic mechanism of the kidney and heart. The peptide degrading property of NEP is found to be involved in the pathophysiology of various diseases such as renal and heart disease. In the context of increased prevalence of interlinked diseases such as heart failure and chronic kidney disease, the new NEP inhibitor could be an attractive target for treating kidney and heart diseases[2]. NEP inhibition mediated, upregulation of natriuretic peptide showed impressive results in cardiovascular and renal disease alone and in combination with ARB.

Now a days, sacubitril, thiorphan and phosphoramidon are used as NEP inhibitors. But various drawbacks are associated with the present treatment strategies. Similarly, available treatment options for NEP inhibition are also limited. Hence drug repurposing for NEP inhibition using existing FDA approved drug can be an attractive strategy to overcome this problem. Use of *in-silico* tools like molecular docking can aid in speeding up the process of drug repurposing for the development of new treatment options for heart and renal diseases. Hence in the present study molecular docking of FDA approved drugs was performed with NEP, to find out new NEP inhibitors.

2 METHODS AND RESULTS

Molecular docking study was performed by maestro molecular modelling platform (Version 11.9) by Schrödinger. The structure of human substrate-free NEP was retrieved from the protein data bank (PDB ID: 6GID). The catalytically essential triad of HIS587, HIS583 and ZN805 was present in selected protein structure. Similarly, the structure of FDA approved drugs were downloaded from Zinc 12 database. The protein preparation module was used for protein preparation. During this process, using the prime tool missing back chains and side chains were included. Further water molecules occupying protein structure were removed. Finally, optimization and minimization processes were done. In the selected protein structure, the ligand was not present. Hence to detect the drug-able site of NEP, sitemap tool was used. The site score values were used to rank sites (Table 1). The receptor cavity was generated, as per the previously described set of amino acids[3]. In the sitemap module, the identification of the potential binding site of a particular protein was done on the basis of site-points which are linked together.

Table 1 Selected active site based on site score and residues involved using SiteMap tool of Maestro

Predicted site	Site score	D score	Volume(Å^2)	Residue number (Chain A)
Site 1	1.020	0.998	1034.488	102, 104, 105, 106, 107, 110, 201, 203, 206, 208, 217, 220, 221, 222, 223, 225, 226, 227, 228, 229, 231, 232, 280, 283, 284, 286, 287, 288, 289, 290, 292, 293, 295, 298, 346, 529, 533, 535, 542, 543, 544, 545, 546, 547, 548, 549, 558, 579, 580, 583, 584, 587, 590, 591, 592, 593, 596, 636, 637, 639, 641, 642, 643, 644, 645, 646, 692, 693, 694, 695, 696, 697, 706, 707, 708, 709, 710, 711, 712, 713, 714, 715, 717, 805, 1289

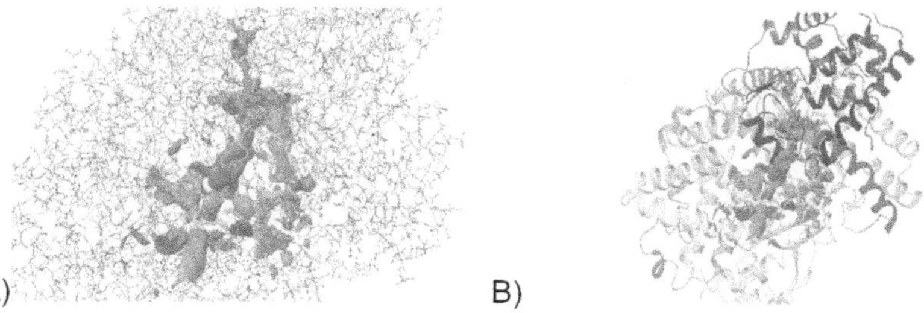

A) B)

Figure 1 A) Selected active site of NEP B) Ribbon plot of the selected active site of NEP

The LigPrep module was used to optimise all the FDA approved drugs, at pH 7.4 under OPLS3e force field. To determine protein-ligand interaction, all the FDA approved drugs were docked in the active site of NEP in high throughput virtual screening (HTVS) mode. These were reanalysed by standard precision (SP) and extra precision (XP). The lowest glide score and best docked pose of each ligand were recorded.

The molecular mechanics/ generalised Born and surface area (MM/GBSA) method from the prime module, was used to estimate binding free energies of a protein-ligand complex of FDA approved drugs from Zinc 12 data base. The MM/GBSA method is the most commonly used method for the estimation of the free binding energy of small ligand to biological macromolecules using MMGBSA approach. The free binding energy was calculated for all XP docked ligands. The selected FDA approved drugs such as ZINC000100071256 and ZINC000001530948 showed negative binding energy score which indicates that stronger and stable formation of the ligand-protein complex. The binding energy of selected FDA approved drugs is described in Table 2.

Based on the docking score and estimated binding free energy, we were able to find out 2 inhibitors ZINC000100071256 and ZINC000001530948. Both of the drugs showed interaction with catalytically important residues of NEP such as HIS587, HIS583 and ZN805 as well as the other residues important of NEP inhibition as demonstrated in Figure 2 and Table 2.The interactions of these selected FDA approved drugs in catalytic pocket of NEP seems to result in stable transition state of Zn^{2+} leads to decrease catalytic activity of NEP as per previous report[4]. Both selected compounds are showing polar interactions with HIS711, which has been shown to be involved in the stabilization of the transition state of NEP[5]. Previous studies reported that GLU584 is also contributing to the catalytic activity of NEP[6]. In our study, both of the selected drugs showed interaction with amino acid residue GLU584. Similarly, finding also suggests that ALA543 and ASN542 are catalytically important residues for NEP inhibitor [7]. In our study, ZINC00000153094 and ZINC000100071256 also showed hydrophobic, hydrogen bond and polar interaction with ALA543 and ASN542.

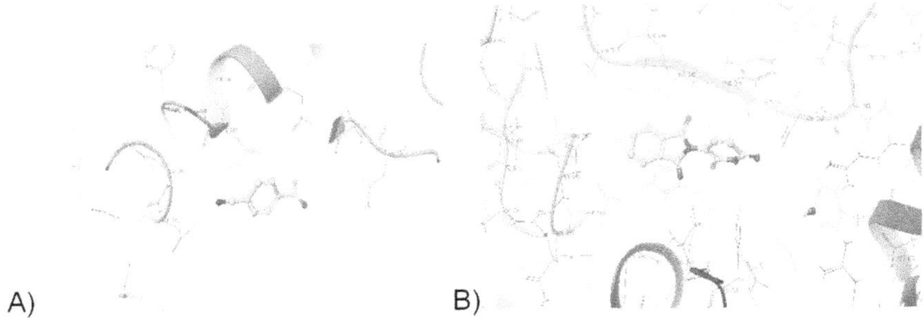

Figure 2 The ligand interaction diagram of the selected drugs based on molecular docking and binding free energy A) Ligand interaction diagram of ZINC000100071256 with NEP B) Ligand interaction diagram of ZINC000001530948 with NEP

Table 2 Docking score, Binding energy and ligand interactions of selected drugs

Compound ID	Docking Score	Binding energy	Hydrophobic interactions	Other interactions
ZINC000100071256	-4.982	-9.80	TYR545, PHE 544, ALA543	Polar Interactions: HIS711, HIS583, HIS587, SER547, ASN592 Salt bridge and metal co-ordination: ZN805 Charged Negative: GLU646, GLU584
ZINC000001530948	-6.631	-15.01	TYR545, ALA543, VAL710	Polar Interactions: HIS711, ASN592, HIS583, HIS587, ASN542 Salt bridge and metal co-ordination: ZN805 Charged Negative: GLU646, ASP590, GLU584

3 CONCLUSION

In the present study, the concept of drug repurposing of FDA approved drugs were used first time to develop NEP inhibitors against the pathogenesis of kidney and heart disease. In this study, we have identified some potential NEP inhibitors based on the molecular docking study. Selected drugs, ZINC000100071256 and ZINC000001530948 showed a docking score of -4.982 and -6.631 and retained the crucial H-bond interactions with amino acid residues such as GLU584, ALA543, ASN542, HIS587, HIS583 and ZN805 as reported in the earlier works. In future, the effect of the selected drug should be tested in *in-vitro* and *in-vivo* system of kidney and heart diseases. The identification of a new NEP inhibitor also opens new avenues to increase the treatment options in kidney and heart disease by modulating NEP levels.

References

1. A. Bayes-Genis, J. Barallat and A. M. Richards, *J. Am. Coll. Cardiol.,* 2016, **68**, 639–653.
2. V. Malek and A. B. Gaikwad, *Biomed. Pharmacother.,* 2017, **90**, 752–759.
3. N. Schiering, A. D’Arcy, F. Villard, P. Ramage, C. Logel, F. Cumin, G. M. Ksander, C. Wiesmann, R. G. Karki and M. Mogi, *Sci. Rep.,* 2016, **6**, 27909.
4. C. Oefner, A. D’Arcy, M. Hennig, F. K. Winkler and G. E. Dale, *J. Mol. Biol.,* 2000, **296**, 341–349.
5. C. Marie-Claire, E. Ruffet, G. Tiraboschi and M.-C. Fournie-Zaluski, *FEBS Lett.*, 1998, **438**, 215–219.
6. A. J. Barrett, J. F. Woessner and N. D. Rawlings, *Handbook of Proteolytic Enzymes*, Elsevier, 2012.
7. S. Voisin, D. Rognan, C. Gros and T. Ouimet, *J. Biol. Chem.,* 2004, **279**, 46172–46181.

ANTICANCER POTENTIAL OF PHYTOCONSTITUENTS MODULATING NA/ K ATPASE PUMP; A NOVEL REPURPOSING STRATEGY

S. Jubie,[a] M. Suresh Kumar[b] and Niladri Saha[a]

[a]Department of Pharmaceutical Chemistry JSS College of Pharmacy, JSS Academy of Higher Education & Research, India; [b]Tifac core.Research in herbal drugs, JSS College of Pharmacy, JSS Academy of Higher Education & Research, India

1. INTRODUCTION

Na/K ATPase is a heteromeric protein complex located on the plasma membrane of eukaryotic cells that uses ATP to regulate sodium and potassium transport. It is composed of three subunits, α-subunit, β-subunit and γ-subunit. The α-subunit has four isoforms (α1, α2, α3 and α4). Report says that the α1 sub unit is up regulated in certain cancer types such as renal cell carcinoma, glioma and melanoma[1]. It is well evidenced that there is an elevation of activity observed in malignant cells. Drugs targeting Na/K ATPase induces apoptosis and autophagy in cancer cells. Ouabain, a well-known cardenolide, inhibitor of Na/K ATPase, has also been reported for anticancer potential in neuroblastoma[1]. As most of non-cancer drugs have little or tolerable side effects in human, repositioning of non-cancer drugs for anticancer therapy is an excellent strategy for future anticancer drug development[2]. Based on these facts, the present study was aimed to identify novel inhibitors of Na/K ATPase pump, assuming that the identified inhibitors would have anticancer potential through the inhibition of Na/K ATPase pump.

2 METHOD AND RESULTS

2.1. Selection of phytoconstituents and target

A library of forty three phyto terpenes and phyto tannins were selected for the study. The α-subunit of NA/ K ATPASE with ouabain (PDB id: 3A3Y) was chosen for our study.

2.2. *In-silico* ADMET profile

In-silico Absorption, Distribution, Metabolism, Excretion and Toxicity of all the forty three compounds had been performed by using QikProp (v5.3) module of Schrodinger. It was observed that all the forty three compounds were predicted to be safe and having drug like properties.

2.3. Molecular docking protocol

The molecular docking studies were carried out by using Glide v7.6 integrated Maestro v11.3 and binding free energy calculation were carried out by using

Prime/MM-GBSA approach (v4.9, Schrödinger 2017-3)[3]. The designed molecules were docked along with the native ligand Ouabain which served as reference standard also. G- score was used to assess the binding affinities of the studied ligands to the target receptor. The G score of Ouabain is -7 and it was kept as cut off point. The fourteen compounds which had G-scores above 7 (Table 1) were selected for further binding energy studies. The key amino acid residue Arginine 270 was found to be very important for the interaction and it was shown to interact with all the fourteen compounds. Along with arginine, the other residues such as Phenyl alanine 150, Isoleucine 264, Leucine 226 (were also in alignment with the standard ouabain.

The compounds 2,3-dihyroxy-5-carboxyphenoxyacetic acid, 2,3,4-trihydroxy benzylidine anthranilic acid and 3-benzoyl gallic acid were top scored compounds. We also carried out the binding free energy of top fourteen compounds by MM-GBSA approach to get a better estimate of the binding strengths and relative predicted potencies and the results are depicted in Table 2.

Table 1 *G- Scores from the glide XP docking run of phyto constituents in the active site of 3A3Y.*

S. No	Name of the Phytoconstituent	G Scores
1.	n-butyl gallate	-7.61
2.	2,3-dihyroxy-5-carboxyphenoxyacetic acid	-10.11
3.	Methyl 2,3-dihydroxy-5-carbomethoxyphenoxyacetate	-7.41
4.	3,4,5-trihydroxybenzoyl-glycolamide	-7.42
5.	3-benzoyl gallic acid	-9.11
6.	3,5-dibenzoyl gallic acid	-8.54
7.	2,3-dihyroxyphenoxyacetic acid	-7.32
8.	Methyl 2,3-dihyroxyphenoxyacetate	-7.10
9.	2,3,4-trihydroxybenzylidine-2'-methyl-5'-sulfo- aniline	-8.67
10.	2,3,4-trihydroxybenzylidine-sulfanilic acid	-8.57
11.	2,3,4-trihydroxybenzylidine anthranilic acid	-9.39
12.	[6-(ethoxymethyl)-tetrahydro-2H-pyran-2,3,4,5-tetraol with 1-isopropyl-4-methoxybenzene	-8.43
13.	Isopropyl gallate	-7.15
14.	Gallic acid	-7.46
15.	Oubain	-7.36

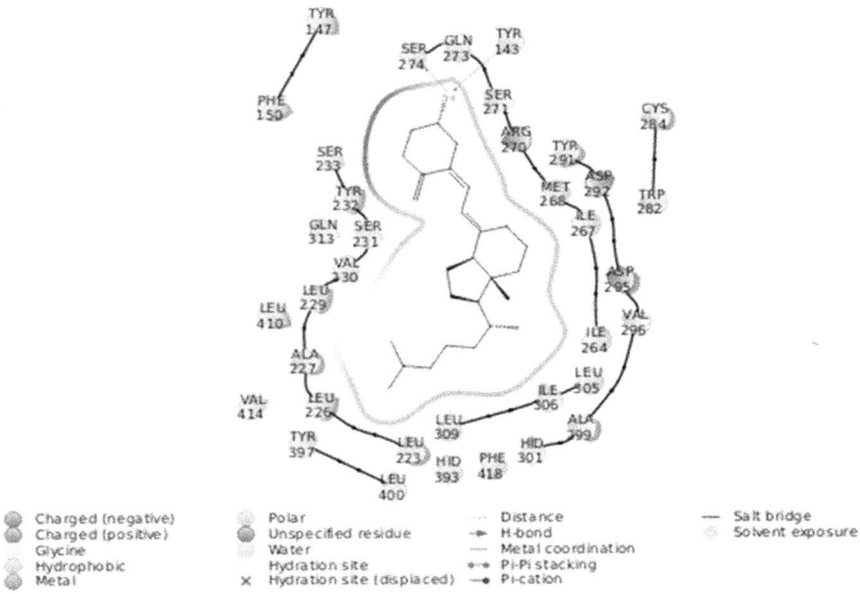

Figure 1 *2D ligand interactions ouabain with 3A3Y.*

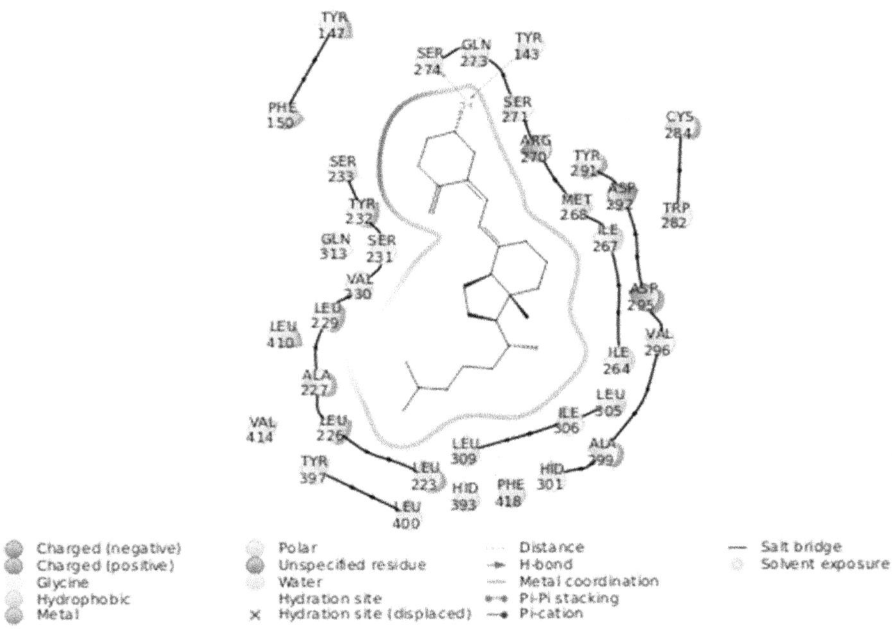

Figure 2 *2D ligand interactions of phyto constituent with 3A3Y.*

Table 2 *Binding free energy calculation using Prime/MM-GBSA approach.*

S. No	Name of the Phytoconstituent	Binding free energy (kcal/mol)
1.	n-butyl gallate	10.74
2.	2,3-dihyroxy-5carboxyphenoxyacetic acid	14.48
3.	Methyl 2,3-dihydroxy-5-carbomethoxyphenoxyacetate	32.64
4.	3,4,5-trihydroxybenzoyl-glycolamide	29.67
5.	3-benzoyl-gallic acid	15.26
6.	3,5-dibenzoyl-gallic acid	53.35
7.	2,3-dihyroxyphenoxyacetic acid	20.77
8.	Methyl 2,3-dihyroxyphenoxyacetate	15.57
9.	2,3,4-trihydroxybenzylidine-2'-methyl-5'- sulfo-aniline	33.90
10.	2,3,4-trihydroxybenzylidine-sulfanilic acid	10.97
11.	2,3,4-trihydroxybenzylidine-anthranilic acid	20.47
12.	[6-(ethoxymethyl)-tetrahydro-2H-pyran-2,3,4,5-tetraol with 1-isopropyl-4- methoxybenzene	35.16
13.	Isopropyl gallate	35.26
14.	Gallic acid	39.56
15.	Ouabain	41.98

4. CONCLUSION

Based on docking study and MM-GBSA calculations, the two compounds namely 3-benzoyl-gallic acid and 2,3,4-trihydroxybenzylidine-sulfanilic acid were selected. Further *in vitro* screening against cancer cell lines and enzyme inhibition studies are in progress in our laboratory. The identified compounds, may be repositioned for cancer therapy through the inhibition of Na K ATPase after thorough preclinical and clinical studies.

References

1. R. A. Karasneh, L. J. Murray and C. R. Cardwel, *Int. J. Cancer*, 2017, **140**, 1035–1041.
2. X. Hanqing, L. Jie, X. Haozhe and W. Yadong, *Int J Biol Sci.*, 2018, **14**(10), 1232–1244.
3. H. Beard, A. Choolleti, D. Pearlman, W. Sherman and K. A. Loving, *Plos One,* 2013, **8**(12), 1–11

IN SILICO HIT IDENTIFICATION, DRUG REPURPOSING, PHARMACOKINETIC AND TOXICITY PREDICTION OF C-MET KINASE INHIBITORS FOR CANCER THERAPY

Pretisha Flora Cutinho,[a*] Venkataramana C. H. S.[a] and Suma B. V.[a]

[a]Department of Pharmaceutical Chemistry, Faculty of Pharmacy, Ramaiah University of Applied Sciences, Bangalore-560 054, Karnataka, India

1. INTRODUCTION

Cancer is globally one of the leading causes for death and through years it has been emerged to increase in the incidents of morbidity and mortality worldwide.[1] Despite immense research and development in cancer therapy, lack of a feasible, safe and affordable treatment, opens wide scope in this area of study. Though chemotherapy is widely used for the treatment, adverse effects associated with it has brought about a step back in the therapy. Thus, targeted protein tyrosine kinase therapy has gained significance in recent times.[2] Among the tyrosine kinase therapies, c-Met is a promising target. This target is found to be involved in tumurogenesis of different cancers and so has brought our interest towards it. c-Met belongs to the subfamily of receptor tyrosine kinases (RTKs) which are encoded by the Met proto-oncogene. This involvement of HGF/c-Met signalling pathway in oncogenesis has been reported in various types of cancers. Over expression of the target c-Met has brought about transcriptional regulation, gene amplification, autocrine or paracrine ligand stimulation leading to uncontrolled and enhanced tumor cell proliferation, survival, motility, scattering, migration, angiogenesis, invasion, epithelial-mesenchymal transition (EMT), metastatic dissemination.[3,4] This over expression is commonly observed in cancers like gastric, bladder, cervical, endometrial, kidney, liver, lung, head, neck, thyroid, breast, ovarian, prostatic, pancreatic, colorectal cancers and so on. Targeting of HGF/c-Met signalling pathway is thus expected to give promising results, which has put forward our focus on it.[4]

2. METHODOLOGY AND RESULTS

2.1 Methodology

Computational studies were carried out on a machine running with Linux operating system, 16GB RAM, i3 processor with Schrodinger Drug Design Suite and few open access web servers. Advancement of technology, vast information on the crystal structure and active pocket of target protein inspired us for design by structure-based approach. Structure-based pharmacophore model was generated using optimized crystallographic structure of c-Met (PDB ID: 3LQ8) as target protein. The generated pharmacophore model was then screened with publicly available databases (Figure 1) of natural compounds like flavonoids (Fl), chalcones (Ch) and coumarins (Co). In addition, we have screened a database of FDA approved drugs for repurposing. In this study, a comparison was made with the marketed c-Met kinase inhibitors like Cabozantinib and Foretinib to obtain promising hit molecules. The molecules having similar pharmacophoric features with respect to standard were further taken for molecular docking using Glide to investigate binding interactions and conformations at the active pocket. The top five hits were studied for pharmacokinetic and toxicity parameters by Schrodinger's QikProp and some freeware like pkCSM and SwissADME for design of safe and promising hit molecules.[5]

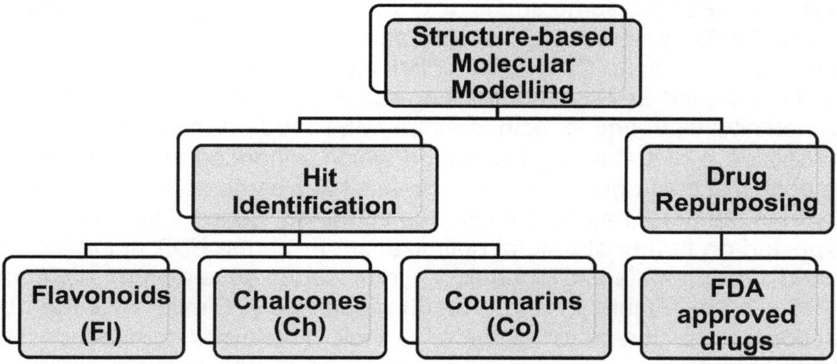

Figure 1 *Flow chart listing databases screened.*

2.2 Results

2.2.1 Structure based pharmacophore modelling

Structure-based approach utilized in the study, focuses on target protein (PDB ID: 3LQ8) to scan and identify active sites with respect to the identified active sites of the co-crystallized ligand Foretinib which is a marketed drug as an anti-cancer agent in particular a c-Met kinase inhibitor. The Phase module of Schrodinger uses this principle to generate a seven featured pharmacophore model (7-ADHHRRR) as shown in Figure 2.

This generated pharmacophore model provides information regarding the features required for a compound to behave as a promising lead c-Met kinase inhibitor. The molecules under study were screened on this pharmacophore model to identify hit molecules, in comparison with reference drugs cabozantinib and foretinib. Interestingly, the compounds have shown a better fitness score than one of the reference drugs, representing their scope as c-Met kinase inhibitors (Table 1). Top twenty molecules were filtered and then carried forward to analyse their binding affinity with the same target protein by performing molecular docking studies.

2.2.2 Molecular Docking

Top twenty compounds docked with the protein (PDB ID: 3LQ8) expressed a very good range of docking scores which were comparable to that of cabozantinib and foretinib. Among the flavonoids, Fl-1 topped the list with a score of -12.297 followed by Fl-7 with -12.160. Both the flavonoids showed similar hydrogen bond interactions with the protein, specifically with amino acid residues ASP1222 and MET1160. The hydroxy groups of the flavonoids were commonly found to participate in the interaction. Further, chalcones Ch-19, Ch-2 also presented promising scores of -11.102 and -10.060 respectively, among the twenty compounds. They have also shown to form hydrogen bonds with amino acid MET1160. The interaction of Ch-19 was observed between its hydroxy group and MET1160 whereas, Ch-12 expresses hydrogen bond formation between the same amino acid MET1160 but, with its carbonyl group. The next database of coumarins also showed good affinity with the target protein that is, coumarin Co-4 gave a docking score of -10.210 following which the next compound Co-2 gave a docking score of -10.184. Likewise, both have shown affinity with amino acid residue MET1160. Additionally, Co-2 has demonstrated a π-π stacking interaction which enhanced the affinity with the active pocket of the target protein. Coumarins have comparatively shown lesser affinity than the other natural compounds screened, hence a better affinity can be expected on further structural optimization. From the FDA approved drugs screened, bicalutamide and diphenidol hydrochloride gave a promising docking score of -13.739 (Figure 2) and -13.463 respectively. Similar hydrogen bond interactions were observed by them as well. The other compounds in the database have found to express a similar pattern of binding to the amino acids of target protein. Surprisingly, all the compounds have shown a close comparison with the reference drugs, thus the study infers that hydrogen bond formation with MET1160 is essential for compounds to demonstrate potential hits as c-met kinase inhibitors. Drug repurposing in this study has also shown the potential of existing drugs as c-Met kinase inhibitors.

2.2.3 Prediction of Pharmacokinetic parameters

The top five hits filtered from each database were checked for their safety and efficacy with the help of some *in silico* techniques. Schrodinger's QikProp explains that, the hit molecules obey's Lipinski rule of five by showing values within the limits. SwissADME gives drug likeness of each molecules in the form of a bioavailability radar, which justifies parameters like lipophilicity (LogP), size (Molecular weight), polarity (TPSA), solubility (LogS), flexibility and saturation to be within their respective range (pink region). Bioavailability radar of a representative molecule Bicalutamide is represented in Figure 3 in comparison with that of reference drug. Additionally, pkCSM predicts other pharmacokinetic and toxicity parameters which further gives the safety and assures to take these hit molecules for further *in vitro* and *in vivo* evaluation of c-Met kinase inhibition.

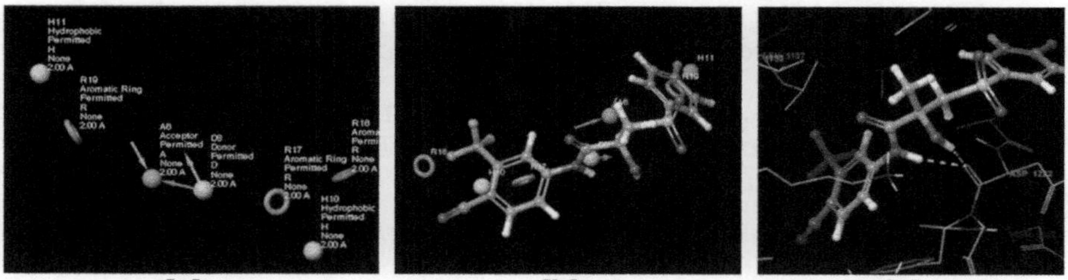

[a] [b] [c]

Figure 2 *[a] Generated seven featured pharmacophore model from target protein (PDB ID: 3LQ8), [b] Overlay of a representative hit molecule (Bicalutamide) on the pharmacophore model and [c] 3D binding interaction of a representative hit molecule (Bicalutamide).*

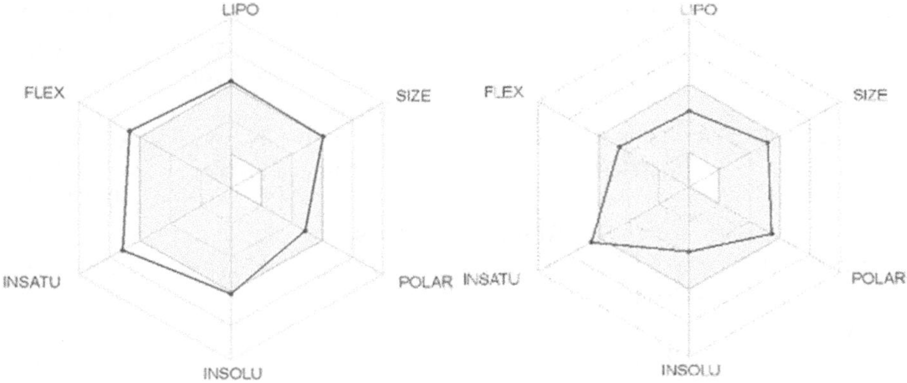

Cabozantinib Bicalutamide

Figure 3 *Bioavailability radar of Reference compound Cabozantinib and a representative hit molecule (Bicalutamide).*

Table 1 *Fitness score, docking score and binding interaction of top two hit molecules from screened databases for hit identification and drug repurposing.*

Top hit molecules	No. of matched features	Fitness Score	Docking Score	Binding Interactions	
				Type	Amino acid residue
Flavonoids					
Fl-1	4-DHRR	1.430	-12.297	H-bond	ASP1222 MET1160
Fl-7	4-DHRR	1.304	-12.160	H-bond	ASP1222 MET1160
Chalcones					
Ch-19	5-AHHRR	1.336	-11.102	H-bond	MET1160
Ch-2	4-HHRR	1.441	-10.060	H-bond	MET1160
Coumarins					
Co-4	4-AHHR	1.329	-10.210	H-bond	MET1160
Co-2	4-AHHR	1.358	-10.184	H-bond π - π stacking	MET1160 PHE1223
FDA approved drugs					
Bicalutamide	6-ADHHRR	1.462	-13.739	H-bond	ASP1222
Diphenidol HCl	5-AHRRR	1.566	-13.436	H-bond	ASP1222 MET1160
Reference drugs					
Cabozantinib	7-ADHHRRR	0.794	-15.752	H-bond	ASP1222 MET1160
Foretinib	7-ADHHRRR	1.796	-17.041	H-bond	ASP1222 MET1160

3. CONCLUSION

A pharmacophore model with seven essential features was built. This pharmacophore model was used to screen various databases, to identify the best twenty compounds. Filtered compounds were then taken for molecular docking studies. Among the screened database of natural compounds, flavonoids demonstrated better affinity than chalcones and coumarins. Computationally identified potential compounds were Fl-1, Fl-7, Ch-19, Ch-2, Co-4 and Co-2. Further, from the FDA approved database of drugs, bicalutamide marketed as an anti-androgenic agent and diphenidol hydrochloride marketed as an anti-emetic agent, have expressed potential to be repurposed as c-Met kinase inhibitors. Safety of selected compounds were also checked by predicting their pharmacokinetic and toxicity parameters. This study suggests for *in vitro* and *in vivo* evaluation of c-Met kinase inhibition of identified molecules. Understanding SAR of molecules and further structural modifications will contribute safe and potent c-Met kinase inhibitors.

References

1. J. Li, W. Gu, X. Bi, H. Li, C. Liao, C. Liu, W. Huang and H. Qian, *Bioorg. Med. Chem.*, 2017, **25**, 6674.
2. W. Shi, H. Qiang, D. Huang, X. Bi, W. Huang and H. Qian, *Eur. J. Med. Chem.*, 2018, **158**, 814.
3. G. Baldanzi and A. Graziani, *Biomedicines,* 2015, **3**, 1.
4. P. K. Parikh and M. D. Ghate, *Eur. J. Med. Chem.*, 2018, **143**, 1103.
5. P. F. Cutinho, J. Roy, A. Anand, R. Cheluvaraj, M. Murahari and H.S.V. Chimatapu, Journal of Biomolecular Structure and Dynamics, 2019, **37**, 1.

DESIGN POTENT TELOMERASE INHIBITORS USING STRUCTURE BASED APPROACHES

Shalini Bajaj,[a] Nilofer Gerald Arakal,[b] Manikanta Murahari[b] and Mayur C. Yergeri[a]

[a]SPP SPTM-School of Pharmacy and Technoslogy Management, SVKM'S NMIMS, Mumbai, India; [b]Department of Pharmaceutical Chemistry, Faculty of Pharmacy, M.S. Ramaiah University of Applied Sciences, Bangalore, India

1 INTRODUCTION

Target based drug design is increasingly gaining more attention for anticancer therapy. Telomere are repetitive nucleotide sequences that are present at the end of chromosomal DNA. Shortening of telomere length has been associated with cellular senescence and the aging process. In cancerous cell telomerase is activated which maintain the length of telomere and leads to cell proliferation. Therefore, telomerase is the major recognised target for the development of better cancer therapeutics. In this study we have designed novel derivatives of telomerase inhibitors as anticancer agents. In continuation of this we have generated structure based pharmacophore model of telomerase and screened ZINC database to identify compounds with similar pharmacophoric features and further subjected to Glide molecular docking and ADMET prediction. The flow chart of entire study is as shown:

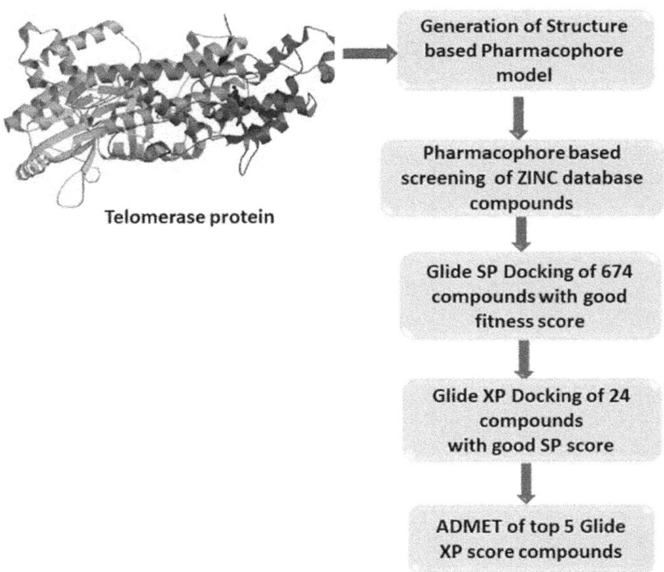

2 METHODS AND RESULTS

1.1. Structure based Pharmacophore modelling

Structure based pharmacophore hypothesis of targeted telomerase protein (PDB ID: 5CQG) was generated using the Phase module of Schrodinger. The generated pharmacophore hypothesis containing four features such as two aromatic ring (R), one hydrophobic (H) and one negative ionic (N) group which are sown in the Figure 1. These four pharmacophore features are assumed to play an essential role for telomerase inhibitory activity.

1.2. Pharmacophore based screening for lead identification

Based on generated structure based pharmacophore hypothesis, ZINC database was screened and molecules which were ranked higher in term of fitness score are selected for Glide Standard Precision (SP) docking on telomerase receptor (PDB:5CQG). Top 24 Glide SP scored compounds were subjected to Glide Extra Precision (XP) docking mode. Among them, top five compounds with good XP score range from -9.562 to -7.698 are shown in Table 1. The 2D binding interaction of top scored ZINC00395553 compounds with ARG486, PHE494 and TYR551 active site residues of telomerase are presented in Figure 2. Reference molecule BIBR1532 also showed similar binding interaction with PHE494 residue of telomerase receptor (Figure 2). Pharmacokinetic and toxicity property of top five Glide XP scored compounds was done using Qikprop module of Schrodinger and reported in Table 2. ZINC00395553 compound showed good value for drug-likeness feature and may use as a lead molecules for the development of potent anticancer agents with potent telomerase inhibitory activity. Predominantly, our assumption of work to identify the novel molecules with potent telomerase inhibitory activity and screen for *in vitro* and *in vivo* biological activity evaluation in future.

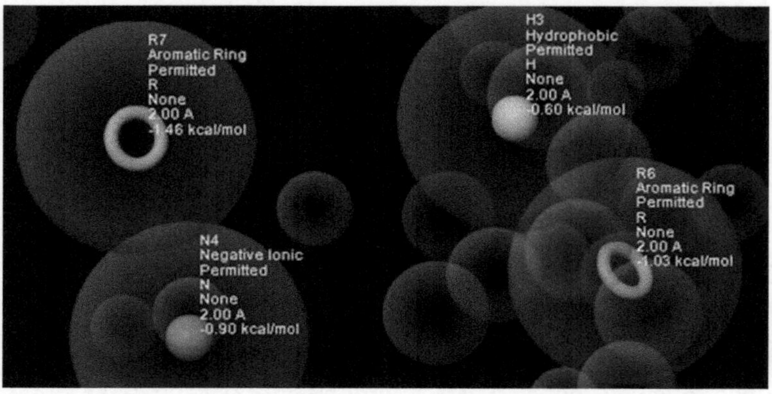

Figure 1 *Structure based pharmacophore model of telomerase.*

Table 1 *XP Score and binding interaction of top five compounds.*

S. No.	*ZINC Comp.*	*XP score*	*Binding interaction*	
			Type of interaction	*Interaction residues*
1.	ZINC00395553	-9.562	H-bond, Salt bridge	ARG486
			Pi-Pi stacking	PHE494
				TYR551
2.	ZINC59443596	-8.796	H-bond, Salt bridge	ARG486
			Pi-Pi stacking	PHE494
				TYR551
3.	ZINC71285419	-8.608	Salt bridge	ARG486
			Pi-Pi stacking	PHE494
				TYR551
			H-bond	ILE550
				GLY553
4.	ZINC71284970	-8.243	Salt bridge	ARG486
			Pi-Pi stacking	TYR551
5.	ZINC06803303	-7.698	H-bond, Salt bridge	ARG486
			Pi-Pi stacking	PHE494
				TYR551
			H-bond	ILE551

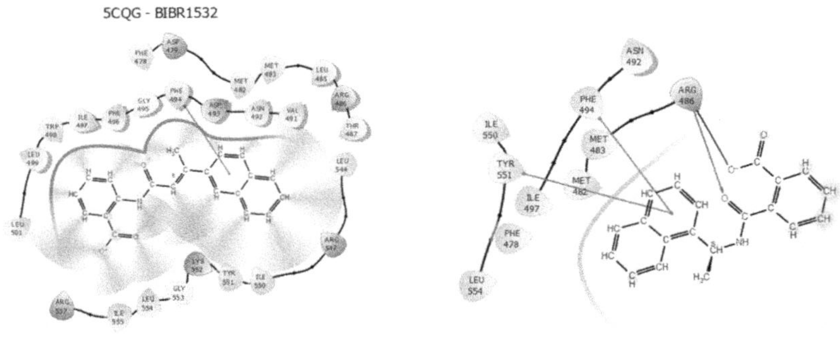

Figure 2 *Docking interaction of BIBR1532 reference molecule and ZINC00395553 compound.*

1.3. ADMET (Pharmacokinetic) and T (toxicity) Predictions
Pharmacokinetic and toxicity prediction plays an important role in the process of drug discovery to identify safe molecules which can be further taken for experimental evaluation and lead optimization. Top five ZINC database compounds showed good ADMET properties. ZINC00395553 compound showed good Caco cell permeability with highest 309.816 QPPCcao value.

Table 2 *Pharmacokinetic property of ZINC database compounds.*

Properties	ZINC00 395553	ZINC59 443596	ZINC71 285419	ZINC71 284970	ZINC06 803303
M.W.	319.36	382.39	351.32	352.35	336.43
SASA	630.81	605.74	633.92	635.57	609.98
QPlogBB	-0.666	-1.723	-2.338	-1.909	0.026
QPPMDCK	177.31	4.59	4.16	9.77	112.50
PSA	70.13	136.61	155.07	142.90	55.86
QPlogPo/w	4.34	3.13	0.82	1.52	2.68
QPPCaco	309.81	6.22	9.63	21.22	185.22
Percent Human Oral Absorption	96.96	59.52	49.35	59.63	83.26

CONCLUSION

In the present study, an endeavour has been made to design telomerase inhibitors with potent anticancer activity. Structure based pharmacophore model of telomerase protein (PDB: 5CQG) was generated and screened ZINC database. Generated ZINC database compounds were subjected to molecular docking, pharmacokinetic and toxicity prediction using Schrodinger Drug Design Suite and pkCSM software to identify lead molecules as potent and safe telomerase inhibitors. Results obtained from each molecular modelling methods were compiled and correlated with telomerase inhibitor which are in clinical trials. Pharmacophore modelling and Glide docking approaches could filter top 5 compounds from ZINC database. Study concludes that ZINC00395553 compound may be used as lead molecule for the development of potent telomerase inhibitors.

References

1. Y. Wang, F.X. Cheng, X.L. Yuan, W.J. Tang, J.B. Shi, C.Z. Liao and X.H. Liua, *Eur. J. Med. Chem.*, 2016, **112**, 231.
2. P. F. Cutinho, J. Roy, A. Anand, R. Shankar, M. Murahari and C. H. S. Venkataramana, *J. Biomol. Str. Dynamics*, 2019, 1–30.
3. Schrödinger Release 2018-3: Maestro, Schrödinger, LLC, New York, NY, 2018; D. R. Koes and C.J. Camacho, *Nucleic Acids Res.*, 2012 **40**, W409–W414.

MOLECULAR DOCKING BASED SCREENING OF "NITROGEN CONTAINING BISPHOSPHONATE CONJUGATE WITH HYDROXYAPATITE" ACTIVE CONSTITUENTS TOWARDS MEVALONATE PATHWAY IN FINDING POTENT INHIBITORS FOR ANTI-OSTEOPOROTIC ACTIVITY

Kalyani A.,[a] Bharath Srinivasan[a] and Parasuraman P.[b]

[a]Department of Pharmaceutics, Faculty of Pharmacy, M.S. Ramaiah University of Applied Sciences, Bangalore, Karnataka, India; [b]Department of Pharmaceutical Chemistry, Faculty of Pharmacy, M.S. Ramaiah University of Applied Sciences, Bangalore, Karnataka, India

1. INTRODUCTION

Osteoporosis is a skeletal disease causing bone fragility by disturbing micro architecture of the bone by decreasing Bone mineral density (BMD). Osteoporosis is seen commonly in post menopausal women and also in patients with long term use of medications like corticoids termed as glucocorticoid induced osteoporosis. Midst of the disease increases fracture risk and decreases the life expectancy and quality of life. Globally there are 323 million people above the age of fifty were beingaffected with osteoporosis, 34 million are at the risk of developing the disease and the number may be increasesup to 1.55 million (310% in men and 240% in women) by 2050[1].Bisphosphonates (BP's) like Residronate, Ibandronate, Pamidronate, Zoledronic acid andEstrogen patchesare the choice of drugs for treatment of osteoporosis. Nitrogen containing Bisphosphonates aid in the effective treatment of osteoporosis by directly acting on osteoclasts cells by inhibiting the Farnesyl pyrophosphate synthase (FPPS) of melvonate pathway and their ability to bind with the mineral component of bone contributing to their potency of BP's. The P-C-P backbone is responsible for strong affinity towards hydroxyapatite and facilitate potent inhibitory effects. These drugs are having high affinity and strong binding sites towards Hydroxyapatite. Hydroxyapatite is the mineral component present in the bone containing calcium and phosphorus in the ratio of 1.67 and also aid in effective treatment against hypercalcemia, multiple myeloma, metastatic cancer.

Cholesterol synthesis of Mevalonate pathway is a bio synthetic pathway responsible for osteoclasts inactivation by inhibiting some of the isoprenoid lipids like HMG CO-A synthase, HMG CO-A reductase, Mevalonate kinase, Human Geranyl Geranyl pyrophosphate synthase which bind to GGPP that areessential for the activation of prenylation of small GTPase isoprenoids to prenylation GTPase signalling protein that helps in normal functioning of proteins. Another protein target Cytokine receptor triad Receptor Activator of Nuclear factor kappa B ligand (Human RANKL/OPG) a bone morphogenic protein (BMP-2) that crucially controls the terminal differentiation step of osteoblasts and osteoclasts also binds to a protein secreted by the cells of osteoblasts lineage known as Osteoprotegerin.

BP's acts on inhibiting the key enzyme Farnesyl pyrophosphate synthase (FPPS) in mevalonate pathway of cholesterol biosynthesis which is crucial in achieving the therapeutic target.

2 METHOD AND RESULTS

AutoDock 4.2 as a molecular docking tool

AutoDock is a molecular modelling simulation software used for protein-ligand docking in the research community. It works on Monte Carlo simulated annealing and Lamarckian genetic algorithm (LGA) to create a set of probable conformations. LGAmethod is used as a global optimizer and energy minimiser.The docking studies was carried out using AutoDock 4.2 suite to predict binding of small molecules to the 3D target protein structure. Apart from generating the binding energies of the docked molecules, nature and the position of the ligand in the binding site can be visualized.

In this docking simulation, Target proteins are set aside as rigid by using semi-flexible docking protocols. Further drug substrate to be docked were kept flexible, in order to discover random number of torsional degrees of freedom and six spatial degrees of freedom spanned by the translational and rotational parameters

In the present work, the comparative docking analysis of newly synthesized Ibandronate–hydroxyapetite conjugate and pure Ibandronate (N-Bisphosphonate drug)ligand with target proteins of Mevalonate pathway enzymes and RANKL/OPG have been performed.

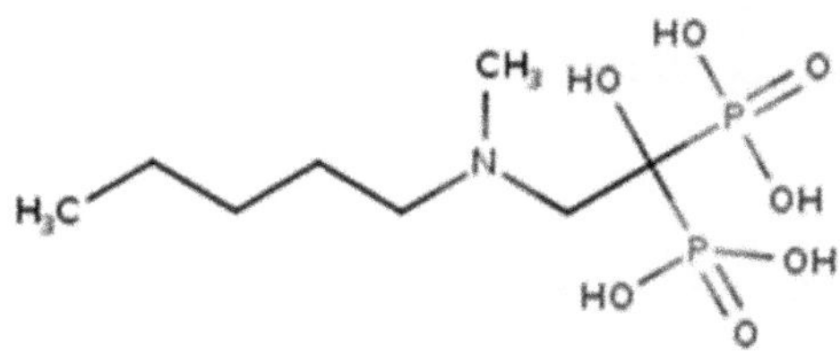

Figure 1 *Two dimensional ligand structure of Ibandronate.*

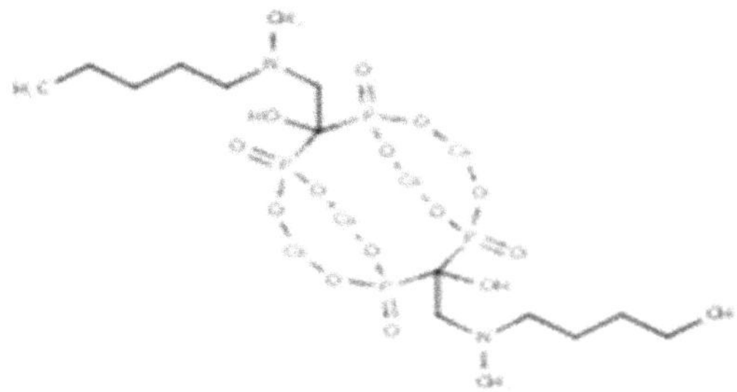

Figure 2 *Two dimensional ligand structure of Ibandronate drug conjugate with Hydroxyapetite.*

Ligand preparation

The ligand molecular structure of Ibandronate and Ibandronate-hydroxyapetite conjugate were drawn using chem draw Ultra 7.0.4. Geometrical refining of the chemical structures were done by using Ligprep model. Ligands were defined for rigid roots of Ibandronate and Ibandronate conjugate hydroxyapatite automatically

Protein preparation

The crystal structure of protein receptors with PBD ID's – 3SQZ (HMG CoA synthase), 1DQ9 (Human HMG-COA Reductase), 2F7M (Human Farnesyl Pyrophosphate synthase (FPPS)), 2HFU (Mevalonate kinase), 2Q80 (Human Geranyl Geranyl pyrophosphate synthase bound to GGPP) of Melvonate pathway and 3URF (Human RANKL/OPG complex) solved using X- ray diffraction method obtained from protein data bank (PDB)[2]. Protein structure were subjected for protein preparation using BOVIA Discovery Studio 2017 R2 Client software.He-atoms and water molecules were removed from the protein structure and subjected to applied force field in order to minimize the energy of the atoms present in the structure and made them stable to react with the ligands. Hydrogen atoms were added, Gasteiger chargers were assigned by merging non polar bonds, followed by assigning of AD4 type to the protein receptor.

Grid generation

In the preparation of grid box, x,y,z dimensions are taken and set at 126,126,126 respectively. Spacing remains 0.375 A°, and centred the grid box on macromolecule. Lamarckian Genetic Algorithm (LGA) were preferred to search for the best conformers.

Docking

For docking, a macromolecule was selected.ligandand protein undergoes 27000 maximum number of generations, 0.02 as mutation rate, 0.2 Å was step size, 2.0 Å as cluster tolerance, adjusting external grid energy to 1000.0 with initial energy as 0.0 and at last exhibit 10 the best conformational runs were performed.

3. ASSESSMENT OF DOCKING

Docking simulation at the active sites of HMG-COA synthase, HMG-COA Reductase, FPPS (Figures 3&4), Mevalonate kinase, GPPS, RANKL/OPG with respect to pure Ibandronate & Ibandronate-hydroxyapatite conjugate were assessed and docking simulation score and pIC50 values was recorded in Table 1. Mainly the oxygen atoms of phosphate groups had undergone conventional hydrogen bonding with Amino acid residues of targets[3].On the basis of minimum energy conception, it was concluded as ΔG binding energy and pIC50 values showing promising potential anti osteoporotic effect for Ibandronate-hydroxyapatite conjugate and it can act as a most powerful binary to FPPS, GPPS & RANKL/OPG in contrast to pure Ibandronate. By using Data warrior the drug- conjugate also complied with Lipinski's rule without any compliance for mutagenecity, irritability and tumarogenecity.

Table 1: Docking results of the parent and final drug molecules with various protein targets

Sl. No.	Protein (PDB ID)	Binding Energy (Kcal/mol)		pIC50 values	
		Pure Ibandronate (Parent)	IBA-HAP Conjugate	Pure Ibandronate (Parent)	IBA-HAP Conjugate
1	3SQZ	-1.84	-0.16	44.58mM	17.47mM
2	1DQ9	-1.28	-6.12	25.12mM	32.49 μM
3	2F7M	-2.07	-6.75	30.31mM	11.37μM
4	2HFU	-2.98	-1.67	6.51mM	7.85mM
5	2Q80	-1.97	-5.33	35.88mM	123.93μM
6	3URF	-3.23	-6.49	4.3nM	17.47 μM

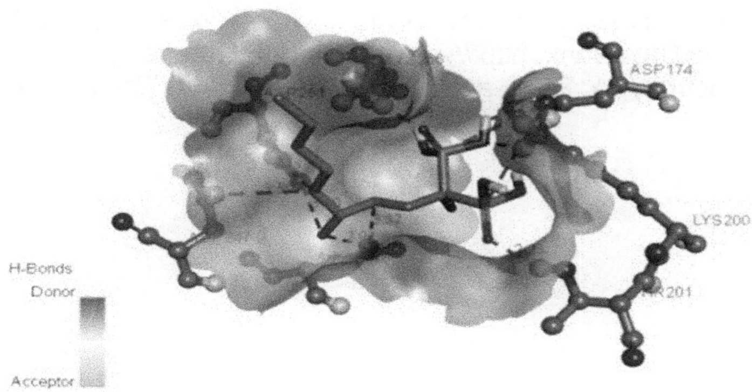

Figure 3 *3D surface interactions of 2F7M with ligand pure Ibandronate drug molecule.*

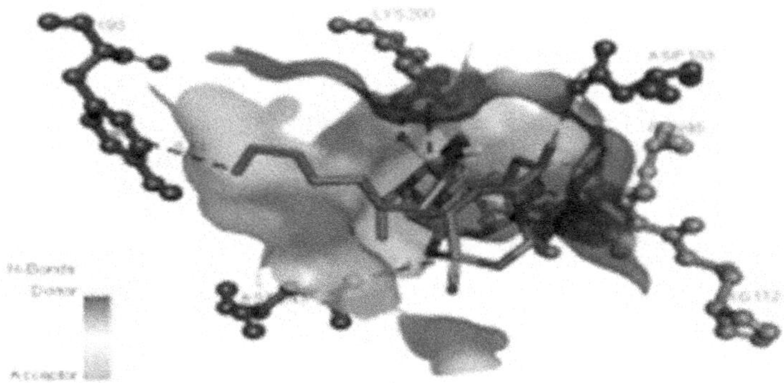

Figure 4 *3D surface interactions of 2F7M with synthesized ligand Ibandronate–Hydroxyapatite conjugate.*

3 CONCLUSION

N-BP's are broadly used as potent inhibitors towards bone as bone resorbing agents. The docking accuracy and scoring reliability of Ibandronate (N-BP's) with Hydroxyapatite conjugate of Human RANKL/OPG proteins and mevalonate pathway enzymes has been represented as highly potent new drug moiety in the treatment of Bone related diseases and osteoporosis. It can be concluded that Ibandronate-Hydroxyapatite conjugates the most powerful inhibitor of Human HMG-COA Reductase, FPPS, GGPP and RANKL/OPG when compared to Pure Ibandronate drug.

The results from the current study has proved to be very useful in designing of new drug-conjugate moietyto treat osteoporosis.

References

1. WHO, *Assessment of fracture risk and its application to screening for postmenopausal osteoporosis. Report of a WHO Study Group*, 1994, World Health Organization technical report series 843, pp. 1–129.
2. Protein Data Bank, http://www.rcsb.org/pdb/home/home.do.
3. J, Choubey, A. Patel and M. K. Verma, *Int. J. Comput. Appl.,* 2013, 66–72.

ZANAPEZIL, A POTENTIAL REPURPOSABLE DRUG AGAINST HUNTINGTON DISEASE: AN *IN SILICO* FORECAST

S. Sumedha Kulkarni,[a] G. N. S. Hema Sree[b] and G.R. Saraswathy*[b,c]

[a]Department of Pharmaceutics, Faculty of pharmacy, M.S. Ramaiah University of Applied Sciences, Bangalore, Karnataka, India; [b] Pharamacological Modelling and Simulation Centre, M.S. Ramaiah University of Applied Sciences, Bangalore, Karnataka, India; [c] Department of Pharmacy Practice, Faculty of Pharmacy, M.S. Ramaiah University of Applied Sciences, Bangalore, Karnataka, India

1. INTRODUCTION

Huntington's Disease (HD), a fatal neurodegenerative disorder, is an autosomal dominant condition characterized by improper involuntary movements of the voluntary muscles (chorea), dystonia, difficulty in speech and cognitive impairment along with psychiatric symptoms. On an average the disease affects about 0.25% (~7000 individuals) of the general population [1]. Pathophysiology of HD involves mutation in Huntingtin (HTT) gene due to repetition of CAG nucleotide, characterized by degeneration or loss of medium spiny neurons resulting in atrophy of striatal region. The early development of the disease also shows a significant slimming of cortical ribbon and proceeds from posterior to anterior portion of the cortical regions. Further, neurodegeneration is attributed to mitochondrial dysregulation leading to decrease in ATP production and loss in calcium buffering capacitance eventually leading to apoptosis. Tetrabenazine, a current FDA approved drug for the treatment of HD, used to treat choreiform movements, inhibits Vesicular Monoamine Transporter (VMAT2) protein which is involved in the storage and release of dopamine [2] [3]. The appropriate treatment for the disease is still under research. Contemporary research is focussed on drug repurposing which explores new therapeutic indications for existing drugs rather than *de novo* drug discovery to circumvent the hiccups involved in the later process. Many repurposing trials have come up based on existing omics data to identify the novel repurposable targets. In fact, handful of repurposing projects reached the heights of clinical trials. Some of the repurposed drugs in clinical trials for HD are Pridopidine under phase 2 trials (NCT01306929) [4]; Cysteamine under phase 2/3 (NCT02101957) and Latrepiridine under phase 3 trials (NCT00920946) [2], [5]. The present research is an *in silico* approach to predict disease specific targets and drugs. In this work, we extracted Differentially Expressed Genes (DEGs) in HD through Gene Expression Omnibus (GEO) dataset, studied their interaction with literature derived reported genes through Protein-Protein Interaction (PPI) network analysis using Search Tool for Retrieval of Interacting Genes (STRING). Finally, the shortlisted targets were subjected to molecular docking with FDA approved drugs from Drugbank database to identify the potential anti-HD drugs.

2. METHODS AND RESULTS

2.1 Selection and extraction of DEGs:

The dataset GSE1767 deposited by Borovecki *et al.*, 2005 in GEO at National Centre for Biotechnology Information (NCBI) comprising of microarray data extracted from peripheral blood in 14 healthy controls and 17 patients analysed on the platform GPL1449 was selected for the present study. Significant DEGs ($p<0.05$) were retrieved using Geo2R based on Benjamini and Hochberg False Discovery Rate (FDR) method. A total of 208 upregulated genes (CASD1, ETNK1, FAM8A1......) and 14 downregulated genes (LAMB2, RBFA, ITGB4......) were found to be significantly expressed in HD (Table 1)

2.2 Functional enrichment analysis and PPI network

The DEGs were processed for functional enrichment analysis by Database for Annotation, Visualization, and Integrated Discovery (DAVID) tool (version 6.7) [6]. DAVID analysis revealed a significant association of 62 DEGs in the cytosol (GO:0005829, $p = 4.0E^{-5}$), 53 DEGs in the nucleoplasm (GO:000565, $p = 1.3E^{-4}$) and 140 DEGs in protein binding (GO:0005515, $p = 9.6E^{-9}$) in the Gene Ontology (GO) category. A total of 63 and 67 DEGs were found to be significantly expressed in the epithelium and placenta respectively. PPI constructed between 6 reported genes obtained through literature mining as well as NCBI and 9 DEGs obtained through DAVID using STRING, exhibited 134 nodes and 86 edges (Figure 1) with PPI Enrichment value of 0.01044381784.

2.3 Selection of targets and drugs for repurposing studies

The proteins exhibiting significant interactions with reported genes were selected for further studies. Caspase 3 (CASP3, PDB ID: 1GFW) which is up regulated (log Fc value 1.684) with node degree 8 was selected as a potential target. Minocycline was identified as a standard drug for Casp3 through Drug Repurposing Hub (a drug repurposing app of Library of Integrated Network Based Cellular Signatures (LINCS) consortium). Drug similarity search resulted in 29 drugs from Drug Bank database.

2.4 Molecular docking

The selected targets and their respective drugs were subjected to molecular docking studies through Extra Precision (XP) mode in Glide tool of Schrodinger drug design suite (version 10.7) (Glide, Schrödinger, LLC, New York, NY, 2018). All the docking procedures were performed on Intel® Core™ i7-6700CPU @ 3.40 GHz *8 of HP with memory of 15.6 GiB and OS Type -64 bit. Docking scores and glide energies of the above 29 drugs ranged between -10.303 to -3.339 and -20.399 to – 66.121 kcal/mol respectively (Table 2). The standard Minocycline was found to possess -5.384 docking score, -36.517 kcal/mol glide energy and exhibited hydrogen bond interaction with Arg207 and Phe250 in B chain (Figure 1B). Among 29 similar drugs, 17 drugs were found to score higher than Minocycline. Plicamycin, an antibiotic used in testicular cancer, was found to possess highest docking score (-10.303) and glide energy (-66.121 kcal/mol). The next was found to be Lymecycline, used to treat infections and acne, with docking score (-10.126), glide energy (-58.131 kcal/mol). Zanapezil, an investigational anti-cholinesterase inhibition proposed for the treatment of Alzheimer's treatment, was found with -5.528 docking score, -39.529 Kcal/mol glide energy and possessed interaction with Asn208 in B chain (Figure 1A). Among 17 drugs, Zanapezil was the only drug found to possess CNS permeability.

Table 1 *DEGs extracted from microarray data in HD.*

Up regulated genes	*log Fc value*	*Down regulated genes*	*log Fc value*
CASD1	1.928	LAMB2	-1.194
ETNK1	2.603	RBFA	-0.944
FAM8A1	1.946	ITGB4	-0.999
RASA2	2.215	DSTN	-1.675
SLC26A2	1.618	ALPPL2	-1.374
CCNT2	2.324	G6PC3	-1.444
DOCK11	1.888	DOCK6	-1.648
RICTOR	2.414	SBF1	-1.197
BAG4	1.992	LIMS2	-1.275
CASP3	1.684	GNMT	-1.588

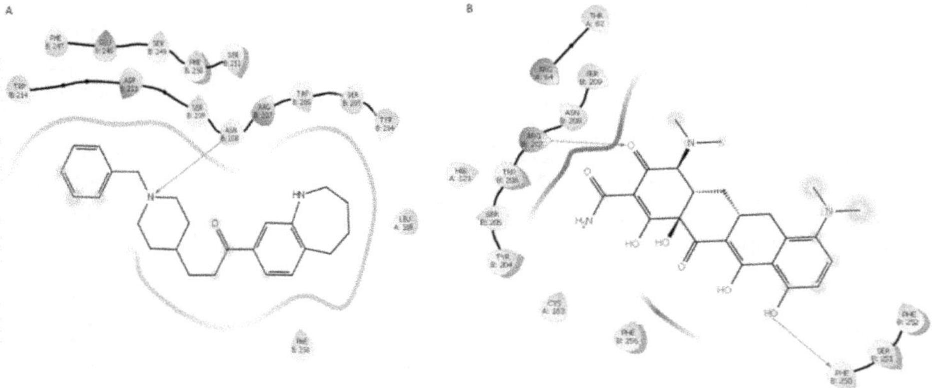

Figure 1 *2D images of Zanapezil (A) and Minocycline (B) with CASP3.*

3. CONCLUSION

Based on our observations, Zanapezil was found to be a potential drug which can be targeted for the Caspase 3 in the treatment of HD. Although *in silico* studies support that role of the Zanapezil in the treatment of HD, further *in vitro* and *in vivo* experimentations should be carried out to validate this *in silico* report.

Table 2 *Docking results of CASP3*

Name of the ligand	*Docking score*	*Glide energy*	*Glide evdw (kcal/mol)*	*glide ecoul (kcal/mol)*	*Xph bond (kcal/mol)*
Plicamycin	-10.303	-66.121	-43.241	-22.88	-6.097
Lymecycline	-10.126	-58.131	-44.668	-13.464	-5.334
Methacycline	-7.914	-32.889	-17.506	-15.383	-3.803
Oxytetracycline	-7.807	-38.162	-29.315	-8.848	-3.862
Sabarubicin	-7.763	-49.095	-40.852	-8.243	-4.415
Doxycycline	-7.345	-33.334	-23.934	-9.4	-3.509
Incyclinide	-7.184	-39.851	-29.693	-10.158	-3.456
Clomocycline	-7.098	-43.109	-34.159	-8.95	-3.881
Rolitetracycline	-6.437	-40.35	-26.898	-13.451	-3.397
Chlortetracycline	-6.376	-36.141	-27.19	-8.951	-3.474
Idarubicin	-6.225	-44.345	-33.263	-11.082	-2.665
Demeclocycline	-6.044	-39.103	-26.156	-12.947	-3.048
Amrubicin	-6.006	-45.338	-34.96	-10.378	-3.454
Tetracycline	-5.77	-34.602	-25.315	-9.288	-3.03
Daglutril	-5.761	-50.608	-41.528	-9.08	-1.553
Tigecycline	-5.578	-36.419	-32.768	-3.651	-3.034
Zanapezil	-5.528	-39.529	-35.508	-4.021	-0.544
Minocycline(Std)	-5.384	-36.517	-23.798	-12.719	-3.1
Pholcodine	-5.109	-33.323	-29.592	-3.731	-1.597
Reserpine	-4.602	-44.112	-40.126	-3.986	-1.033
Nalorphine	-4.588	-28.344	-18.69	-9.653	-1.44
Hydromorphone	-4.453	-25.416	-16.591	-8.825	-1.31
Oxymorphone	-4.391	-25.361	-16.867	-8.495	-1.31
Naloxone	-4.280	-28.095	-19.219	-8.876	-1.31
Naltrexone	-4.250	-28.798	-18.684	-10.113	-1.288
Ajmaline	-4.196	-24.961	-19.895	-5.067	-0.911
Methylnaltrexone	-4.177	-28.736	-19.035	-9.701	-1.359
Oxycodone	-3.706	-22.59	-19.046	-3.544	-0.7
Hydrocodone	-3.492	-20.399	-17.191	-3.208	-0.7
Deserpidine	-3.339	-37.962	-35.565	-2.397	-0.289

References

1. C. Kay, M. R. Hayden, B. R. Leavitt, in *Handbook of Clinical Neurology*, 2017.
2. A. Deb, S. Frank, C. M. Testa, in *Handbook of Clinical Neurology*, 2017.
3. G. Zheng, L. P. Dwoskin and P. A. Crooks, *AAPS J.*, 2006, E682–E692.
4. A. McGarry, K. Kieburtz, V. Abler, I. D. Grachev, S. Gandhi, P. Auinger, S. Papapetropoulos and M. Hayden, *J. Huntingtons. Dis,* 2017, 189–199.
5. E. R. Dorsey, *Arch. Neurol.*, 2013, 25–33.
6. D. W. Huang, B. T. Sherman and R. A. Lempicki, *Nat. Protoc.,* 2009, **4**, 44–57.

DRUG REPURPOSING FOR HEPATOCELLULAR CARCINOMA VIA VIRTUAL SCREENING INTEGRATED WITH BIOINFORMATICS ANALYSIS

Gouri Nair,[a] Hema Sree G. N. S.,[b] Saraswathy Ganesan Rajalekshmi,[b] Subeesh K. Viswam[c] and Jayaraman Anbu[a]

[a]Department of Pharmacology, Faculty of Pharmacy, M. S. Ramaiah University of Applied Sciences, Bangalore, Karnataka, India; [b]Pharmacological Modelling and Simulation Centre, M. S. Ramaiah University of Applied Sciences, Bangalore, Karnataka, India; [c]Department of Pharmacy Practice, Faculty of Pharmacy, M. S. Ramaiah University of Applied Sciences, Bangalore, Karnataka, India

1 INTRODUCTION

Intratumoral hetrogenecity is an emerging issue which holds back the major developments in clinical oncology, as it leads to cancer progression and resistance to therapy. Hepato Cellular carcinoma (HCC) is one of the most common malignant hepatic diseases worldwide. However; the molecular mechanism is still unclear. In majority of patients, chronic inflammation of liver and associated wound healing process is closely linked to the development of fibrosis which further leads to cirrhosis and HCC.[1] Intratumoral heterogeneity in the morphological, histological and molecular level from chronic inflammatory liver disease to dysplastic lesions hampers the therapeutic progress. Studies stated that the cellularity, vascularity and gene expression of invasive border cells in the tumor microenvironment is distinctive from the central part of tumor.[2,3] It is important to study different patterns of gene expression in central tumor as well as the invasive border region to evaluate the cancer progression and treatment options.

As of now, Sorafenib is the only FDA (Food and Drug Administration) approved drug for HCC. Drug repurposing notion is a pioneering drug discovery method to identify the off-label indication of an existing drug. This helps to bypass the hurdles of conventional drug discovery process in terms of dearth of novel moieties and financial overburden. The study aimed to identify common potential targets of tumor and invasive border region, followed by screening of FDA approved drugs against the druggable targets with drug repurposing concept.

2. METHODS AND RESULTS

2.1 Dataset accession and identification of DEGs

The Gene Expression Omnibus (GEO) dataset GSE84598 containing microarray data of 74 samples from 2 different regions of hepatocellular carcinoma (37 from tumor tissues and 37 from border region) analysed on the platform GPL10558 was downloaded from the National Centre of Biotechnology Information (NCBI). The Differentially Expressed Genes

(DEGs) list was analysed with GEO2R, using bioconductor package and characterised by *Benjamini & Hochberg false discovery rate method* ($p < 0.05$). A total of 916 genes (443 down-regulated and 473 up-regulated) in the tumor region and 431 genes (267 up-regulated and 164 down-regulated) in invasive border region were identified.

2.2 Protein–Protein Interaction (PPI) network construction

PPI network of tumor cells and border cells was constructed separately using Search Tool for the Retrieval of Interacting Genes (STRING) with high confidence score (0.7) to predict the significant interactions. Cytoscape software was used to visualize the common DEGs (Figure 1) among tumor cells and border cells. Another PPI network (Figure 2) was constructed between the common DEGs (visualized by cytoscape) and reported genes (obtained from Gene database from NCBI). Among them, the gene CDC20 showed highest node degree (11) interactions with the reported genes of HCC and was selected as a potential target.

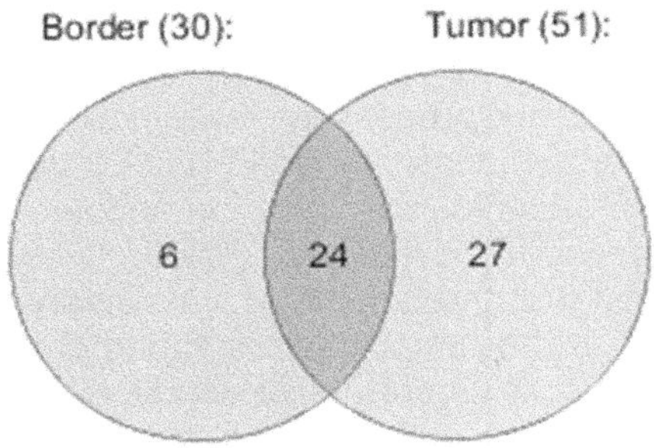

Figure 1 *Venn diagram represents the number of common DEGs obtained from tumor and invasive border region.*

2.3 Molecular docking

The docking simulations of the potential target were performed using Glide tool[4] in Schrödinger drug design suite (version 10.7). The ligand structures (all the FDA approved drugs from the Drugbank) were prepared for docking by LigPrep application. The 3D structure of selected protein (X-ray crystallographic structure) CDC20 (PDB code: 4N14) was taken from Protein Data Bank (PDB), and was prepared by protein preparation wizard of Schrödinger suite.

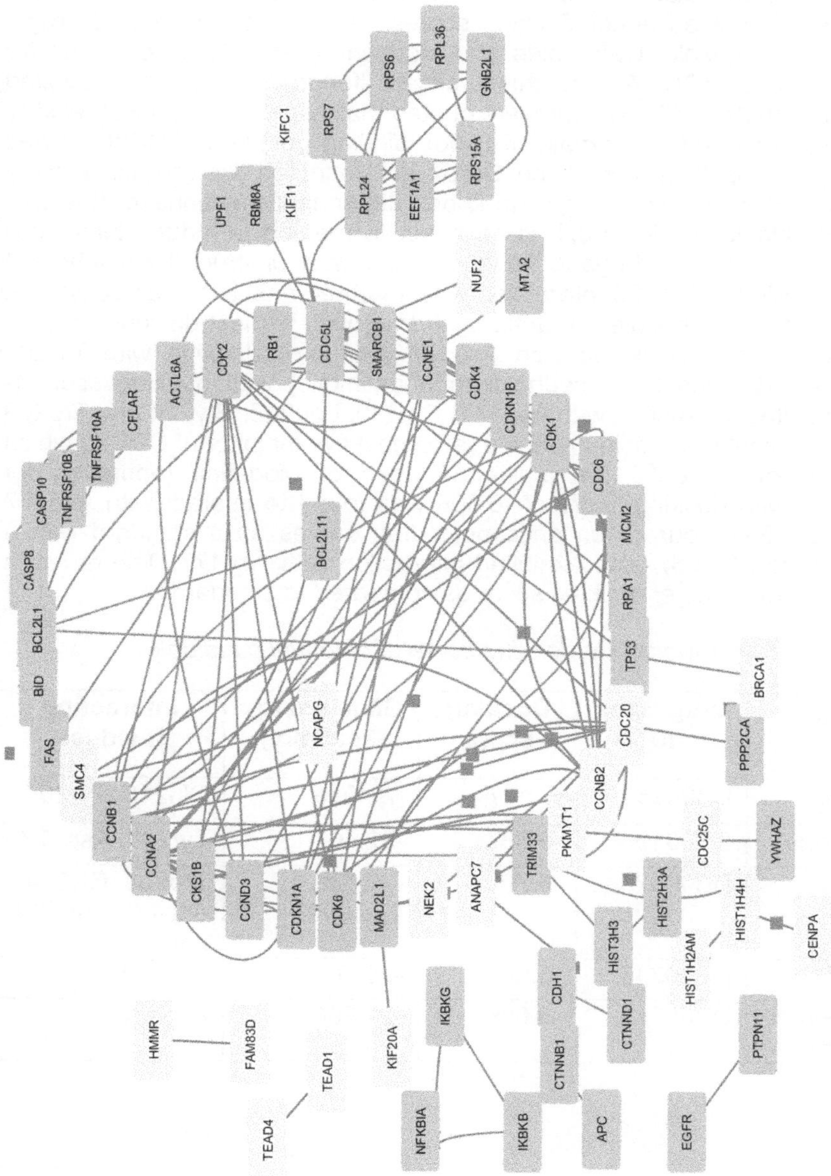

Figure 2 *PPI of common DEGs and reported genes. Common DEGs are represented in light grey and reported genes are represented in dark grey.*

CDC20: All FDA approved drugs were subjected to molecular docking against CDC20 by using High Throughput Virtual Screening (HTVS) mode followed by XP mode. Docking results showed that among the 2388 FDA approved drugs, 7 drugs like Pitavastatin, Ifenprodil, Fluvoxamine, Fluconazole, Ceforanide and Modafinil showed better docking scores. Of the aforementioned drugs: Ifenprodil, Fluconazole, Ceforanide and Modafinil (Table 1) were not so far explored against HCC. Among them, except Ifenprodil all others exhibited interaction with Asp177. Researchers quoted that Asp177 is a crucial residue for binding to the WD40 domain of D-box binding pocket in CDC20. It was reported that substrate containing D-box are recruited for ubiquitilation by APC/C-CDC20 complex through the D-box dependent mechanism. It is well known that blocking mitotic exit is an effective method to induce tumor cell death. Upon binding of drugs to WD40 domain, may obstruct the mitotic exit by blocking APC/C-CDC20 interaction which further inhibits the ubiquitylation of substrates.[5] Fluconazole (an antifungal drug) with highest docking score (-6.887) possesses interaction with Asp177 and pi-pi stacking with Tyr207 Figure 3(A). Ceforanide (a cephalosporin antibiotic) with docking score (-6.709) exhibited interaction with Asp177, Glu180, Leu202, Gly214 and Trp209 Figure 3(B). Similar to Apcin molecule (a competitive inhibitor of CDC20 which prevents ubiquitylation[5]), the aminal nitrogen of Modafinil (approved for narcolepsy) with docking score of -6.608 was found to interact with Asp177 Figure 3(C). 5-fluorouracil an anticancer drug serves as a standard with a docking score (-3.656), also exhibited interaction with Asp177. The drugs in question showed higher docking score as compared to standard.

Table 1 *Molecular docking score of potential candidate.*

Drug	Drug Bank ID	Docking score	Glide energy (kcal/mol)	Interacting residues
Ifenprodil	DB08954	-7.127	-29.019	Gln218, Tyr207
Fluconazole	DB00196	-6.887	-32.467	Tyr207, Asp177
Ceforanide	DB00923	-6.709	-57.588	Trp209, Gly214, Asp177, Glu180, Leu202
Modafinil	DB00745	-6.608	-26.765	Asp177
5-Fluorouracil (Standard)	DB00544	-3.656	-21.806	Asp177

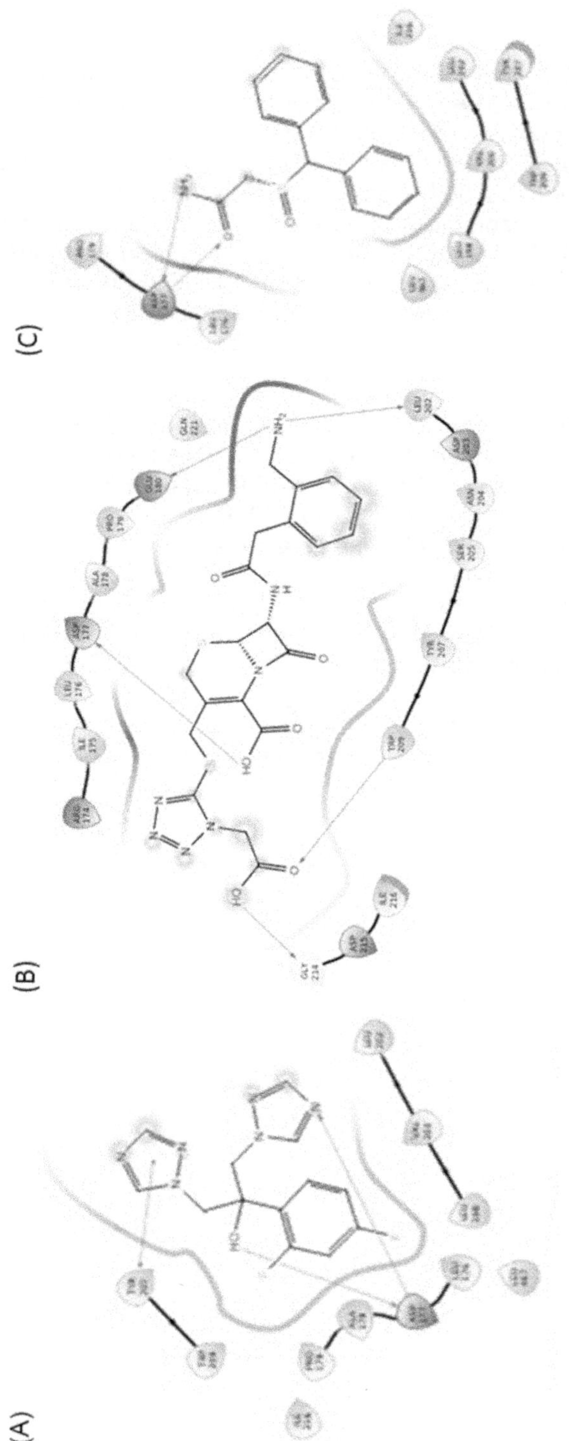

Figure 3 *2D structure of fluconazole (A) ceforanide (B) and Modafinil (C) exhibiting interaction with CDC20.*

3 CONCLUSION

It was observed that invasive tumor border cells and central part of tumor shared some common genes, which illustrates that the gene expression signature of invasive border cells might be helpful to predict the development of HCC. We identified CDC20 as a potential target for both tumor and invasive border cells as it exhibited high node degree interaction with the reported genes for HCC. Fluconazole, Ceforanide and Modafinil showed significant interaction with Asp177 of CDC20 and hence they can be considered as potential drugs that could be repurposed against HCC. Further *in vitro* and *in vivo* studies are required to strengthen these findings.

References

1. D. Castven, M. Fischer, D. Becker, S. Heinrich, J. B. Andersen, D. Strand, M. F. Sprinzl, S. Strand, C. Czauderna, S. Heilmann-Heimbach, S. Roessler, *Oncotarget*, 2017, **8**, 48688.
2. G. Stanta and S. Bonin, *Front. Med.*, 2018, **6**, 85.
3. T. Luedde and R. F. Schwabe, *Nat. Rev. Gastroenterol. Hepatol.*, 2011, **8**, 108.
4. R. A. Friesner, J. L. Banks, R. B. Murphy, T. A. Halgren, J. J. Klicic, D. T. Mainz, M. P. Repasky, E. H. Knoll, M. Shelley, J. K. Perry, D. E. Shaw, P. Francis and P. S. Shenkin, *J. Med. Chem.*, 2004, **47**(7), 1739–1749.
5. K. L. Sackton, N. Dimova, X. Zeng, W. Tian, M. Zhang, T. B. Sackton, J. Meaders, K. L. Pfaff, F. Sigoillot, H. Yu and X. Luo, *Nature*, 2014, **514**, 646.

AN *IN SILICO* TARGET SPECIFIC DRUG REPURPOSING APPROACH FOR MULTIPLE SCLEROSIS

V. Lakshmi Prasanna Marise,*[a] Surekha,[b] G. N. S Hema Sree[c] and Saraswathy G. R.[a,c]

[a]Department of Pharmacy Practice, Faculty of Pharmacy, M.S.Ramaiah University of Applied Sciences, Bangalore, Karnataka, India; [b]Department of Pharmaceutics, Faculty of pharmacy, M.S. Ramaiah University of Applied Sciences, Bangalore, Karnataka, India; [c] Pharmacological Modelling and Simulation Centre, Faculty of Pharmacy, M.S. Ramaiah University of Applied Sciences, Bangalore, Karnataka, India

1. INTRODUCTION

Multiple sclerosis (MS) is a multifocal demyelinating disease with progressive neurodegeneration caused by an increased auto immune response to self-antigens. Based on the ability of remyelination, MS is characterized into 4 subtypes: Relapsing Remitting MS (RRMS) (most commonly found ~87%), Primary Progressive MS (PPMS), Secondary Progressive MS (SPMS), and Progressive Relapsing MS (PRMS).[1] MS is termed as T cell-mediated autoimmune disorder with a predominant expression of CD8+ cells, (involved (MHC-I) activation), CD4+ T cells and microglia/macrophages. The pathophysiology of MS involves focal demyelination (myelin sheath destruction), oligodendrocyte destruction and astrocyte proliferation.[2] Majority of people experience symptoms like muscle spasms, constipation, weakness, urinary problems and alterative sessions of relapses and remissions between the ages of 20 and 40.[3] Currently, Disease Modifying Therapies (DMTs) were known to decrease the frequency and severity of relapses, exacerbations and development of new lesions. To date, U.S FDA approved drug for the treatment of PPMS is Ocrelizumab. Nevertheless, many drugs exhibited success in the preclinical phases and successfully reached clinical trials. Examples of such drugs are: (i) Daclizumab, under phase 2 study (NCT00870740);[3] (ii) Peginterferonbeta-1α, under phase 3 (NCT00906399);[4] (iii) Glatiramer acetate under phase 3 study (NCT00451451);[5] and (iv) Cladribine under phase 3 study (NCT00725985).[6] Although drugs are available to treat MS, there's always a research gap in the exploration of genetic mechanisms and pathways involved in progression and re-attacks of the disease. In our present study we made an attempt to identify the disease specific targets and drugs that can be repurposed for the identified targets through an *in silico* approach using gene expression dataset available in Gene Expression Omnibus (GEO) database. This approach could potentially bypass the extensive work involved in initial stages of de novo drug discovery.

2. METHODS AND RESULTS

2.1 Selection and extraction of DEGs

The dataset GSE21942 deposited by Kemppinen *et al.*,[7] in Gene Expression Omnibus (GEO), National Centre for Biotechnology Information (NCBI) comprising of microarray data extracted from peripheral blood mononuclear cells in 15 healthy controls and 14 MS patients analyzed on platform GPL570 was selected for the present study. Significant DEGs ($p<0.05$) were retrieved using Geo2Enrichr based on Benjamini and Hochberg False Discovery Rate (FDR). A total of 108 upregulated genes (CCL5, HBB, RPL8...) and 92 downregulated genes (PSAP, S100A4, LTA4H...) were found to be significantly expressed in MS (Table 1)

2.2 Functional enrichment analysis and Protein-Protein Interaction (PPI) network

The DEGs were processed for functional enrichment analysis by Database for Annotation, Visualization, and Integrated Discovery (DAVID) tool (version 6.7).[8] The PPI was constructed between common DEGs obtained in DAVID and reported genes obtained through literature mining and NCBI using Search tool for Retrieval of Interacting Genes (STRING) with a threshold of 0.7-0.9 [9] and visualized by Cytoscape software (version 3.6.1).[10]

2.3 Selection of targets and drugs for repurposing studies

Based on the criteria laid, two up regulated genes, Ras-related C3 botulinum toxin substrate 1(RAC1, PDB ID: 3TH5) with node degree 12 and Nuclear Factor Kappa B Subunit 1(NFκB1, PDB ID: 2O61) with node degree 7 were selected as potential targets in MS. Dextromethorphan (RAC1) and Aspirin (NFκB1) were taken as standard drugs from LINCS consortium Drug Repurposing Hub (a drug repurposing app of Library of Integrated Network Based Cellular Signatures (LINCS) consortium. Drug similarity search was performed for Dextromethorphan (5 drugs) and Aspirin (4 drugs) from Drug Bank database. These 9 drugs were subjected to docking against RAC1 and NFκB1.

2.4 Molecular docking

The selected targets and their respective drugs were subjected to molecular docking studies through Extra Precision (XP) mode in Glide tool of Schrodinger drug design suite (version 10.7) (Glide, Schrödinger, LLC, New York, NY, 2018). All the docking procedures were performed on Intel® Core™ i7-6700CPU @ 3.40 GHz *8 of HP with memory of 15.6 GiB and OS Type -64 bit. The standard Dextromethorpan and Aspirin were found to possess -1.158, -5.565 docking score, -4.354, -34.822 kcal/mol glide energy respectively. The standard Dextromethorphan has a pi-pi stacking with Tyr139 in B chain (Figure 1A) whereas Aspirin has a hydrogen bond interaction with His170 and Leu173. (Figure 2A). Similar drugs for both the standards were found to exhibit higher docking scores except Carbaspirin calcium. Butorphanol a synthetic opioid agonist-antagonist analgesic and Salsalate a non-steroidal anti-inflammatory drug were found to possess highest docking scores (-3.505, -7.168) and glide energies (-22.523, -42.09 kcal/mol) than standards Dextromethorphan and Aspirin. Amongst the similar drugs for Dextromethorphan - Dezocine, Levorphanol and Butorphanol were found to possess CNS permeability.

Table 1 *DEGs extracted from microarray data in MS.*

GSE series	Up regulated genes	log Fc value	Down regulated genes	log Fc value
GSE21942	CCL5	0.146388	PSAP	-0.144177
	HBB	0.622313	S100A4	-0.118115
	RPL8	0.296145	LTA4H	-0.111731
	LYZ	0.124881	RPS8	-0.103688
	RPS26	0.124498	HLADRA	-0.092662
	RGS2	0.11344	COX1	-0.081268
	CD74	0.098308	TMSB10	-0.080381
	RPL19	0.07794	IGJ	-0.076606
	RPS18	0.073896	TXNIP	-0.076479
	RPL36	0.072369	HLA-DPA1	-0.07503

Table 2 *Docking results of RAC1 and NFkB1.*

Name of the Ligand	Docking Score	Glide Energy	Glide evdw (kcal/mol)	Glide ecoul (kcal/mol)	Xph Bond (kcal/mol)
Dextromethorphan(STD)	-1.158	-4.354	-4.949	0.595	0
Levallorphan	-3.265	-18.882	-18.882	-3.445	-0.35
Butorphanol	-3.505	-22.523	-22.523	-5.039	-0.7
Pentazocine	-2.303	-16.966	-16.966	-6.971	-0.7
Levorphanol	-1.882	-16.003	-16.003	-3.22	0
Dezocine	-2.507	-15.85	-15.85	-3.539	-0.491
Aspirin (STD)	-5.565	-34.822	-27.146	7.676	-0.921
Salsalate	-7.168	-42.09	-37.566	-4.523	-1.44
Triflusal	-6.733	-31.816	-26.254	-5.562	-0.96
Phenyl salicylate	-7.046	-39.851	-34.025	-5.825	-0.96
Carbaspirin calcium	-5.544	-34.82	-27.11	-7.71	-0.917

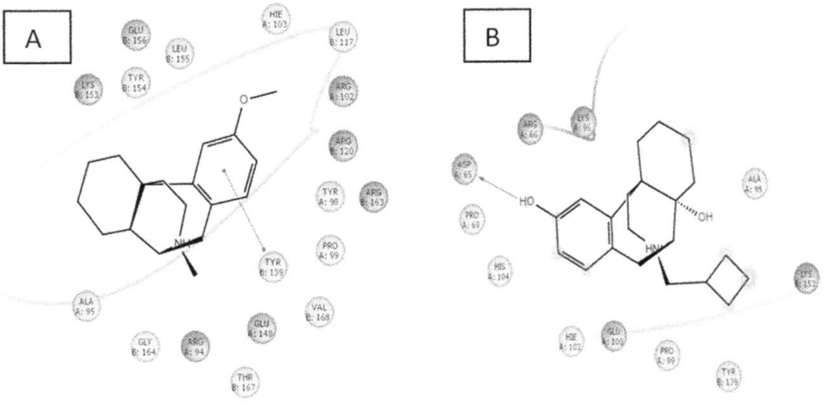

Figure 1 *2D images of Dextromethorphan (A) Butorphanol (B) with RAC1.*

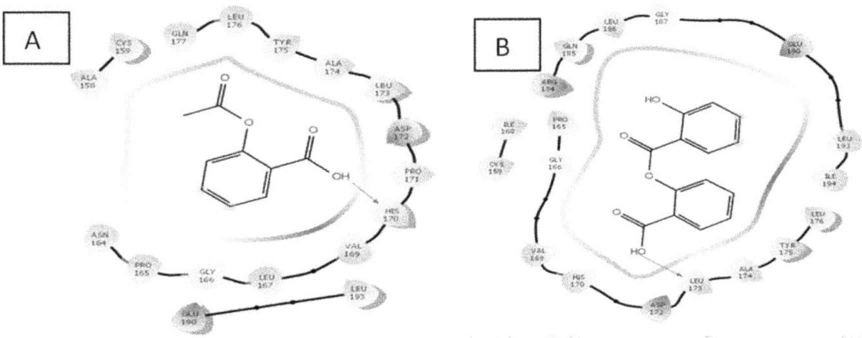

Figure 2 *2D images of interaction of Aspirin(A) and Salsalate(B) with NFκB1.*

CONCLUSION

All the drugs possess similar docking scores and no violations in Lipinski rule. Butorphanol was the only drug to possess CNS permeability. Hence, Butorphanol can be proposed for repurposing to treat MS. Besides the *in silico* results supporting the efficacy of Butorphanol, further *in vitro* and *in vivo* experimentations must be performed to validate the outcomes.

References

1. N. Ghasemi, S. Razavi, and E. Nikzad, Multiple sclerosis: pathogenesis, symptoms, diagnoses and cell-based therapy, *Cell J.*, 2017, **19**(191), 1–10.
2. B. Bielekova, *et al.,* Expansion and Functional Relevance of High-Avidity Myelin-Specific CD4 + T Cells in Multiple Sclerosis, *J. Immunol.*, 2004, **172**(6), 3893–3904.

3. P.-P. Axisa and D. A. Hafler, Multiple sclerosis, *Curr. Opin. Neurol.*, 2016, **29**(3), 345–353.
4. T. F. Scott, Improvement in relapse recovery with peginterferon beta-1a in patients with multiple sclerosis, *Mult. Scler. J.*, 2016, **2**, DOI: 10.1177/2055217316676644.
5. M. Hutchinson, R. J. Fox, E. Havrdova, N. C. Kurukulasuriya, S. P. Sarda, S. Agarwal, M. K. Siddiqui, A. Taneja and B. Deniz, Placebo-Controlled Phase 3 Study of Oral BG-12 or Glatiramer in Multiple Sclerosis, *New Engl. J. Med.*, 2012, **367**(12), 1087–1097.
6. T. P. Leist, G. Comi, B. A. Cree, P. K. Coyle, M. S. Freedman, H. P. Hartung, P. Vermersch, F. Casset-Semanaz and M. Scaramozza, Effect of oral cladribine on time to conversion to clinically definite multiple sclerosis in patients with a first demyelinating event (ORACLE MS): A phase 3 randomised trial, *Lancet Neurol.*, 2014, **13**(3), 257–267.
7. A. K. Kemppinen, J. Kaprio, A. Palotie and J. Saarela, Systematic review of genome-wide expression studies in multiple sclerosis, *BMJ Open*, 2011, **1**(1), e000053.
8. D. W. Huang, B. T. Sherman, R. Stephens, M. W. Baseler, H. C. Lane and R. A. Lempicki, DAVID gene ID conversion tool, *Bioinformation*, 2008, **2**(10), 428–430.
9. D. Szklarczyk, J. H. Morris, H. Cook, M. Kuhn, S. Wyder, M. Simonovic, A. Santos, N. T. Doncheva, A. Roth, P. Bork, L. J. Jensen, C. von Mering, The STRING database in 2017: quality-controlled protein–protein association networks, made broadly accessible, *Nucleic Acids Res.*, 2017, **45**(D1), D362–D368.
10. P. Shannon, A. Markiel, O. Ozier, N. S. Baliga, J. T. Wang, D. Ramage, N. Amin, B. Schwikowski and T. Ideker, Cytoscape: A Software Environment for Integrated Models of Biomolecular Interaction Networks, *Genome Res.*, 2003, **13**, 2498–2504.

Part 2: Structure Based Drug Design and Biological Evaluation

DESIGN, SYNTHESIS AND BIOLOGICAL EVALUATION OF 3-PHENYLISOXAZOLO[5,4-D]PYRIMIDINE DERIVATIVES AS ANTICANCER AGENTS

Nikhil Gaikwad and Y. V. Madhavi*

Department of Pharmaceutical Technology and Process Chemistry, National Institute of Pharmaceutical Education and Research (NIPER), Hyderabad, Telangana, India. 500037.

1 INTRODUCTION

Cancer is the leading cause of the death since last few decades leading to the death of millions of people worldwide. Cancer is a generic term for a large group of diseases characterized by the growth of abnormal cells beyond their usual boundaries that can then invade adjoining parts of the body and/or spread to other organs. Also as the medication available is leading to the resistance, thus the continuous supply of new and novel anticancer drugs is the demand of the current situation. Derivatizing the existing anticancer drugs is an important strategy for the development of new anticancer agents. Lapatinib is 4-anilinoquinazoline derivative a dual tyrosine kinase inhibitor, acts by disrupting the epidermal growth factor receptor and HER2/neu pathways. Many derivatives of the lapatinib were previously reported by modification on quinazoline ring such as furanopyrimidine[1] thiophenopyrimidine[2] were found to be promising anticancer agents. Many isoxazole and isoxazolines derivatives have been reported for their anticancer potential3-5, but the fused ring of isoxazole and pyrimidine is not yet explored for anticancer potential.[2]

2 METHODS AND RESULTS

Synthesis of 3-Phenylisoxazolo [5,4-d] pyrimidine

The synthesis of 3-Phenylisoxazolo[5,4-d]pyrimidine was started with benzaldehyde to get benzaldehyde oxime on treatment with hydroxylamine sulphate and sodium hydroxide in first step. Benzaldehyde oxime on reacting with N-chlorosuccinamide in acetonitrile gave compound **2**, which upon reaction with malononitrile resulted in cyano-aminoisoxazole (**4**)[6] using click chemistry. This intermediate which is the key for the synthesis of isoxazolopyrimidine was fully characterised by ^{1}H and ^{13}C NMR and HRMS. Compound **4** upon hydrolysis using Conc H_2SO_4 at room temperature for 3 hours[7], resulted in the amide compound **5** which was converted to 3-phenylisoxazolopyrimidine **6** by refluxing with $(EtO)_3CH$ (Triethyl orthoformate) using the reported procedure of Garcia *et al* with a slight modification where in acetic anhydride[7] was not used. Further **6** was chlorinated by using phosphoryl chloride to get 4-chloro-3-phenylisoxazolo [5,4-d] pyrimidine **7**. Compound **7** was then refluxed with different amines in DCE in presence of trimethylamine to give the desired final products (**8-10**).

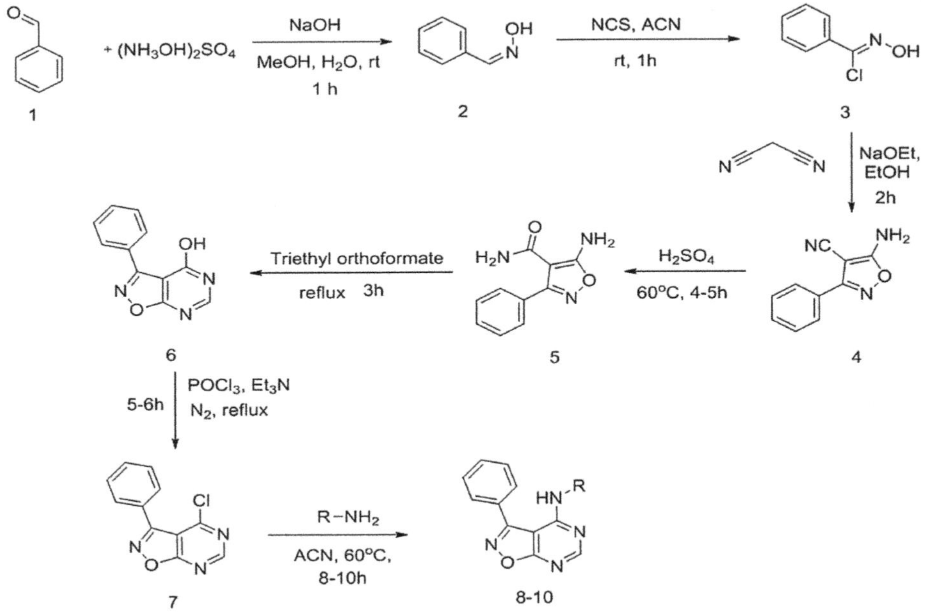

Scheme 1 *Synthesis of 3-phenylisoxazolo[5,4-d]pyrimidin-4-amine derivative.*

The substituted benzyloxyanilines and phenoxyanilines were synthesized by using 4-fluoronitrobenzene, substituted benzyl alcohol and phenols respectively to generate a series of substituted benzyloxynitrobenzene and phenoxynitrobenzene which on further reduction with iron and acetic acid gave benzyloxyanilines and phenoxyanilines. These anilines were refluxed in DCE in presence of catalytic triethylamine to give the final compounds (**8-10**)

2.1 Biological evaluation by NCI-60 cell line:

Three series (8-10) encompassing 40 isoxazolopyrimidine derivatives were synthesized, characterised, and assessed for their cytotoxic activity against a panel of 60 human cancer cell lines at National Cancer Institute (NCI), U.S.A. The selected 40 compounds (III a–f) were subjected to preliminary *in vitro* screening at a single dose (single dose study; 10 µM) against NCI-60 cell-line group representing 9 humanoid cancer types such as melanoma, leukemia and cancers of brain, colon, ovary, breast, kidney, lung and prostate. One-dose data will be reported as a mean graph of the percent growth of treated cells and will be similar in appearance to mean graphs from the 5-dose assay. The number reported for One-dose assay is growth relative to the no-drug control, and relative to the time zero number of cells. This allows detection of both growth inhibition (values between 0 and 100) and lethality (values less than 0). (The same as for the 5-dose assay, described below. For example, a value of 100 means no growth inhibition. A value of 40 would mean 60% growth inhibition.

A value of 0 means no net growth over the course of the experiment. A value of -40 would mean 40% lethality. A value of -100 means all cells are dead.) Thus according to their standard screening procedure[8] all the compounds were screened. Single dose study results given as mean graph of the % growth of the treated cells verses untreated control cells. Compound 10h exhibited promising activity while preliminary screening which is carried out at 10 µM on all 60 cell lines and was selected for the further 5 dose study regime which is carried out at five different concentrations of 0.01, 0.1, 1, 10 and 100 µM. The results for the 5 dose studies were given as three response parameters *i.e* GI_{50}: LC_{50} and TGI for each cell line, Results of one dose study and five dose study obtained for the compound **10h** were presented in Table 1.

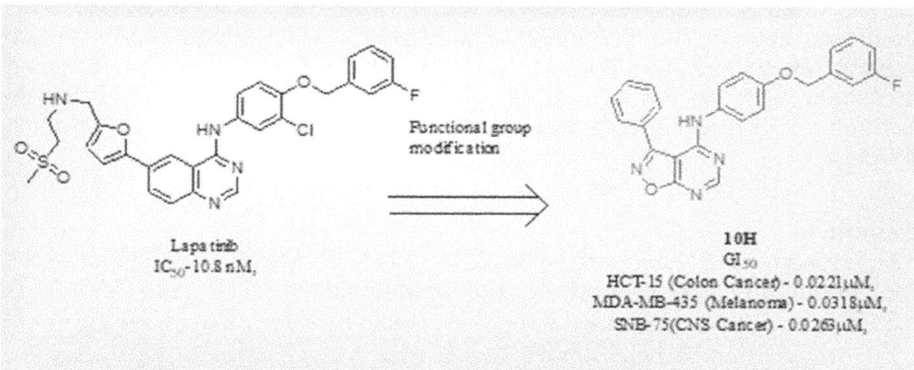

Table 1 *%Growth, GI_{50}, TGI and LC_{50} of the tested compounds against NCI-60 human cancer cell line panel.*

S. No.	Cell line	Growth %	GI_{50}	TGI	LC_{50}
1	Leukemia- K-562	12.08	0.0947	>100	>100
2	Leukemia- RPMI-8226	12.49	0.0582	>100	>100
3	Leukemia- SR	26.06	0.0671	>100	>100
4	Non-Small Cell Lung Cancer-NCI-H460	2.61	0.0502	28.4	>100
5	Colon Cancer-HCT-116	8.15	0.0577	>100	>100
6	Colon Cancer-HT29	3.86	0.0528	11.6	>100
7	Colon Cancer-KM12	9.57	0.08	14.8	>100
8	CNS Cancer-SNB-75	-34.93	0.0263	0.386	>100
9	Melanoma- LOX IMVI	23.07	0.949	>100	>100
10	Melanoma- MDA-MB-435	-54.67	0.0318	-	-
11	Breast Cancer-MCF7	14.05	0.0372	>100	>100

*Units of GI50, TGI and LC50 are expressed in µM

3 CONCLUSION

We have developed the 3 new series of N-3-diphenylisoxazolo [5,4-d] pyrimidin-4-amine, N-(4-(benzyloxy)phenyl)-3-phenylisoxazolo [5,4-d] pyrimidin-4-amine and N-(4-phenoxyphenyl)-3-phenyl-isoxazolo[5,4-d] pyrimidin-4-amine from condensation of 4-chloro-3-phenylisoxazolo[5,4-d] pyrimidine and substituted aniline derivatives. Out of all synthesized analogues many have shown good to excellent anticancer activity on 9 different human cancer cell lines at 10 μM. Series **8** and **9** have shown weak anticancer profile. Series **10** molecules have shown better cytotoxicity than other two series and compound **10h** was found to show promising cytotoxicity to lines like HCT-15 (Colon Cancer) - 0.0221μm, MDA-MB-435 (Melanoma) -0.0318μm, SNB-75(CNS Cancer) - 0.0263μm. The 3-fluorobenzyloxyanline attachment to the 3-phenylisoxazolo[5,4-d]pyrimidine was found to be crucial for the activity. This was then proven by the molecular docking studies of the compound **10h** and Lapatinib which was the structural basis for the design, have similar binding interactions with the kinase protein. The outcome of this study will help in development of novel anti-cancer agents based on isoxazolo[5,4-d]pyrimidine scaffold.

References

1. M. Hossam, D. S. Lasheen, N. S. M. Ismail, A. Esmat, A. M. Mansour, A. N. B. Singab and K. A. M. Abouzid, *Eur. J. Med. Chem.*, 2018, **144**, 330–348.
2. S. R. Abd El Hadi, D. S. Lasheen, M. A. Hassan and K. A. M. Abouzid, *Arch. Pharm.*, 2016, **349**(11), 827–847.
3. J. P. Yong, C. Z. Lu and X. Wu, *Anti-Cancer Agents Med. Chem.*, 2015, **15**(1), 131–136.
4. T. L. Dadmal, K. Appalanaidu, R. M. Kumbhare, T. Mondal, M. J. Ramaiah and M. P. Bhadra, *New J. Chem.*, 2018, **42**(19), 15546–15551.
5. B. Çalışkan, E. Sinoplu, K. İbiş, E. Akhan Güzelcan, R. Çetin Atalay and E. Banoglu, *J. Enzyme Inhib. Med. Chem.*, 2018, **33**(1), 1352–1361.
6. E. M. Beccalli, A. Manfredi and A. Marchesini, *J. Org. Chem.*, 1985, **50**(13), 2372–2375.
7. E. C. Taylor and E. E. Garcia, *J. Org. Chem.*, 1964, **29**(8), 2116–2120.
8. A. Monks, D. Scudiero, P. Skehan, R. Shoemaker, K. Paul, D. Vistica, C. Hose, J. Langley, P. Cronise, A. Vaigro-Wolff, M. Gray-Goodrich, H. Campbell, J. Mayo and M. Boyd, *J. Natl. Cancer Inst.*, 1991, **83**(11), 757–766.

QSAR AND DOCKING STUDIES ON TRIAZOLE BENZENE SULFONAMIDES WITH HUMAN CARBONIC ANHYDRASE IX INHIBITORY ACTIVITY

Gopinath P.[a] and Kathiravan M. K.[b]

[a] Department of Pharmaceutical Chemistry, SRM College of Pharmacy, SRMIST, Kattankulathur, India
[b] 209, Dr. APJ Abdul Kalam Research Lab, SRM College of Pharmacy, SRMIST, Kattankulathur, India

Corresponding Author E-Mail: gopi.pharma@gmail.com

1 INTRODUCTION

According to WHO reports,[1] Cancer is the second leading cause of death worldwide and is responsible for an estimated 9.6 million deaths in 2018. Breast cancer accounted for 2.09 million cases in the year 2018. Carbonic anhydrases are involved in lowering tumor pH by converting carbon dioxide and water to carbonic acid. It is known that transcription of carbonic anhydrase IX (CA IX) is activated by hypoxia and suppressed in normoxia. 4-Hydroxy-3-(3-(phenyl ureido) benzenesulfonamides as SLC-0111 is currently in clinical trials for CA-IX inhibition activity in hypoxic tumors.[2] One of the rational and successful methods in drug design is Quantitative structure-activity relationship (QSAR), playing a key role in optimizing leads and thereby improving their biological activity. The aim of this work is to design better leads for carbonic anhydrase IX inhibition, a major hypoxic target for breast cancer.

2 METHODS AND RSULTS

'QSARINS'[4,5] allows developing MLR-QSAR models by ordinary least squares, attentively verified and in detail validated according to the chemometric approach. Dataset compounds were calculated for 3D-MoRSE molecular descriptor values from ChemDes - Chemopy servers.[3] Genetic algorithm is used for descriptor selection relational to biological activities of molecules along with Friedman's 'lack-of-fit' (LOF) function, to evaluate the fitness of models. Internal validation by cross validation leave-one-out ($Q2_{LOO}$), cross validation leave-many-out ($Q2_{LMO}$), root mean squared error (RMSE), Y-Scrambling and external validation by Q^2_{F1}, Q^2_{F2}, Q^2_{F3}, and CCC were applied on selected models. To evaluate reproducibility of models, concordance correlation coefficient (CCCext) is analysed. In combination with the developed QSAR model, affinities and interactions of carbonic anhydrase inhibitors were estimated using AUTODOCK V4.2.6.[6]

Out of many experimental models, only the final model is included having with molecular descriptors positively correlated for bioactivity. 3D-MoRSE descriptors meaning molecular representation of structure by electron diffraction weighted by charges, polarizability, masses and volume have played a key role compared with other weighted and unweighted descriptors. Scattering parameter of 3D-MoRSE descriptor provide information about the structure of molecules.

pKi = 2.0573 + 3.3553 (MoRSEV22) + 16.4033 (MoRSEC17) + 0.2339 (MoRSEV1) + 2.0175 (MoRSEC4) + 0.1158 (MoRSEE2)

n_{tr} = 56, n_{pred} = 10, R^2 = 0.7852, LOF = 0.2278, $RMSE_{tr}$ = 0.3920, CCC_{tr} = 0.8797, s=0.4149, F=36.5464, Q^2_{LOO} = 0.7237, $Q2_{LMO}$ = 0.7071, R^2_{Yscr} = 0.0889, Q^2_{Yscr} = -0.1471, $RMSE_{cv}$ = 0.4446, CCC_{cv} = 0.8472, R^2_{ext} = 0.7894, $RMSE_{ext}$ = 0.4095, CCC_{ext} = 0.8784, Q^2_{F1} = 0.7255, Q^2_{F2} = 0.7254, Q^3_{F3} = 0.7656.

Figure 1 displays Kxy vs. R^2_{Yscr} and Q^2_{Yscr} graph of final model from which scramble models can be differentiated from normal ones, clearly stating model is not obtained by chance correlation.

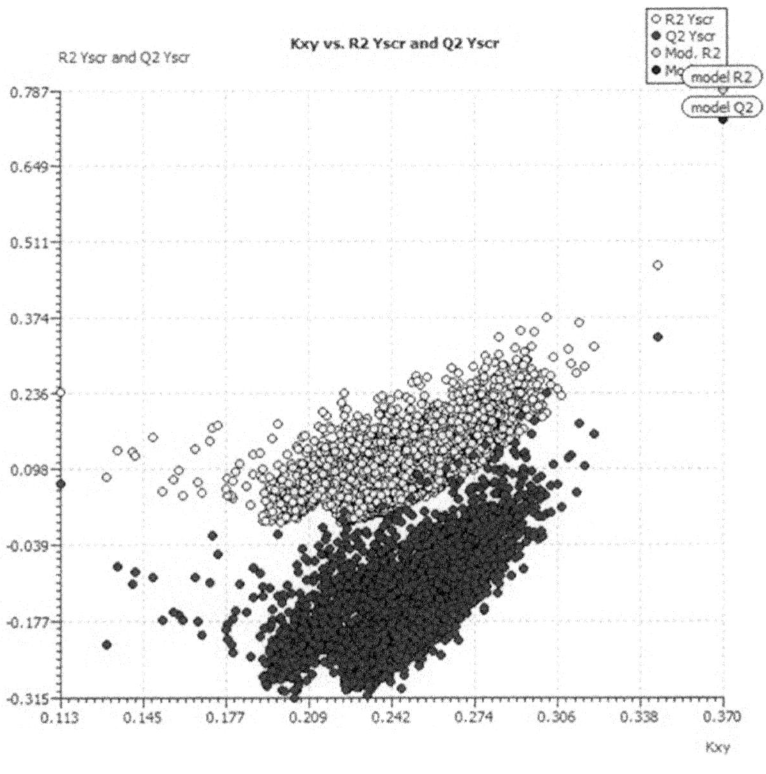

Figure 1 *Kxy vs.* R^2_{Yscr} *and* Q^2_{Yscr} *of model 4.*

It is evident that, the final model shows better fitting criteria, internal validation parameters and external predictive parameters, particularly Q^2_{LOO}, LOF, Q^2_{LMO}, CCC, Q^2_{F1}, Q^2_{F2}, Q^2_{F3}, R^2 and RMSE. Results state that, the model obtained is not by chance and truly there is a relationship between structures of triazole benezene sulphonamide analogs with corresponding carbonic anhydrase IX inhibitory activity.

Compound 3e interacted well in the active site of CA IX with Zn, Thr200, His96 and His119. Glu106 and Val121 also participated in H-bonding stabilizing the core of benzene sulfonamide derivative. Compound 6f showed interactions at both hydrophilic and hydrophobic region of active site. The results were favorable giving importance to model equation descriptors supporting QSAR studies. Figure 2 display interactions of compound 3e with 5FL4 from Ligplot$^+$ program.[7]

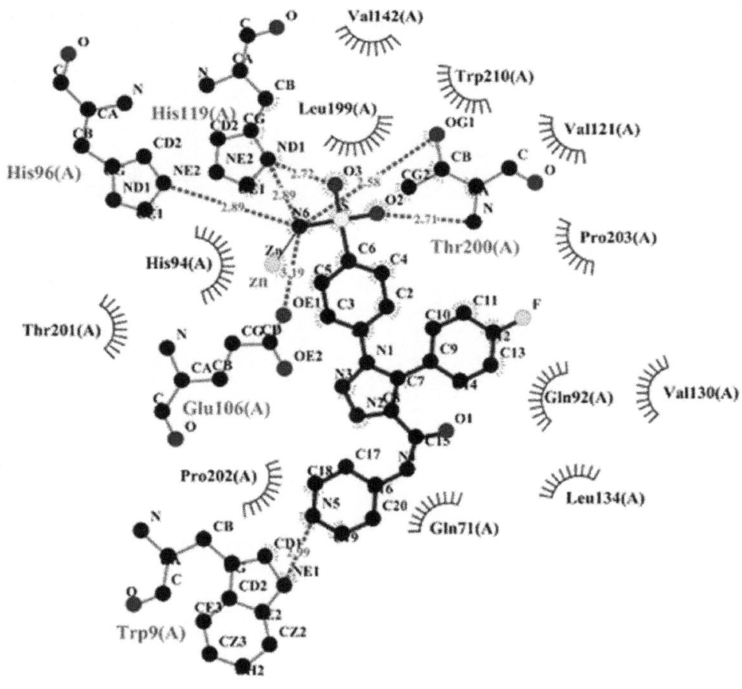

Figure 2 *Interaction of Compound 3E with Carbonic anhydrae IX (5FL4).*

3. CONCLUSION

The objective of using QSAR study conducted by 3D-MoRSE descriptors resulted in a robust and stable model by which better leads have been designed and studied by Autodock. The model is validated both internally and externally using software statistical parameters. The model is tested on designed compounds for predicting bioactivity and subjected to molecular docking studies for structural and interaction information which gave compounds 3e and 6f as potential leads and will be considered for future studies.

References

1. https://gco.iarc.fr/today/data/factsheets/cancers/20-Breast-fact-sheet.pdf.
2. F. Pacchiano, F. Carta, P. C. McDonald, Y. Lou, D. Vullo, A. Scozzafava, S. Dedhar, and C. T. Supuran, *J. Med. Chem.,* 2011, **54**, 1896.
3. J. Dong, D. S. Cao, H. Y. Miao, S. Liu, B. C. Deng, Y. H. Yun, N. N. Wang, A. P. Lu, W. B. Zeng and A. F. Chen, *J. Cheminf.*, 2015, **7**, 60.
4. P. Gramatica, N. Chirico, E. Papa, S. Cassani and S. Kovarich, *J. Comput. Chem.,* 2013, **34**, 2121.
5. P. Gramatica, S. Cassani and N. Chirico, *J. Comput. Chem.,* 2014, **35**, 1036.
6. D. Santos-Martins, S. Forli, M. J. Ramos and A. J. Oslon, *J. Chem. Inf. Model.,* 2014, **54**, 2371.
7. R. A. Laskowski and M. B. Swindells, *J. Chem. Inf. Model.,* 2011, **51**, 2778.

DESIGN AND SYNTHESIS OF SOME NOVEL ERLOTINIB DERIVATIVES AS POTENTIAL EGFR KINASE INHIBITORS

V. Ajeesh,[a] S. Shubham[b] and M. Ruchi[b]

[a] Department of Pharmaceutical Sciences, Al-Shifa College of Pharmacy, Perinthalmanna, Kerala, 679325, India; [b] Department of Pharmacy, School of Chemical Sciences and Pharmacy, Central University of Rajasthan, Bandarsindri, Ajmer Dist.-305817, India

1. INTRODUCTION

Despite many significant improvements in cancer therapies over the past 50 years, metastatic solid cancers remain largely incurable. The survival for patients with these malignancies is often measured in months. Substantial efforts are being made in the development of targeted therapies for a specific type of cancer. These efforts have been spurred by few successes, such as imatinib for Chronic Myelogenous Leukaemia (CML); Trastuzumab for breast cancer with amplification of ERBB2 (also known as HER2); and erlotinib and gefitinib for lung cancer that regulates mutant epithelial growth factor receptor (EGFR). In the era of targeted drug discovery tyrosine kinases (TKIs) are the most studied pathways. The targeted drug discovery is preferable over conventional method and is used to reduce the maximum resistance and side effects. Especially with the EGFR mutation cause dramatic changes in patients. The mutations in the EGFR signalling pathway is one of the main reasons for one-third of non-small cell lung cancer.

2. METHOD AND RESULTS

2.1 *In silico* study

The drug design technique has been implemented to design molecules possessing better selectivity, ADMET etc. In this study, we have identified a binding pocket followed by an e-pharmacophore generation which gave us an insight about the binding pose and structural requirements essential for selectivity and activity (Figure 1). After e-pharmacophore prediction, potential T790 mutant EGFR inhibitors have been designed by the de-novo method and it has been designed by the de-novo method and it has been validated by docking and scoring function as well as predicted descriptor values suitable for predicting ADME properties.

Docking study using Schrodinger was performed. Schrodinger maestro is a genetic algorithm-based docking program that allows a wide range of confirmations and flexibility for the ligand in the binding pocket of the desired protein. This has been implemented to predict and find out the binding modes and orientation of the newly synthesized compounds at the active binding site of EGFR binding protein. The crystal structure and coordinates of the T790 mutated EGFR structure along with their inhibitor was obtained from the protein data bank, (PDB ID: 1M17 and 2JIU). The root means square difference (RMSD) between the top docking confirmation and original crystallographic geometry was found 0.6Å and 0.8Å for WTEGFR and T790M

EGFR respectively. The e-pharmacophore hypothesis (AARRRH) which consists of two acceptors (A), three rings (R), one hydrophobic (H). the pharmacophore hypothesis was based on interaction energetics provided by ligand docking of erlotinib in resistant type EGFR. The docking protocol of developed lead molecule, erlotinib has been validated in the active site of both wild and T790 mutated EGFR kinase, (PDB ID: 1M17 and 2JIU), binding pattern of erlotinib in the active site of the X-ray crystal structure of EGFR kinase has been retrieved. This showed protein (PDB ID 2JIU), where the N-1 atom of quinazoline heterocycle interacts with the backbone of Met769 via H-bond. 4-anilinoquinazoline based inhibitors resemble the binding pattern with the proposed design towards the protein. The best-scored pose of compound RM-AV-09, the main interactions between the ligand and receptor has been observed and identified as following: (i) the N- 1 atom of quinazoline interacted by H-bond with backbone NH of Met769 (ii) the N-3 made water-mediated H-bond to the water molecule H10, (iii) the substituted 4-anilinoquinazoline was shown remarkable fitting pattern in the deep pocket back of the ATP binding site. The chlorine group in the derived molecule found that it has the capability of forming an extra hydrogen bonding with Cys-775 which was absent in erlotinib, considered as the reference molecule. This was predicted which could enhance the binding affinity and fix the ligands in its binding pocket.

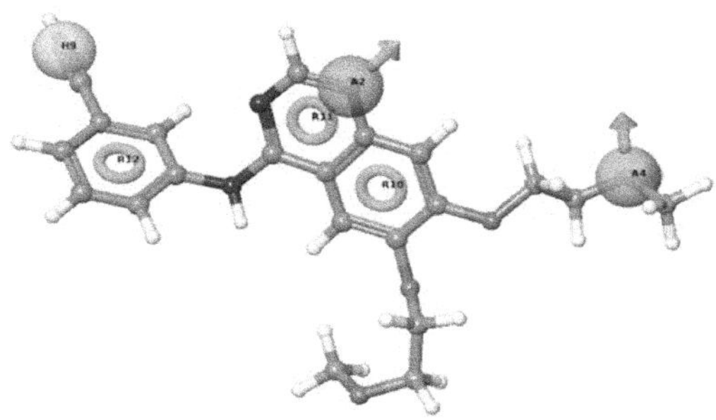

Figure 1 *Erlotinib e-pharmacophore in T790 mutant EGFR.*

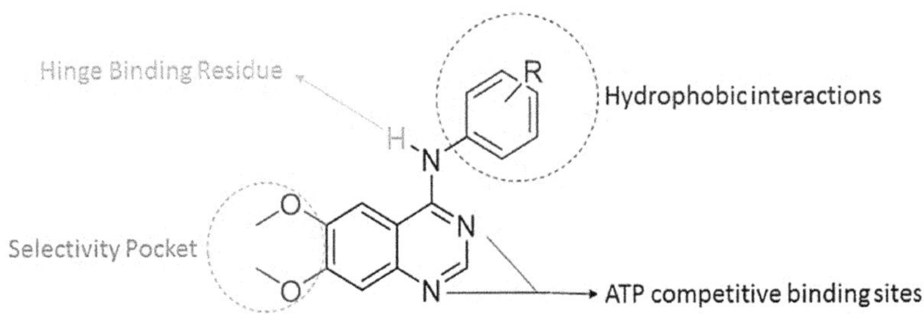

Figure 2 *Design approach.*

The molecules have been designed as per figure 2, in which the hydrophobic interaction part has been modified with the diversity of anilines as well as other heterocyclic diversities. The linker NH between quinazoline and phenyl ring binds to the hinge binding region which is the most important residue for an inhibitory activity for EGFR. The two methoxy groups were implemented for selectivity towards receptor additionally the nitrogen's in quinoline scaffold acting as an ATP competitive binding pocket.

Table 1: *e-pharmacophore hypothesis.*

Wild type EGFR (1M17)			T790M EGFR (2JIU)		
Rank	**Feature**	**Score**	**Rank**	**Feature**	**Score**
1	R11	-2.88	1	R11	-2.23
2	A2	-2.17	2	A2	-2.18
3	R10	-0.98	3	R10	-1.07
4	A1	-0.67	4	R12	-1.02
5	R12	-0.49	5	A4	-0.48
			6	H9	-0.25

2.2 Synthesis of Erlotinib derivative

The synthesis of erlotinib derivative had been started from a key intermediate 4-chloro-6,7-dimethoxyquinazoline. This was prepared from 3,4-dimethoxy benzoic acid in a modified and feasible way. The spectroscopic techniques, IR, 1H NMR, 13C NMR, and MASS spectrometry used to confirm the structure of the synthesized molecule.

In the synthesis of erlotinib derivatives, it was found that efficient formation of an intermediate ethyl-2-amino-4,5-dimethoxy benzoate (RM-AV-03) from 4,5-dimethoxy-2-nitro benzoate (RM-AV-02), and 6,7-dimethoxy quinazoline-4(3H)-one (RM-AV-04) from ethyl-2-amino-4,5-dimethoxy benzoate (RM-AV-03) was difficult. To overcome this problem an efficient and feasible method was developed.

Previously reported methods for synthesis of quinazoline scaffold derivatives were either lengthy or had purity issues. Herein, synthetic route optimization was focused on reducing the time of reaction and to get better yield. Formation of 6,7-dimethoxy quinazoline-4(3H)-one was the crucial step. Previous reports on quinazoline ring formation were an either futile or tedious process. To overcome this problem, reaction conditions were optimized and its time of reaction had been reduced to 4 hrs as well as the yield was increased up to 73%. So as in results 6,7-dimethoxy quinazoline-4(3H)-one (RM-AV-04) from ethyl-2-amino-4,5-dimethoxy benzoate (RM-AV-03) were obtained in good yields.

2.3 Descriptor values of the derivatives

The predicted descriptor values were calculated computationally by Qik prop version 4.6 application of Schrodinger maestro. Most of them will be useful in predicting ADMET properties.

Fifteen descriptor values calculated include, predicted central nervous system activity on a –2 (inactive) to +2 (active) scale (CNS), total solvent accessible surface area (SASA) in square angstroms using a probe with a 1.4 Å radius (SASA). Donor hydrogen bond (donrHB) predicted polarizability in cubic angstrom (PolrZ), predicted aqueous solubility (Log S), predicted octanol/water partition coefficient (log Po/w), Caco-2 permeability (PCaco), predicted skin permeability (Log Kp), predicted blood/brain partition coefficient (Log BB), Human ether-a-go-go-related gene (log HERG), acceptor hydrogen bond (accpt HB), predicted qualitative human oral absorption (HOA value ranges from 1 for low, 2 for medium and 3 for High), number of violation of rule of five, Madin Darby canine kidney cell line permeability (MDCK), Van der Waals surface area of polar nitrogen and oxygen atoms and carbonyl carbon atoms (PSA), and Predicted human oral absorption on 0 to 100% scale. The prediction is based on a quantitative multiple linear regression model. This property usually correlates well with Human Oral- Absorption, as both measures the same property.

CONCLUSION

We have implemented a hybrid approach to of structure-based drug design and ligand-based drug design techniques to overcome the resistance problems of the present therapy available. The results obtained from the drug design aspect were implemented to design novel compounds that could preferably overcome T790 mutation in EGFR inhibition. Substitution in R12 ring position can make a remarkable binding affinity towards EGFR receptor.

The synthesis of these developed molecules has been done with the different chemical process infeasible manner and the characterisation has been done by using different chemical tools like NMR, melting point, mass etc. In this present study, the compound RM-AV-09 and RM-AV-10 showed good binding affinity towards T790 mutant EGFR also. The compound RM-AV-09 possesses an additional hydrogen bond which was not present in erlotinib.

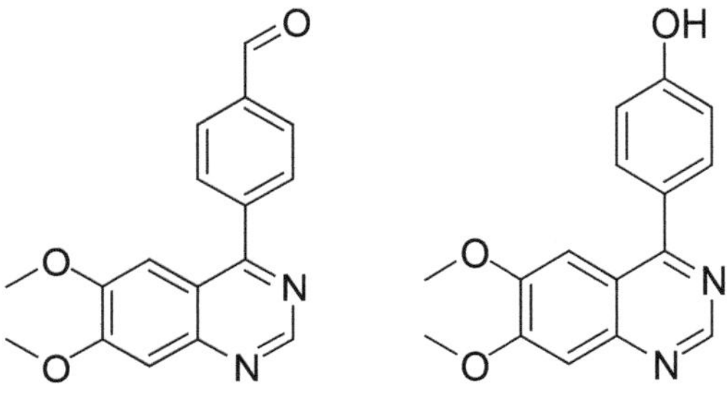

RM-AV-09 RM-AV-10

References

1. M. Scaltriti and J. Baselga, *Clin. Cancer. Res.*, 2006, **12**, 18.
2. L. Zhang, Y. Yang, H. Zhou, Q. Zheng, Y. Li, S. Zheng, S. Zhao, D. Chen and C. Fan, *Eur. J. Med. Chem.,* 2015, **102**, 445.

DESIGN, SYNTHESIS AND CHARACTERIZATION OF NOVEL SERIES OF 1,3-THIAZOLE CONTAINING HETEROCYCLES AS INHIBITORS OF DHFR FOR USE AS ANTICANCER AGENTS

Sunil Dilip Pawar, Mithila N. S., Abhishek Yadav, Shraddha Bhat, Roshni Varghese, Anjali Sharma, Sharvari Somayaji, Sanjukta G. P., Vibha S. and Anil Kumar Agarwal

Department of Chemistry, CHRIST (Deemed to be University), Bengaluru, India

1 INTRODUCTION

Dihydrofolate reductase (DHFR) is an indispensable cellular enzyme which catalyzes the reduction of 7,8- dihydrofolate to tetrahydrofolate (THF) [1]. Tetrahydrofolate is used as a cofactor during the formation of serine and thymidine (required for DNA replication) via a one carbon transfer reaction pathway involving pyrimidine, purine, and amino acid biosynthesis, resulting in cell growth and proliferation. It is also used in the methylation of dUMP to dTMP [2]. DHFR is a highly targeted, pivotal enzyme in human therapy as it is the only source of tetrahydrofolate [3]. DHFR inhibitors are potential pharmacological agents showing promising results as novel anti-cancer, anti-bacterial, anti-malarial therapeutics. Methotrexate is a popular DHFR inhibitor whose biological potency is offered by 1,3-thiazole, which exists in many fused heterocyclic systems. Several compounds bearing thiazole heterocycle such as tiazofurin, 4-phenyl- thiazole-1,3,5-triazines, netropsin and thia-netropsin were reported as nonclassical DHFR inhibitors [4]. Molecular modelling studies have concluded that recognition with key amino acid Leu4, Glu30, Arg22 and Val115 are essential for DHFR binding and the thiourea moiety is seen as an anchoring group to enhance the DHFR inhibition [5]. In view of these facts, the present study targeted the design, synthesis and biological evaluation of novel series of 1,3-thiazole heterocycles containing electron withdrawing groups acting at the DHFR enzyme active site. The obtained model could be useful for the development of a new class of DHFR inhibitors.

2 METHODS AND RESULTS

2.1 Molecular modeling

Docking studies were performed for all the designed ligands using human DHFR X-ray crystal structure bound to MTX (1U70.pdb) via Glide docking method from Schrodinger. **Compound-3** showed highest docking score as the electron withdrawing nitrile group was able to make a hydrogen bond interaction with tryptophan-24 (TRP24) along with other pi-pi and cation-pi interactions with rest of the molecule.

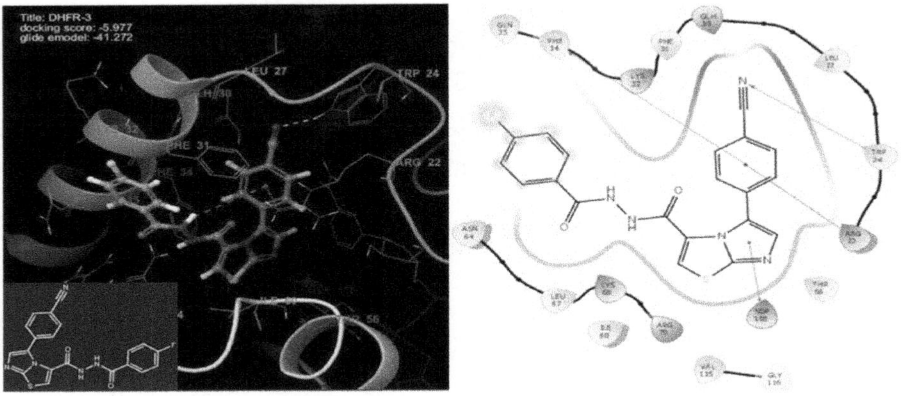

Figure 1 *Docked pose of* ***Compound-3*** *showing interactions with protein in 3D and 2D.*

2.2 Chemistry

The target molecules were synthesized as per the scheme-1.

Ethyl 2-aminothiazole-4-carboxylate was taken in a round bottom (RB) flask and dissolved in ethanol. 2- bromo-1-(4-substitutedphenyl) ethanone was then added to the RB flask and stirred under reflux for 24h. Ethyl 5- (4-substitutedphenyl) imidazo[2,1-b]thiazole-3-carboxylate was then treated with hydrazine hydrate using ethanol as solvent and refluxed for 18h to form 5-(4-substitutedphenyl) imidazo[2,1- b]thiazole-3-carbohydrazide. Intermediate 4 was dissolved in pyridine and 4-substituted benzoyl chloride was added and stirred for 24h at room temperature to give the final compound.

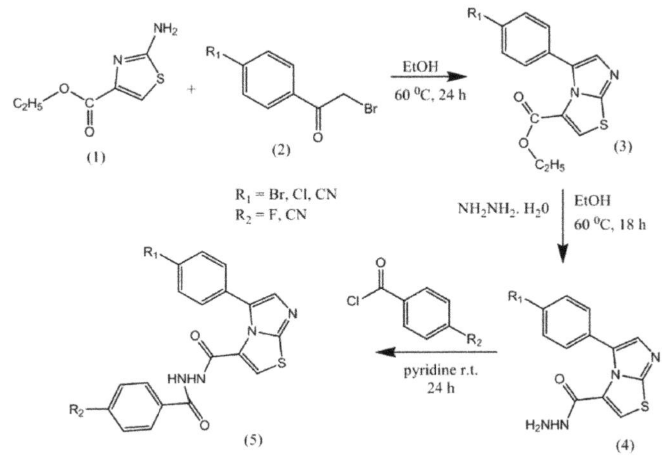

Scheme 1 *Synthesis of target compound.*

Substituting R_1 (Br, Cl, CN) and R_2 (F, CN), six compounds were synthesized as shown in Figure1. The synthesized compounds were characterized by FT-IR, ^{1}H NMR, and Mass spectral analysis (LCMS).

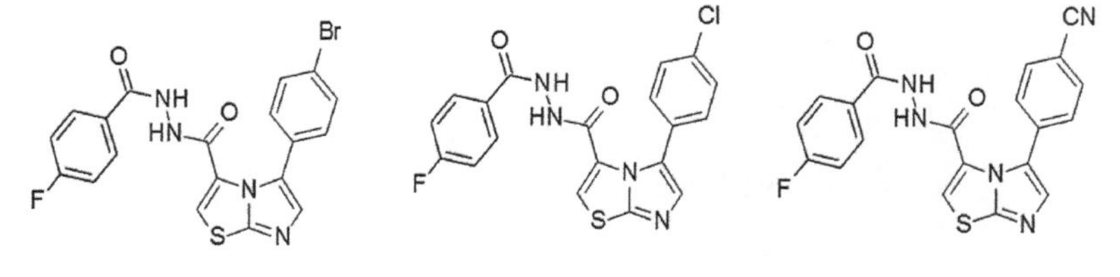

1 2 3

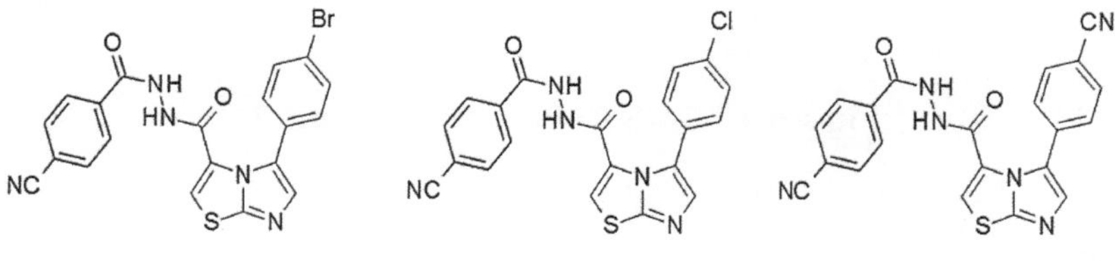

Figure 2 *Structures of synthesized compounds.*

2.3 Analytical data

The ^{1}H NMR and LCMS data for **Compound-3** is shown below.

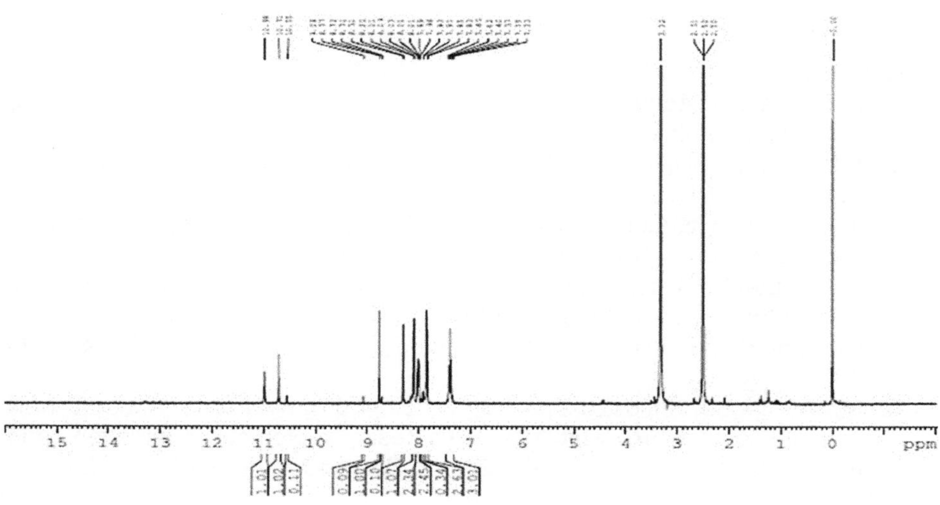

Figure 3 *^{1}H NMR Spectra of* ***Compound-3****.*

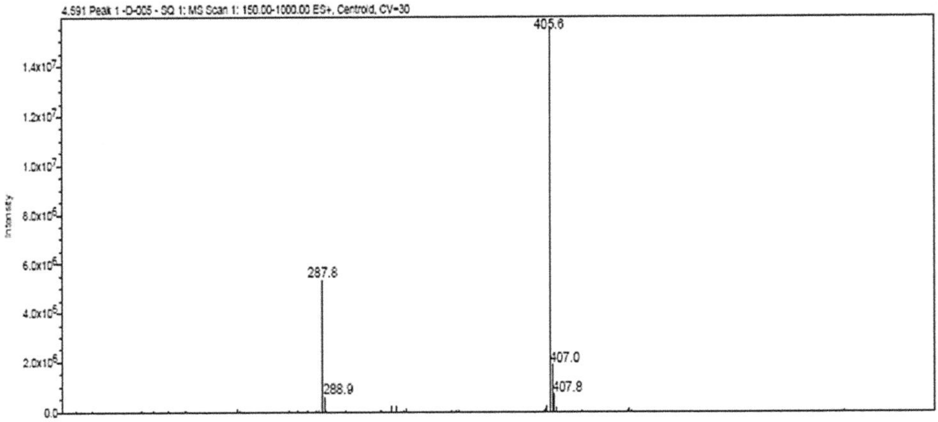

Figure 4 *LCMS Data of* ***Compound-3****.*

3 CONCLUSION

Thiazoloimidazole based heterocycles with electron-withdrawing groups were designed and synthesized for screening against DHFR. All the compounds were characterized using FT-IR, ^{1}H NMR, and Mass spectral analysis (LCMS). Biological evaluation for the synthesized compounds for in vitro and in vivo DHFR inhibition will help us in better understanding of SAR for design of safe and potent anticancer compounds.

References

1. R. L. Blakley, *Advances in Enzymology and Related Areas of Molecular Biology*, Memphis, Tennessee, 1995, vol. 70, pp. 23–102.
2. I. M. Kompis, DNA and RNA synthesis: Antifolates, *Chem. Rev.*, 2005, **105**, 593–620.
3. J. R. Schnell, Structure, dynamics and catalytic function of dihydrofolate reductase, *Annu. Rev. Biomol. Struct.*, 2004, **33**, 119–140.
4. M. A. Ewida, Thiazolo[4,5-d]pyridazine analogues as a new class of dihydrofolate reductase (DHFR) inhibitors: Synthesis, biological evaluation and molecular modeling study, *Bioorg. Chem.*, 2017, **74**, 228–237.
5. M. A. Ewida, Imidazo[2′,1′:2,3]thiazolo[4,5-d]pyridazinone as a new scaffold of DHFR inhibitors: Synthesis, biological evaluation and molecular modeling study, *Bioorganic Chemistry*, 2018, **80**, 11—23.

IN SILICO AND *IN VITRO* STUDIES OF BENZOHYDRAZIDE ANALOGUES AS POTENT ANDROGEN RECEPTOR ANTAGONIST

H. A. Arjun,[a] Ravi Kumar Rajan,[b] R. Elancheran,[a] K. Lakshmithendral,[a] M. Ramanathan,[b] Atanu Bhattacharjee[c] and S. Kabilan[a*]

[a] Drug Discovery Lab, Dept. of Chemistry, Annamalai University, Annamalainagar - 608002, India; [b] Dept. of Pharmacology, PSG College, Coimbatore - 641004, Tamil Nadu, India; [c] Dept. of Bioinformatics, North Eastern Hill University, Shillong, India

1 INTRODUCTION

In the late 1950s, the Prostate Cancer (PCa) is commonly diagnosed in men. The growth of PCa is largely due to the stimulation of the AR by androgens. Currently, hormonal therapy, surgery, and oral chemotherapeutic drugs are the main therapeutic for the treatments of prostate cancer. About 1,74,650 men were diagnosed and 31,620 deaths due to PCa in the U.S reported by the American Cancer Society.[1] Androgens, testosterone, and Dihydrotestosterone binds with the ligand binding domain (LBD) of the androgen receptor (AR) and incite the growth of prostate cancer. There are few available drugs for androgen receptor antagonists such as bicalutamide, flutamide, and enzalutamide are used to prevent the activation of the androgen receptor. Currently, Sipuleucel-T is used as a drug for the immunotherapy to inhibit PCa. To reduce toxicity, lower the doses, and minimize drug resistance, numerous studies are enduring for the development of target based drugs.[2] Recently, benzohydrazide derivative was reported for antifungal–antibacterial, anti-inflammatory, cruzipain inhibitor. Entamoeba histolytic, antimalarial, anti-tuberculosis activities, and as EGFR Tyrosine Kinase Inhibitors. Thus, benzohydrazides are the important moieties which play a vital role in effective inhibition against different cancer cell line HepG2, HeLa, A549, and MCF-7.

All these facts made us screen new substituted benzohydrazide derivatives as a potent androgen receptor antagonist. Structure-based drug design approach revealed the findings of substituted benzohydrazide derivatives, which shows the possible binding modes of these compounds with the androgen receptor (AR PDB ID: 3V49). Further, the compounds *(4a-d)* were synthesized based on the highest glide score and analyzed well by spectroscopic techniques (NMR, LC-MS, IR). The anticancer studies for the compounds *(4a-d)* were tested against LNCaP & PC3 cell lines and further confirmed by DNA Methyl Transferase enzyme inhibitor assay (ELISA).[3] MD simulation was carried out for the potent molecule of the series. Pharmacokinetic properties for all the compounds *(4a-d)* were predicted.[4] These applications and results prompted us to synthesize new series of benzohydrazide derivatives.

2 METHOD AND RESULTS

2.1 In *silico* molecular docking

This approach was carried out using Schrodinger software. The protein was obtained from the PDB (3V49) and prepared using a protein preparation module. The ligands were prepared by Ligprep wizard and docking was done using GLIDE XP.[5] The compounds were synthesized as shown in (Scheme 1) and characterized by spectroscopic IR, ^1H & ^{13}C NMR and LC-MS methods.

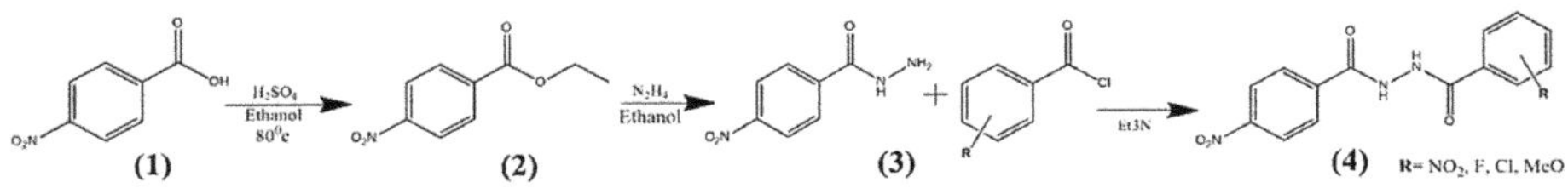

Scheme 1 *Synthesis of compounds (4a-d).*

The MTT assay for the compounds (4a-d) was carried out by LNCaP & PC-3 cell lines, which are obtained from NCCS Pune, India. In RPMI medium with 10% FBS, the cells were cultured at 35°C with 6% CO_2. In 96 well plates, around 5,000 cells/well were seeded and incubated for 24 hrs to evaluate the antiproliferative activity. DNA Methyl Transferase enzyme inhibitor assay was done using EpiQuik, DNMT activity/inhibition screening assay kit from Epigentek, Brooklyn, USA. Further, Molecular Dynamic stimulations studies were carried out for 50.098 ns using Desmond module for a potent molecule 4b which showed inhibition against prostate cancer. The pharmacokinetic properties were determined by using Qikprop-4 module of Schrödinger software. Drug-likeness properties of the compounds were analyzed by Lipinski's rule of three and five.

2.2. Molecular docking studies

They revealed the highest glide score of -9.332 kcal/mol for the compound *4b* and all the compounds*(4a-d)* were in the range of (-9.33 to -7.84 kcal/mol). The amino acids Leu707, Leu704, Met895, Leu873, Met742, Trp741, Phe764, and Met749 are having hydrophobic interactions with AR receptor are shown in Figure. 1. As shown in scheme 1, the molecules were synthesized and were confirmed by ^1H & ^{13}C NMR, IR, and LC-MS. The purity of compound *4b* and *4c* were analyzed using LC.

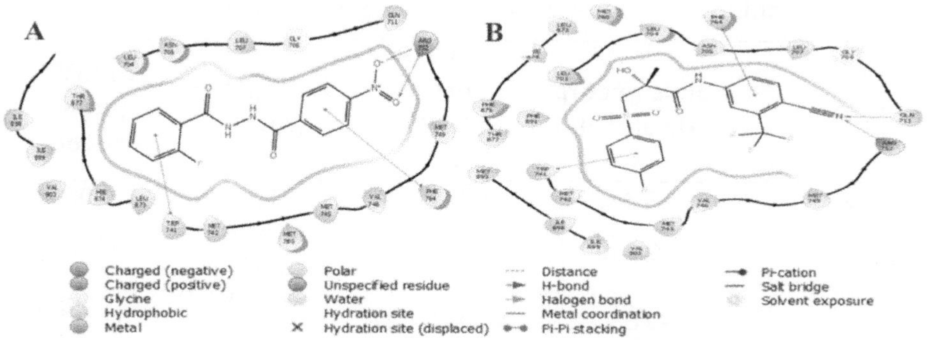

Figure 1 *Binding modes of compound 4b (A) and Std. drug Bicalutamide (B) in AR LBD 3V49.*

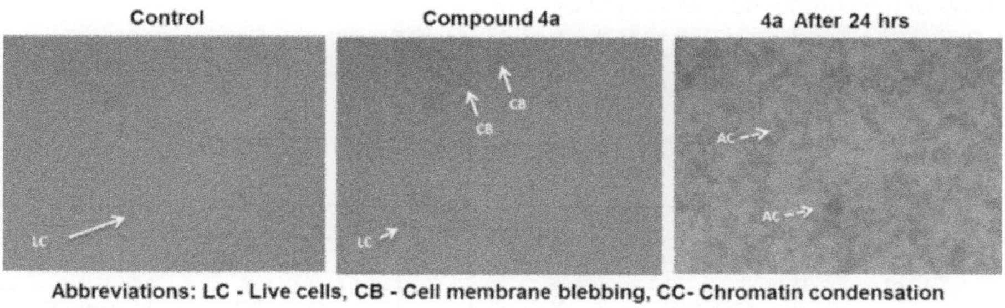

Figure 2 *The untreated cells control & morphological changes in 4b and after 24 hrs in LNCaP cell line.*

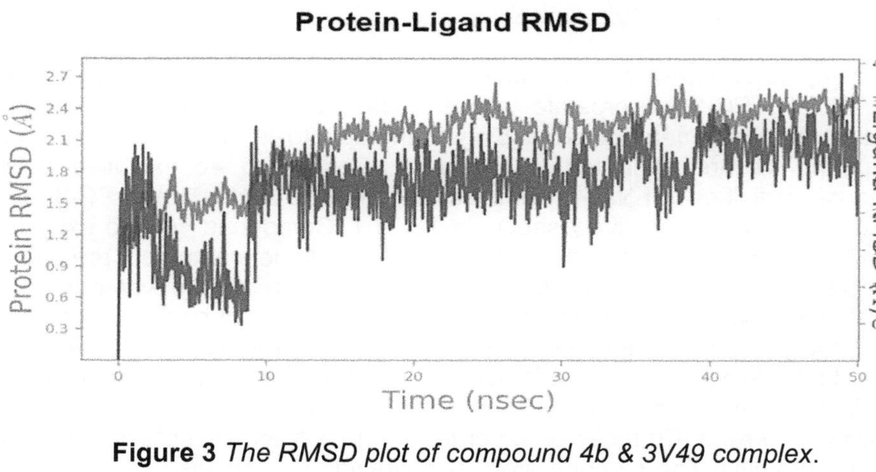

Figure 3 *The RMSD plot of compound 4b & 3V49 complex.*

Table 1 *MTT assay of (4a-d) and standard compounds.*

Ligand	glide gscore (kcal/mol)	LNCaP (μM)*,a (T877A AR+)	PC3 (μM)a (AR -)	3T3 (μM)a (non-cancer)
4a	-7.841	48.57± 2.39	>100	>100
4b	**-9.219**	**14.56 ± 2.01**	**31.60 ± 2.57**	**>100**
4c	-9.333	37.57± 2.27	72.14± 1.25	88.31 ± 1.78
4d	-9.213	28.48 ± 3.64	40.41 ± 3.88	74.81 ± 1.43
Bicalutamide[b]	-11.064	26.3 ± 0.07	58.26 ±1.86	62.55 ± 0.32

*IC50 of the compounds stimulated by 1nM DHT, [b] Positive control

[a] The values are the mean ± standard deviation (SD) of three independent experiments in triplicate

Table 2 *DNMT inhibition assay of compound 4b.*

DNMT 1 Inhibition assay				
Test Compound	1 μM		100 μM	
	OD ± SD	% Inhibition	OD ± SD	% Inhibition
Control	1.519 ± 0.062			
5-Azacytidine	1.289 ± 0.022	11.8	1.022 ± 0.196	38.6
4b	1.298 ± 0.810	14.1	1.257 ± 0.246	24.3

The compounds *(4a-d)* were evaluated for the antiproliferative activity against the LNCaP, PC-3 of prostate cancer. The live cells having chromatin condensation, cell membrane blebbing during apoptosis, were observed in the compound *4b* are illustrated in Figure. 2. Compound *4b* exhibited dose-response, activities in LNCaP cell lines IC_{50} 14.56 ± 2.01 μM and PC-3, IC_{50} 31.60 ± 2.57 μM respectively shown in Table 1. Compound *4b* also exhibited DNMT1 enzyme inhibition activity at 1μM and 100μM concentration, are shown in Table. 2. In the enzyme inhibition assay, gradient response was perceived and showed 24.3% of inhibition. So, the compound *4b* can be further used for DNA Methyl Transferase enzyme inhibitor isoform targeted therapies to treat PCa. The protein RMSD started from 0.9 Å and later stabilized at 1.7 Å. In ligand RMSD, stimulation started from 2.0 Å and later stabilized at around 1.8 Å are shown in Figure. 3. Changes of the order of 1 to 3 Å are acceptable for small, globular proteins. The nitrogen atom of hydrazide ring shows hydrogen bond interaction with THR877 (53%), LEU704 (18%) and Pi-Pi stacking interaction with TRP741 (37%) and PHE764 (27%) respectively. In ADME, the compound (4a-d) obeys the suggested drug-likeness properties.

3 CONCLUSION

In the current study, a series of compounds *(4a-d)* were intended using structure based molecular docking approach for the androgen receptor antagonist. The compounds having good glide score with AR receptor were further synthesized. The structures of compound *(4a-d)* were confirmed by NMR, IR, LC-MS and purity of the compound *4b* were 99.23 by liquid chromatography. Further, (4a-d) demonstrate anti-prostate cancer activity. Among the series, the compound *4b* showed reduced cell viability against LNCaP & PC3 cell lines with the IC_{50} of 14.56 ± 1.54 µM & 31.60 ± 2.57 µM respectively. Moreover, it is evident from the above results that the fluorine at R- substitution has enhanced binding affinities and has prominent anticancer activities comparing to OCH_3 & Cl substitutions. The *4b* also exhibited 24.3% in the enzyme inhibition assay in DNA. From MD stimulation the compound 4b-3V49 complex reached stable form at 1.7 Å & 1.8 Å for protein and ligand. With all these facts and figures, we intend to examine the molecules in *vivo* model in which new chemical entities can be developed further.

References

1. American Cancer Society, Facts & Figures 2019, Atlanta, GA, USA.
2. H. A. Arjun, R. Elancheran, N. Manikandan and K. Lakshmithendral, *Front. Chem.*, 2019, **7**, 474.
3. S. Divakar, K. Saravanan and P. Karthikeyan, *Chem.-Biol. Interact.,* 2017, **275**, 22.
4. S. Lakshmana Senthil, R Chandrasekaran and H. A. Arjun, *Carbohydr. Polym.*, 2019, **209**, 350.
5. Maestro, Schrödinger, 2016, LLC, New York, NY.

DESIGN, MOLECULAR DOCKING, SYNTHESIS AND ANTI-FUNGAL ACTIVITY OF NOVEL BENZIMIDAZOLE DERIVATIVES

R. Deepa,[a] M. S. Lakshmi,[a,b] Parasuraman P.[a] and C. F. Sharon[c]

[a]Department of Pharmaceutical Chemistry, Faculty of Pharmacy, M.S. Ramaiah University of Applied Sciences, Bengaluru, Karnataka, India; [b]Pharmacological Modelling and Simulation Research centre, M.S. Ramaiah University of Applied Sciences, Bengaluru, Karnataka, India; [c]Department of Pharmaceutics, Faculty of Pharmacy, M.S. Ramaiah University of Applied Sciences, Bengaluru, Karnataka, India

1 INTRODUCTION

Benzimidazole is a heterocyclic compound with an extensive range of biological activities. Benzimidazole and its derivatives are the most resourceful classes of molecules against microorganisms. The increase in anti-microbial resistance to existing drugs necessitated the search for new molecules for the treatment of fungal infections[1]. Hence, aim of the present study is to design benzimidazole derivatives by molecular docking approach against selected fungal targets to understand binding conformation and interactions at the active pocket of respective targets and binding affinity of compounds was evaluated in terms of docking score. Selected benzimidazole derivatives were synthesized and screened for anti-fungal activity by Zone of inhibition method.

2 METHOD AND RESULTS

2.1 Molecular docking and synthesis

Twenty five benzimidazole derivatives (DAGBI 1 to 25) were designed and molecular docking (Glide module) was carried out against 14 different fungal targets. Compounds have shown good binding scores with three fungal targets namely sterol 14α-demethylase (PDB: 1E9X), N-myristoyl transferase (PDB: 1NMT) and oxysterol binding protein (PDB: 1ZHX) (Table 1 and Figure 2). Ten compounds (DAGBI 1 to 5, DAGBI 14 -16, DAGBI 20 and 22, Figure 1) were selected for synthesis based on the docking score and binding interactions.

Compounds were synthesized by sequence of two reactions- Condensation and Schiff base formation. All the synthesized compounds were purified by recrystallization using absolute ethanol and purity of compounds was ensured by TLC. Further structures were confirmed by IR, ^{1}H NMR, ^{13}C NMR and Mass spectral analysis.

1H-Benzoimidazol-2—ylmethyl)-benzylidene-amine (DAGBI 1): IR (KBr, cm^{-1}) at 2926 (20 N-H), 2830 (C-H), 1505 (C=N); ^{1}H NMR (300 MHz, $CDCl_3$) δ: 8.50 (s, 1H, NH of imidazole), 7.78–7.72 (m, 2H, Ar–H), 7.51–7.45 (m, 3H, Ar–H), 7.44–7.33 (d, 2H, Ar–H), 7.28-7.09 (d, 2H, Ar–H), 5.51 (s, 3H,CH); ^{13}C-NMR (300 MHz, $CDCl_3$) δ: 153 (C-3'), 134.68 (C–Ar), 133.57 (C–Ar), 132.44 (C–Ar),

131.99 (C–Ar), 130.62 (C–Ar), 129.58(C–Ar), 128.74 (C–Ar), 127.39 (C–Ar), 126.59 (C–Ar), 125.95 (C–Ar), 124 .01 (C–Ar), 122 (C–Ar), 118.96 (C–Ar), 40.27 (C-1') and LC-MS: m/z 234 $(M-H)^+$

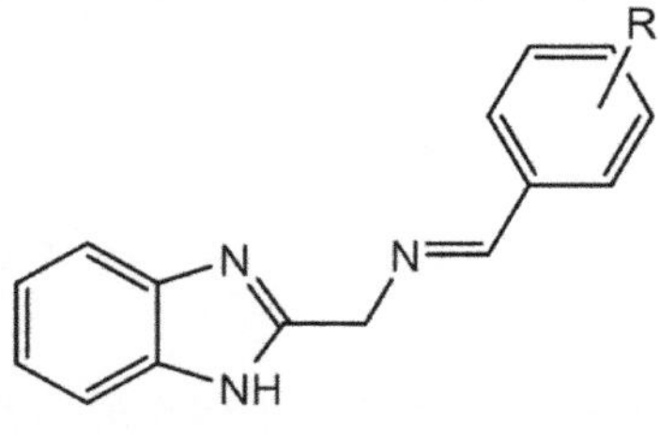

DAGBI

DAGBI 1 : R = H
DAGBI 2 : R= *o*-OH
DAGBI 3 : R= *m*-OH
DAGBI 4 : R= *p*-OH
DAGBI 5 : R= *o*-OCH3
DAGBI 14 : R= *o*-NO2
DAGBI 15 : R= *m*-NO2
DAGBI 16 : R= *p*-NO2
DAGBI 20 : R= *o*-Cl
DAGBI 22 : R= *p*-Cl

Figure 1 *Synthesized molecules with compound code.*

Table 1 *Docking Score of benzimidazole derivatives against fungal targets.*

Compound Code	Sterol 14αdemethylase (PDB:1E9X)	N-Myristoyl Transferase (PDB:1NMT)	Oxysterol Binding Protein (PDB:1ZHX)
DAGBI 1	-7.99	-8.18	-7.72
DAGBI 2	-8.26	-8.58	-7.15
DAGBI 3	-7.58	-6.89	-7.89
DAGBI 4	-7.6	-6.34	-7.73
DAGBI 5	-7.79	-9.17	-7.72
DAGBI 6	-7.22	-5.30	-7.63
DAGBI 7	-6.77	-5.35	-8.19
DAGBI 8	-6.99	-8.65	-7.21
DAGBI 9	-8.18	-5.17	-8.26
DAGBI 10	-8.23	-5.01	-7.98
DAGBI 11	-6.64	-7.69	-7.62
DAGBI 12	-6.76	-7.23	-7.41
DAGBI 13	-7.27	-5.64	-7.06
DAGBI 14	-7.90	-5.87	-8.04
DAGBI 15	-8.19	-5.28	-7.27
DAGBI 16	-6.76	-5.71	-7.27
DAGBI 17	-7.88	-8.77	-8.39
DAGBI 18	-7.55	-8.72	-8.40
DAGBI 19	-7.89	-6.49	-8.07
DAGBI 20	-7.00	-8.78	-8.44
DAGBI 21	-8.41	-6.70	-8.42
DAGBI 22	-8.50	-6.24	-7.99
DAGBI 23	-6.67	-6.26	-8.36
DAGBI 24	-8.31	-5.95	-7.85
DAGBI 25	-8.25	-7.25	-7.36

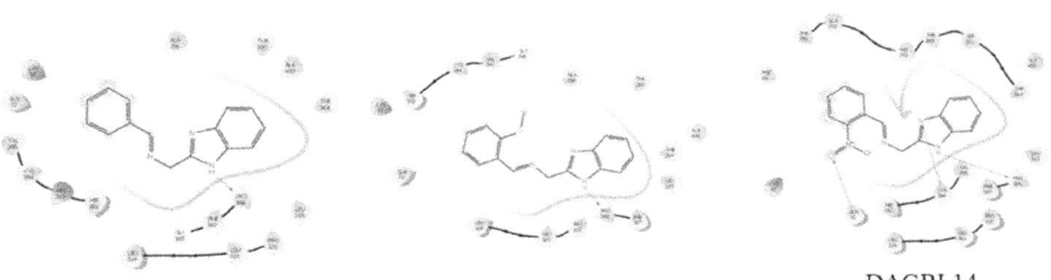

Figure 2 *Docked poses of DAGBI 1, 5 & 14 with Sterol 14α-demethylase (PDB:* 1E9X*).*

2.2 *In vitro* anti-fungal activity

Anti-fungal activity of synthesized benzimidazole derivatives was tested against the fungal microorganisms *Candida albicans* and *Aspergillus niger*. Ketoconazole was used as standard drug. All the ten compounds have shown moderate to potent zone of inhibition[2] for *C. albicans,* whereas no inhibition was observed for *A. niger* except DAGBI 1 and DAGBI 22 which have shown moderate inhibition (Table 2). Based on zone of inhibition results, three compounds (DAGB 1, 5 & 14) have been selected to determine Minimum Inhibitory Concentration (MIC)[3] against *C. albicans* (Table 2).

Table 2 In vitro *anti-fungal activity.*

S.No	*Compound Code	R	***C.albicans***		***A.niger***
			Zone of Inhibition[a] (mm)	***Minimum Inhibitory[b] Concentration (µg/mL)***	***Zone of Inhibition (mm)***
1	DAGBI 1	H	29	32	18
2	DAGBI 2	*o*-OH	20	NT	NI
3	DAGBI 3	*m*-OH	17	NT	NI
4	DAGBI 4	*p*-OH	NI	NT	NI
5	DAGBI 5	*o*-OCH_3	21	64	NI
6	DAGBI 14	*o*-NO_2	18	64	NI
7	DAGBI 15	*m*-NO_2	20	NT	NI
8	DAGBI16	*p*-NO_2	24	NT	NI
9	DAGBI 20	*o*-Cl	18	NT	NI
10	DAGBI 22	*p*-Cl	20	NT	17
Std	Ketoconazole	--	28	64	17

[a]Concentration: 100 µg/mL: NI: No inhibition
[b]Concentration 128-2 µg/mL NT: Not Tested

2.3 Discussion

The compound DAGBI 1 exhibited potent activity against the fungus *C. albicans* at an MIC (32 μg/mL) lesser compared to the standard drug ketoconazole (64 μg/mL). MIC value of other two compounds DAGBI 5 and DAGBI 14 was 64 μg/mL, which was observed similar to that of standard. Among these three compounds, DAGBI 1 and DAGBI 5 have also shown good docking of -7.72 to -9.17 with three targets sterol 14α-demethylase, 1E9X, N-myristoyl transferase, 1NMT and oxysterol binding protein, 1ZHX. However the docking score of compound DAGBI 14 was good with two targets 1E9X (-7.90) and 1ZHX (-8.04) but less for the target 1NMT (-5.87).

3 CONCLUSIONS

In the present study, 25 benzimidazole derivatives (DAGBI 1- 25) were designed and *in silico* screening was performed to investigate their protein-ligand interactions with 14 fungal targets by molecular docking study. Based on the docking scores, 10 compounds (DAGBI 1 to 5, DAGBI 14 -16, DAGBI 20 and 22) were selected for synthesis and evaluated for their *in vitro* anti-fungal activity against *C. albicans* and *A. niger*. The anti-fungal activity of these ten compounds indicated that three compounds DAGBI 1, 5 and 15 had exhibited good inhibition with MIC values observed were comparable to that of standard drug ketoconazole against the fungus *C. albicans*. Experimental results correlated well with the docking scores observed for two fungal targets sterol 14α-demethylase, 1E9X and oxysterol binding protein, 1ZHX. Further *in vitro* screening of ten synthesized compounds against sterol 14α-demethylase and oxysterol binding protein will help us to correlate the molecular docking and respective target inhibition studies. Also for better understanding about the structure activity relationship for the design of safe and potent benzimidazole derivatives as anti-fungal agents.

References

1. S. Tahlan, S. Kumar and B. Narasimhan, *BMC Chem.*, 2019, **13**, 18.
2. M. Kandhasamy, K. D. Arunachalam and A. J. Thatheyus, *Afr. J. Microbiol. Res.*, 2008, **2,** 202.
3. N. Altanlar, *Eur. J. Med. Chem,* 2009, **44**, 1024.

SYNTHESIS, ANTIOXIDANT ACTIVITY AND COMPUTATIONAL STUDIES 2-SUBSTITUTED-L-THIAPROLINES

Romit R. Palei,[a] Priyanka Mohanty,[a] Chandra Sekhar Barik[b] and Sidhartha S. Kar[a]

[a] Department of Pharmaceutical Chemistry, Institute of Pharmacy & Technology, Salipur, Cuttack 754202, Odisha, India; [b] Department of Pharmacology, Institute of Pharmacy & Technology, Salipur, Cuttack-754202, Odisha, India

1 INTRODUCTION

Supplementation of exogenous antioxidants or boosting endogenous antioxidant defences of the body is a promising way of combating the undesirable effects of reactive oxygen species (ROS) induced oxidative damage. Thiaprolines are considered to be the prodrug of cysteine and they have established to be safe and effective antioxidants. Thiaprolines have been reported to affect the formation and scavenging of reactive oxygen species (ROS) and influence free radical-mediated oxidative damage. Thus the development of new synthesized thiaproline molecules is promising, especially regarding their antioxidant activities.[1] However, reports on development of druggable thiaproline derivatives are limited. In this context, the antioxidant capacities of thiaprolines with different chemical substitutions at its thiazolidine heterocycles have been envisaged. The major focus was to develop thiaproline based antioxidants having better druggability and to understand the molecular mechanism of free radical scavenging effect.

2 METHOD AND RESULTS

2.1 Basis of design and synthesis

The thioprolines are extremely variable in structure, due to the various types of substitutions in their basic structure, which can influence their biological activity. A careful structure-system-activity-relationship study of the thioprolines was conducted. The basic structure thiazolidine-4-carboxylic acid was derived from L-cysteine. Our studies started from the design of the thioproline nucleus (Figure 1) with enhanced druggability and antioxidant activities. We approached to increasing of antioxidant activities based on the long conjugated system in the synthesized thioprolines.

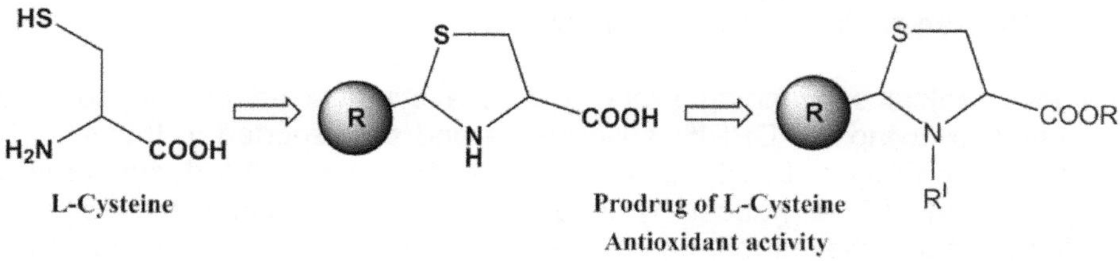

Figure 1 *Design of compounds synthesized in Figure 2.*

2.2 Synthesis of 2-substituted-l-thiaproline derivatives (**1a-e** and **2a-b**)

In view of our aim of building thiazolidine scaffold i.e., thioproline (**1a**) was synthesised from a reaction of formaldehyde with L-cysteine in the presence aqueous ethanol solvent system (Figure 2). Various aryl aldehydes with L-cysteine were used to synthesize compounds **1b-e** in ethanol as the solvent with stirring at room temperature to afford 70-90% yield. Compound **2a** was obtained by using thionyl chloride-methanol as the esterification reagent to esterify compound **1b**. N-acylation of compound **1b** was attempted in acetyl chloride at anhydrous environment. Synthesized compounds were characterized by UV, FTIR and 1H NMR.

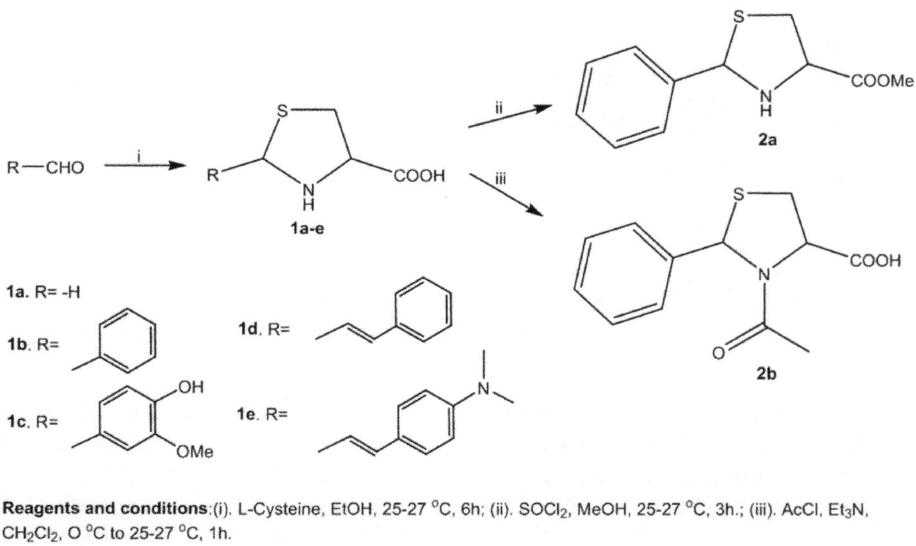

Reagents and conditions:(i). L-Cysteine, EtOH, 25-27 °C, 6h; (ii). $SOCl_2$, MeOH, 25-27 °C, 3h.; (iii). AcCl, Et_3N, CH_2Cl_2, O °C to 25-27 °C, 1h.

Figure 2 *Synthesis of compounds **1a-e** and **2a-b**.*

2.3 DPPH free radical scavenging activity

The free radical scavenging property of the synthesized compounds was evaluated following the DPPH• absorbance and are reported in table-1. The reduction capability of DPPH radicals was determined by the decrease in its absorbance at 517 nm induced by antioxidants. Ascorbic acid was used as the standard. Most of the compounds prepared in Figure 2 (**1a-c**) were found to be poor free radical scavengers in comparison to ascorbic acid. However compounds **1d** and **1e** showed a concentration dependant DPPH free radical scavenging property which could be attributed to the influence of cinnamoyl groups in their moieties. Compound **1e** had a higher radical consumption, indicating that this sample has high reactivity with the DPPH• radical. Compound **1e** exhibited 86.231% inhibition at 400 µg/mL which is comparable to the inhibitory property of ascorbic acid (90.310%). Hence it could be observed that extension of electronic conjugation improves the antioxidant activity.

Table 1 *DPPH free radical scavenging activity of compounds **1a-e** and **2a-b**.*

	% Inhibition					
Comp.	25 µg/mL	50 µg/mL	100 µg/mL	200 µg/mL	300 µg/mL	400 µg/mL
1a	-	-	-	-	-	6.671
1b	-	5.344	7.652	10.336	14.804	15.105
1c	-	8.125	13.895	16.403	23.300	29.367
1d	33.293	51.184	62.056	68.332	76.010	77.3
1e	40.452	55.672	74.569	79.056	80.919	86.231
2a	-	-	-	3.534	13.291	15.544
2b	-	-	-	-	-	-
Ascorbic acid	54.290	63.147	75.176	79.389	84.622	90.310

H_2O_2 is highly important because of its ability to penetrate into biological membranes. H_2O_2 itself is not very reactive, but it can sometimes be toxic to cell because it may give rise to hydroxyl radicals in the cells. Hydrogen peroxide free radical scavenging activity of compounds **1a-e**, and **2a-b** was showed in table 2. Ascorbic acid was used as the standard. Compounds **1c**, **1d** and **1e** showed promising H_2O_2 scavenging effect. Compound **1c** was found to be the most active with 70.107% inhibition at 400 µg/mL which may be attributed to its phenolic group.

Table 2 *Hydrogen peroxide free radical scavenging activity of compounds **1a-e** and **2a-b**.*

Comp.	% Inhibition 25 μg/mL	50 μg/mL	100 μg/mL	200 μg/mL	300 μg/mL	400 μg/mL
1a	-	-	-	-	6.121	11.366
1b	-	-	5.004	9.436	15.564	21.203
1c	-	16.467	38.078	53.175	57.389	70.107
1d	15.227	24.015	33.600	42.390	53.405	65.149
1e	18.088	26.629	35.501	40.204	48.019	61.536
2a	-	-	-	-	-	5.133
2b	-	-	-	-	7.667	13.700
Ascorbic acid	-	4.733	10.184	25.013	36.529	50.130

2.4 Computational study

DFT calculations[2] were carried out in order to associate the free radical scavenging activity of the synthesized compounds with their electronic surfaces through ArgusLab. The single-point energy setting was opened on the toolbar. Single-Point Energy was calculated using PM3 force field. SCF eigen values (au) (MO energy) for HOMO, LUMO, final SCF energy and heat of formation were determined. DPPH free radical scavenging property of synthesized compounds has been correlated with energy of their HOMO and LUMO. Most promising compound **1d** and **1e** showed lowest H-L gap (table 3) which could be attributed to the presence of extended pi electronic conjugation system. It was also observed that compounds with more stabilized LUMO orbital have better free radical scavenging property.

Table 3. *DFT studies compounds **1a-e** and **2a-b**.*

Comp.	SCF eigen values (au) E_{HOMO}[a]	E_{LUMO}[a]	H-L Gap (au)	Final SCF Energy kcal/mol[b]	Heat of Formation kcal/mol[b]
1a	-0.349733	-0.158106	0.191627	-30480.7662	2324.2784
1b	-0.307526	-0.146484	0.161042	-47495.0685	1725.1198
1c	-0.285698	-0.128897	0.156801	-64386.4865	1740.3146
1d	-0.221214	-0.157338	0.063876	-52916.6169	1775.2860
1e	-0.200211	-0.140967	0.059244	-62220.5161	2397.8295
2a	-0.283061	-0.110866	0.172195	-49839.1586	2116.8870
2b	-0.310959	-0.145594	0.165365	-61236.6314	1540.6492

[a]HOMO and LUMO energy were calculated using AurgusLab. [b]Final Self-Consistent Field (SCF) method energy was calculated using PM3 force filed.

3 CONCLUSION

In our research, thioproline derivatives were designed, synthesized and explored for their potential as antioxidants. As a result of the DPPH assay employed in this study, we were able to demonstrate that compound **1d** and **1e**, whose structure present cinnamoyl moiety, proved to be the most effective antioxidant. On the other hand compound **1c** was found to be the most active hydrogen peroxide radical scavenger due to the presence of phenolic group in its moiety. The result of DFT study of compounds revealed that the nature of LUMO and HOMO-LUMO gap significantly influences the free radical scavenging property.

References

1. J. Liu, X. Meng, W. Chan, *J. Agric. Food Chem.*, 2016, **64**, 1361.
2. J. Wang, H. Tang, B. Hou, P. Zhang, Q. Wang, B. L. Zhang, Y. W. Huang, Y. Wang, Z. M. Xiang, C. T. Zi, X. J. Wang and J. Sheng, *RSC Adv.*, 2017, **7**, 54136.

DESIGN, SYNTHESIS, CHARACTERISATION AND ANTIFUNGAL ACTIVITY OF NOVEL PYRIMIDONE DERIVATIVES

P. Vijaybhanu, D. R. Harish Kumar, L. Ramya and C. H. S. Venkataramana

Department of Pharmaceutical Chemistry, Faculty of Pharmacy, Ramaiah University of Applied Sciences, Bengaluru-560 054, Karnataka, India

1. INTRODUCTION

Pyrimidones and oxadiazoles possess a wide range of biological activities such as antibacterial, antifungal, analgesic, anti-inflammatory and anticancer activities.[1-3] The prevalence of fungal infections has progressively increased in recent years and hence it is necessary to continue the search for new, less toxic and more efficacious antifungal agents.[4] As vast number of reports are available pertaining to the antifungal activities of pyrimidones and oxadiazoles, combination of these two moieties may result in additive effect towards antifungal activity. The aim of the present study was to design and synthesize some hybrid molecules containing both pyrimidone and oxadiazole moieties in order to evaluate their antifungal activity. Virtual screening was carried out through molecular docking studies against potential antifungal target glycolipid 2α-mannosyl transferase. Ten novel pyrimidone derivatives were synthesized and characterized using UV, IR, ^{1}HNMR and Mass spectroscopy analysis. The antifungal activity of newly synthesized pyrimidone derivatives was evaluated by Zone inhibition method against two antifungal strains *Aspergillus niger* and *Candida albicans.*

2 METHOD AND RESULTS

2.1 Molecular Docking Studies

In silico screening was carried out for fungal target glycolipid 2α-mannosyl transferase (PDB ID: 1S4N) to identify the specificity of novel pyrimidone derivatives using Schrödinger Drug Design Suite. Energy minimization of title compounds was carried out using Ligprep. Selected target protein was optimized and minimized, a three dimensional grid was generated at the active site, and molecular docking was carried out. The docking calculations were performed using extra precision (XP) mode of GLIDE program. Ten pyrimidone derivatives (3a-3j) have demonstrated hydrogen bonding and Pi-Pi stacking interactions similar to standard drug at the active site of protein with docking scores at a range of -6.3 to -10.325 kcal/mol. Docking score of the standard fluconazole was found to be -7.279 kcal/mol (Figure 1 and Table 1).

2.2 Synthesis of Pyrimidone Derivatives

Cyclocondensation of benzaldehyde or its derivatives, ethyl acetoacetate and urea in presence of chloroacetic acid and alcohol resulted in the formation of ethyl-6-methyl-2-oxo-4-phenyl-1, 2, 3, 4-tetrahydro pyrimidine-5-carboxylate

(1a-1j). The product when refluxed with hydrazine hydrate formed the corresponding hydrazides (2a-2j). Further reaction of these compounds with chloramine-T in presence of 2-chloro-5-nitro benzoic acid resulted in the formation of 5-[5-(2-chloro-5-nitrophenyl)-1, 3, 4-oxadiazol-2-yl]-6-methyl-4 -phenyl-3, 4-dihydro-1*H*-pyrimidin-2-one (3a-3j). The progress of the reaction was monitored by TLC using n hexane: ethyl acetate (7:3) as mobile phase. All the synthesized compounds were purified by recrystallization using absolute ethanol and characterized by NMR and Mass spectral methods.

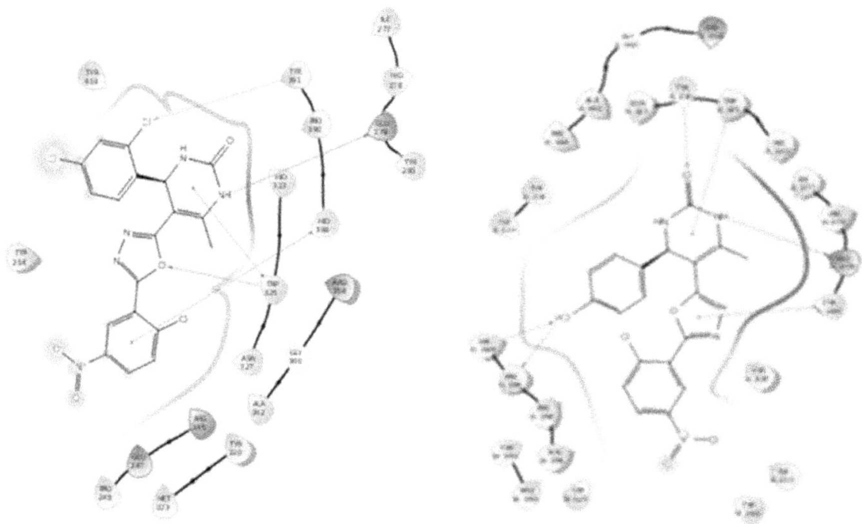

Figure 1 *Ligand interaction of **3j** and **3i** with target protein 1S4N.*

2.3 Characterisation of compounds (3a-3j)

The structures of the synthesised compounds were characterised based on their melting point, TLC and spectral studies.

5-[5-(2-chloro-5-nitrophenyl)-1, 3, 4-oxadiazol-2-yl]-6-methyl-4-phenyl-3,4 -dihydro-1*H*-pyrimidin-2-one (3a): white solid, yield-76%, mp 112-114^{0}C, R_f :0.84; IR (KBr) cm^{-1} C=O(s) 1699, N-H(s) 3232, C-N (s) 1290.29, C-O-C (s) 1222.79; ^{1}H NMR: 7.191-7.741(8H, aromatic), 6.005 (1H, NH), 2.335-2.342 (3H, CH_3), MS: m/z 410 (M-1).

5-[5-(2-chloro-5-nitrophenyl)-1, 3, 4-oxadiazol-2-yl]-4-(2-chlorophenyl)-6-methyl-3, 4-dihydro-1*H*-pyrimidin-2-one (3b): yellow solid, yield-62%, mp 96-98^{0}C, R_f :0.76; IR (KBr) cm^{-1} C=O(s) 1705, N-H(s) 3519.05, C-N (s) 1365.51, C-O-C (s) 1222.79; ^{1}H NMR: 7.197-7.741(7H, aromatic), 2.339 (3H, CH_3), MS: m/z 446 (M+).

5-[5-(2-chloro-5-nitrophenyl)-1, 3, 4-oxadiazol-2-yl]-4-(4-chlorophenyl) -6-methyl -3, 4-dihydro-1*H*-pyrimidin-2-one (3c): yellow solid, yield-69%, mp

135-137^{0}C, R_f :0.60; IR (KBr) cm^{-1} C=O(s) 1730, N-H(s) 3618, C-N (s) 1301, C-O-C (s) 1400; ^{1}H NMR: 7.197-7.754(7H, aromatic), 5.861 (1H, NH), 2.341-2.355 (3H, CH_3), MS: m/z 446 (M+).

5-[5-(2-chloro-5-nitrophenyl)-1, 3, 4-oxadiazol-2-yl] -6-methyl-4-(2-nitro-phenyl)-3, 4-dihydro-1*H*-pyrimidin-2-one (3d): reddish brown solid, yield-28%, mp 96-98^{0}C, R_f :0.70; IR (KBr) cm^{-1} C=O(s) 1731, N-H(s) 3440, C-N (s) 1300, C-O-C (s) 1176; ^{1}H NMR: 7.210-7.767(7H, aromatic), 5.791 (1H, NH), 2.353-2.366 (3H, CH_3), MS: m/z 455 (M-1).

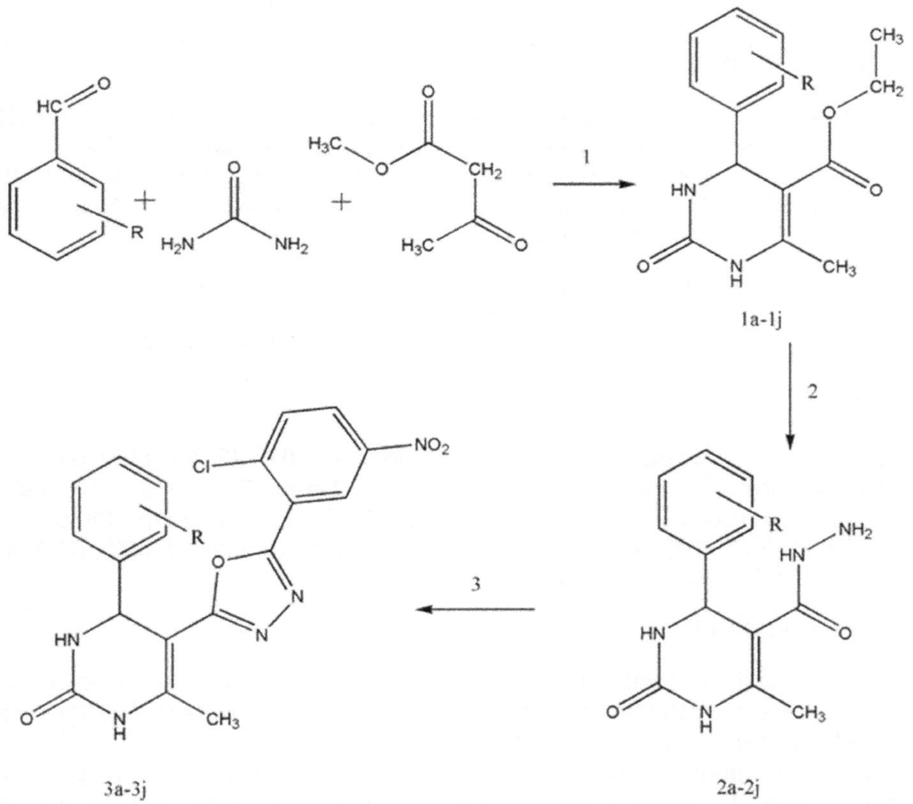

Scheme 1 *Synthetic Scheme of pyrimidone derivatives. Reagents and conditions- 1. Chloroacetic acid, ethanol reflux for 5-10 hrs, 2.Hydrazine hydrate, ethanol reflux for 4 hrs, 3. Chloramine-T, 2-chloro-5-nitro benzoic acid, ethanol reflux for 7-9 hrs.*

5-[5-(2-chloro-5-nitrophenyl)-1, 3, 4-oxadiazol-2-yl]-6-methyl-4-(4-nitro-phenyl)-3, 4-dihydro-1*H*-pyrimidin-2-one (3e): reddish brown solid, yield-54%, mp 94-96^{0}C, R_f :0.52; IR (KBr) cm^{-1} C=O(s) 1645, N-H(s) 3433, C-N (s) 1400, C-O-C (s) 1184; ^{1}H NMR: 7.216-7.739(7H, aromatic), 5.590 (1H, NH), 2.339 (3H, CH_3), MS: m/z 457 (M+1).

5-[5-(2-chloro-5-nitrophenyl)-1 ,3, 4-oxadiazol-2-yl]-6-methyl-4-(3, 4, 5-trimethoxyphenyl)-3, 4-dihydro-1*H*-pyrimidin-2-one (3f): yellow solid, yield-87%, mp 146-148^{0}C, R_f :0.85; IR (KBr) cm^{-1} C=O(s) 1805, N-H(s) 3313, C-N (s) 1172.6, C-O-C (s) 1250; ^{1}H NMR: 7.155-7.699(5H, aromatic), 6.097 (1H, NH), 2.299 (3H, CH_3), 3.711-3.826(9H, OCH_3)MS: m/z 501 (M+).

5-[5-(2-chloro-5-nitrophenyl)-1, 3, 4-oxadiazol-2-yl]-4-(4-dimethylamino-phenyl)-6-methyl -3, 4-dihydro-1*H*-pyrimidin-2-one (3g): reddish brown solid, yield-47%, mp 88-90^{0}C, R_f :0.63; IR (KBr) cm^{-1} C=O(s) 1731, N-H(s) 3352, C-N (s) 1245, C-O-C (s) 1157; ^{1}H NMR: 7.211-7.769(7H, aromatic), 5.944 (1H, NH), 2.347-2.354 (3H, CH_3), 2.368-3.4 (6H, two methyl groups attached to N)MS: m/z 454 (M+).

5-[5-(2-chloro-5-nitrophenyl)-1, 3, 4-oxadiazol-2-yl]-4-(2-methoxyphenyl)-6-methyl -3, 4-dihydro-1*H*-pyrimidin-2-one (3h): white solid, yield-21%, mp 114-116^{0}C, R_f :0.30; IR (KBr) cm^{-1} C=O(s) 1716, N-H(s) 3436, C-N (s) 1200, C-O-C (s) 1208; ^{1}H NMR: 7.226-7.749 (7H, aromatic), 4.876 (1H, NH), 2.358 (3H, CH_3),3.5 (3H, OCH_3) MS: m/z 442 (M+1).

5-[5-(2-chloro-5-nitrophenyl)-1, 3, 4-oxadiazol-2-yl] -4-(4-hydroxyphenyl)-6-methyl -3, 4-dihydro-1*H*-pyrimidin-2-one (3i): brown solid, yield-85%, mp 119-121^{0}C, R_f :0.92; IR (KBr) cm^{-1} C=O(s) 1795, N-H(s) 3315, C-N (s) 1220, C-O-C (s) 1150; ^{1}H NMR: 7.19-7.73(7H, aromatic), 2.32-2.33 (1H, NH), 0.8-1.2 (3H, CH_3), 5.84(1H, OH) MS: m/z 429 (M+2).

5-[5-(2-chloro-5-nitrophenyl)-1, 3, 4-oxadiazol-2-yl] -4-(2, 4-dichloro-phenyl)-6-methyl -3, 4-dihydro-1*H*-pyrimidin-2-one (3j): white solid, yield-59%, mp 104-106^{0}C, R_f :0.56; IR (KBr) cm^{-1} C=O(s) 1899, N-H(s) 3436, C-N (s) 1225, C-O-C (s) 1357; ^{1}H NMR: 7.218-7.767(6H, aromatic), 5.893 (1H, NH), 2.361-2.369 (3H, CH_3), MS: m/z 479 (M-1).

2.4 Antifungal Activity

All the synthesised compounds (500 µg/ml using dimethyl sulphoxide-DMSO as solvent) were tested for their antifungal activity against *Aspergillus niger* and *Candida albicans* by agar diffusion method using ketoconazole (100 µg/ml) as standard. Compounds have shown poor antifungal activity against *Aspergillus niger* (**Table 1**). Compounds **3e** with p-nitro and **3c** with p-chloro groups showed potent antifungal activity and closer to standard against *Candida albicans.* The results indicate that pyrimidones with electron withdrawing substituents such as nitro and chloro at the para position of phenyl ring showed significant antifungal activity. Substitution of methoxy groups on phenyl ring have diminished antifungal activity against both organisms. Similarly, substitution of chloro group at 2nd position have reduced the fungi inhibition, which was observed for both compounds **3b** and **3j**. Molecular docking scores were in the range of -6.0 to -7.1 for compounds **3a-3h**. Surprisingly, compounds **3i** and **3j** have exhibited good docking scores of -8.682 and -10.325 and experimentally moderate fungi inhibition. Further structural modification of pyrimidone derivatives might help us in design of safe and potent antifungal agents.

Table 1 *Molecular Docking and Antifungal activity of synthesised compounds.*

Compound Code	*R*	*Glide Docking Score (PDB ID: 1S4N)*	*Zone of Inhibition (mm)*	
			Aspergillus niger	*Candida albicans*
3a	H	-6.71	11	13
3b	2-chloro	-6.8	10	16
3c	4-chloro	-6.71	12	20
3d	2-nitro	-6.54	10	18
3e	4-nitro	-6.81	13	21
3f	3,4,5-trimethoxy	-6.3	10	12
3g	4-dimethylamino	-6.91	10	12
3h	2-methoxy	-7.1	12	10
3i	4-hydroxy	-8.682	13	18
3j	2,4-dichloro	-10.325	10	14
Ketoconazole	--	-7.279	25	23

3 CONCLUSIONS

Molecular docking studies of novel pyrimidone derivatives with oxadiazole moiety were carried out to investigate the binding interactions at the active site of target glycolipid 2α-mannosyl transferase. Ten pyrimidone derivatives were synthesized and characterised by NMR and Mass spectral data. The synthesised compounds were screened for antifungal activity. Pyrimidone derivatives were found selective to *Candida albicans*, and compounds with p-nitro, p-chloro groups showed significant antifungal activity, closer to standard ketoconazole. Further screening of large databases and experimental evaluation might give us better understanding about SAR of pyrimidone derivatives and design of glycolipid 2α-mannosyl transferase inhibitors as antifungal agents.

References

1. M. A. Azam, L. Dharanya, C. C. Mehta and S. Sachdeva, *Acta Pharm.*, 2013, **63**(1), 19.
2. N. Y. M. Abdo and M. M. Kamel, *Chem. Pharm. Bull.*, 2015, **63**(5), 369–376.
3. E. Abdelghani, S. A. Said, M. G. Assy, and A. M. A. Hamid, *Arabian J. Chem.*, 2017, **10**, S2926.
4. H. X. Ngo, S. Garneau-Tsodikova and K. D. M. Green, *Med. Chem. Commun.* 2016, **7**, 1285.

Part 3: Chemical Synthesis and Biological Evaluation

MICROWAVE-ASSISTED SYNTHESIS OF 4-BENZYLIDENE-2-(2-FLUOROPHENYL) OXAZOL-5(4*H*)-ONE DERIVATIVES CATALYSED BY EGG SHELL POWDER AND EVALUATION OF THEIR ANTI-MICROBIAL ACTIVITY

P. B. Hiremath,[a] K. Kantharaju[a*] and S. H. Pattanashetty[b]

[a] Department of Chemistry, Peptide and Medicinal Chemistry Research Laboratory, Rani Channamma University, P-B, NH-4, Belagavi-591156, Karnataka, India.
[b] Department of Studies in Chemistry, Karnatak University, Pavate Nagar, Dharwad-580 003, Karnataka, India.
E Mail: kk@rcub.ac.in

1 INTRODUCTION

The goal of the current green chemistry is replacement of hazardous organic solvents, which lead to unavoidable toxic, solvent wastage or emission. The usage of reagents like solid supported inorganic catalyst has gained more attention in organic synthesis, such type of approaches not only avoid the use of toxic solvents to carry out reaction, but also create convenient procedure for the reactions. Heterocyclic synthons occur abundantly in nature and are crucial to life, in one of its kind oxazolones gained remarkable biological scope and emerged as a pharmaceutical potent molecules.[1-4] They have been noticed beneficial precursors for the synthesis of amino acids,[5,6] peptides,[7, 8] and also employed in the preparation of various heterocycles.[9,10] Because of wide range of application of the oxazolones exhibited, huge demand for the establishment of eco-friendly, greener, economic and convenient synthesis of oxazolones are in scope.

2 METHOD AND RESULTS

Chemicals purchased are laboratory grade from commercial sources and used as received without further purification. Melting points determined in open capillaries and are uncorrected. FT-IR spectra were recorded in KBr pellets on a Shimadzu spectrometer. ^{1}H-, ^{13}C-NMR spectra recorded on a Bruker 400 MHz spectrometer using TMS as an internal standard. LC-MS spectra recorded in Waters; Synapt G2 High detection Mass spectrometry of high resolution and HR-MS spectra in Waters; XEVOG2 XS Q-TOF.

2.1 Preparation of catalyst[11]

Chicken eggshells are collected from the houses and initially washed with tap water, after separating the membrane. Then, it was again washed with distilled water and dried at room temperature for about 8 hr. After drying, ground in mortar and pestle into fine powder. The resulted white powder is used directly for the reaction.

2.2 Preparation of 2-flouro benzoyl glycine

Glycine (20 mmol) was dissolved in 8 mL of 2N sodium hydroxide in a 100 ml round bottomed flask and stirred vigorously with mechanical stirrer until it is totally dissolved. 2-Fluoro benzoyl chloride (20 mmol) was added in five portions and stirred vigorously further 1 hr on ice bath by maintaining temperature 0 to 10 °C. The crystallized 2-flouro benzoyl glycine **(2)** was left in refrigerator overnight, crystal filtered, washed with ice cold water and dried at 60 °C. The melting point of the compound found 145–149 °C with isolated yield of 80%.

2.3 General procedure for the synthesis of 4-benzylidene-2-(2-fluorophenyl) oxazol-5(4*H*)-one derivatives:

Aldehyde (1 mmol), 2-fluoro benzoyl glycine (1mmol), acetic anhydride (3.3 mmol) taken in a dry 50 mL round bottomed flask followed by eggshell powder (200 mg). The mixture was stirred under microwave irradiation at 300 W power fitted with a reflux condenser for appropriate time (Scheme 1). The progress of the reaction was monitored by TLC. After completion of the reaction, hot ethanol was added, the catalyst separated by filtration, and final product obtained in pure form by recrystallization using absolute ethanol gave chromatographically pure product.

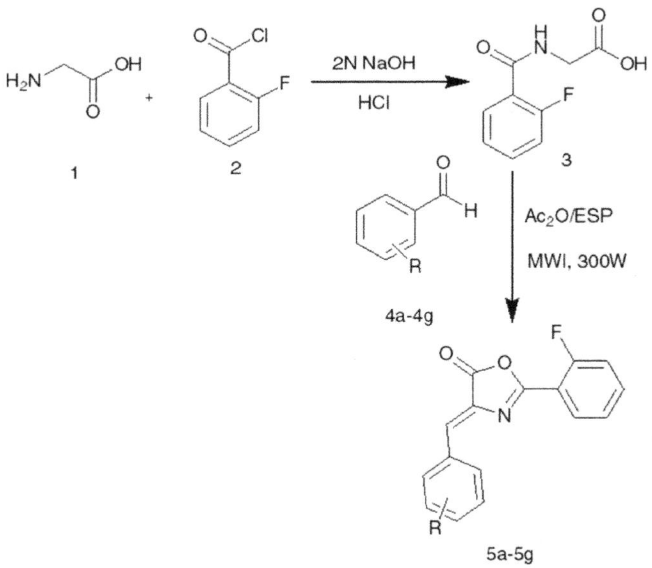

Scheme 1 *Synthetic route of 4-benzylidene-2-(2-fluorophenyl)oxazol-5(4H)-one derivatives synthesis.*

2.4 Discussion

The best optimal condition for the efficient synthesis of azlactone (**5a**) by taking a model reaction of 2-fluoro hippuric acid and benzaldehyde in presence of acetic anhydride by varying reaction condition in terms of catalyst amount loading, solvent effect and advantage of microwave irradiation over conventional heating is carefully studied and tabulated in Table 1. The observation (Table 1) clearly indicated that the best results were obtained for the synthesis of azlactone (**5a**), when 200mg of eggshell powder taken as a catalyst in neat condition under microwave irradiation of 300W, which resulted 90% yield isolation (**5a**). To check the role of the eggshell powder as a heterogeneous catalyst reaction carried out in a model reaction by employing 2-fluoro benzoyl glycine, benzaldehyde as substrates and acetic anhydride in neat condition in the absence of catalyst under microwave at 300W power observed only 18% of product isolation (Table 1, entry 14).

2.5 *In vitro* anti-bacterial and anti-fungal activities

The minimum inhibitory concentration of the compounds **5a** to **5g** was screened by dilution method[12]. The lowest concentration of compounds able to produce bacterial inhibition zone around the disk were considered as MIC values. The synthesized pure compounds are dissolved and diluted to give two-fold serial concentration of the compound was employed to determine the MIC. The MIC value was determined as the lowest concentration of compound inhibiting the visual growth of the microorganism on the agar plate.

Table 1 *Optimization of catalyst load and reaction condition for the synthesis of 4-benzylidene-2-(2-fluorophenyl)oxazol-5(4H)-one (**5a**) using ESP.*

Entry	Solvent	Conditions	Time (min)	Yield[a] (%)
1	EtOH	rt/eggshell (100mg)	30	38
2	CH_2Cl_2	rt/eggshell (100mg)	30	26
3	CH_3CN	rt/eggshell (100mg)	30	24
4	H_2O	60 °C/eggshell (100mg)	30	22
5	EtOH	MWI[c]/eggshell (100mg)	15	42
6	EtOH	MWI[c]/eggshell (200mg)	12	66
7	EtOH	80 °C/eggshell (100mg)	30	44
8	Neat	80 °C/eggshell (100mg)	30	61
9	Neat	MWI[c]/eggshell (50mg)	20	26
10	Neat	MWI[c]/eggshell (100mg)	10	66
11	Neat	MWI[c]/eggshell (200mg)	**5**	**90**
12	EtOH	80 °C	45	Trace[b]
13	Neat	80 °C	60	Trace[b]
14	Neat	MWI[c]	5	18[b]

In vitro antibacterial activities of synthetic compounds were evaluated by disk diffusion method. The experiment revealed that, compounds **5b** and **5c** found to be more effective against the tested bacterial strains and observed high zone of inhibition around 30 mm at higher concentration. Further MIC value of compound **5b** was found to be 0.5 µg/ml for both *E.coli and S. aeruginosa* and 0.25 µg/ml for both *B.subtilis and S. typhi.* The MIC value of compound **5c** was found to be 0.5 µg/mL for *E.coli. P.aeruginosa and S. typhi* 0.25 µg/mL for *B.subtilis.* The diameter of zone of inhibition for all compounds is in between 12 to 16 mm. Further, the MIC of *E.coli* for compounds **5d, 5e, 5g, 5f** and 5a were 4, 8, 2, 4 and 2 µg/mL respectively. The MIC of *S.typhi* for compounds **5d, 5e, 5g, 5f** and 5a were 2, 6, 1, 4 and 8 µg/ml respectively. The compound **5e** was found to be least reactive to the selected pathogens and showed high MIC value. The MIC of *B. subtilis* for compounds **5d, 5e, 5g, 5f** and **5a** were 1, 2, 4, 4 and 2 µg/ml respectively and MIC of *P.aeruginosa* for compounds **5d, 5e, 5g, 5f** and **5a** were 2, 6, 1, 4 and 8 µg/ml respectively. *In vitro* antifungal activities of synthetic compounds were also evaluated against fungal species *C.albicans* and *A.flavus.*compounds **5a, 5b** and **5c** showed good inhibition.

3 CONCLUSION

Herein, we have established a microwave assisted green approach synthesis of 4-benzylidene-2-(2-fluorophenyl)oxazol-5(*4H*)-one derivatives using 2-fluoro benzoyl glycine and various substituted aromatic aldehydes under solvent-free condition catalysed by bio-waste eggshell powder as an efficient reusable catalyst is described. Among the synthesized, compounds **5b** and **5c** show efficient anti-bacterial activity and compound **5a** shows efficient anti-fungal activity.

4 ACKNOWLEDGEMENT

Authors acknowledge UGC-MRP {F.43-181/2014 (SR)}, VGST-SMYSR, DST-FIST for financial support.

References

1. N. Lalitha, J. Annapurna, D. S. Iyengar and U.T. Bhaleras, *Arzeim-Foresh (Drug Research)*, 1991, **41**, 827.
2. C. Rao and N. Lalitha, *Indian J. Chem.,* 1994, **33B**, 3.
3. K. Takenaka, J. Tadakazu, Heterocycl. Chem. 1996, 33, 1367
4. B. S. Holla, R. Gonsalves, B. K. Sarojini. *Indian J. Chem.,* 1997, **36B**, 943.
5. K. Gottwald and D. Seebach, *Tetrahedron,* 1999, **55**, 723.
6. J. Meiwes, M. Schudock and G. Kretzschmar, *Tetrahedron: Asymmetry,* 1997, **8**, 527.
7. E. Bunuel, C. Cativiela and M.D. Diaz-de-Villegas, *Tetrahedron,* 1995, **32**, 8923.
8. F. Cavelier and J. Verducci, *Tetrahedron Lett.,* 1995, **36**, 4425.
9. R. Bossio, S. Marcaccini, R. Pepino and P. Paoli, *J. Heterocycl. Chem.,* 1994, **31**, 729.
10. R. Cannella, F. Clerici, M.L. Gelmi, M. Penso and D. Pocar, *J. Org. Chem.*, 1996, **61**, 1854.
11. E. Mosaddegh and A. Hassankhani, *Catalysis Comm.*, 2013, **33**, 70.
12. C. Valgas, de Souza, S. M. Smânia and E.F.A. Smânia Jr, *Braz. J. Microbiol.,* 2007, **38**, 369.

SYNTHESIS OF OXAZOLOQUINOLONE DERIVATIVES AND EVALUATION OF THEIR POTENCY TO REACTIVATE RAT BRAIN ACETYLCHOLINESTERASE INHIBITED BY CHLORPYRIFOS

S. K. Manjunatha,[a] B. Girish,[b] F. Jennifer,[c] M. Shivalingrao,[d] M. L. Sujatha[a] and A. Rekha[e]

[a]Department of Pharmaceutical Chemistry, Bapuji Pharmacy College, Davangere 577 004, Karnataka, India; [b]Department of Pharmaceutical Chemistry, GM Institute of Pharmaceutical Sciences and Research, Davangere-577004, Karnataka, India; [c]Department of Pharmaceutical Chemistry, N G S M Institute of Pharmaceutical Sciences, Mangalore 574 160, Karnataka, India; [d]Department of Pharmaceutical Chemistry, PES's Rajaram & Tarabai Bandekar College of Pharmacy, Farmagudi-Ponda, 403 401, Goa, India; [e]Department of International Public Health, 226 Grey St, South Bank Campus, Qld 4101, Griffith University, Brisbane, Australia

1 INTRODUCTION

Nowadays organophosphorous (OP) poisoning is one of the most common causes of toxicity throughout the world. Two functionally different types of drugs are used in common to treat such intoxication cases. The first type includes the reactivators of acetylcholinesterase (AChE)- oximes, which have the capability to restore the physiological function of inhibited AChE.The second type includes anticholinergics, such as atropine, which antagonize the effects of excessive ACh by blocking of muscarinic receptors[1]. Alternatively, anticholinergics and reactivators may be co-administered to get synergistic effects. At muscarinic and nicotinic synapses, organophosphorous compounds inhibit AChE release by phosphorylation at enzyme's active site. AChE regenerative process can be accelerated by detaching the OP at -OH group of the enzyme. OP compound combines with AChE enzyme forming a complex making it inactive. After ageing of the inactive state of AChE, it is difficult to break the complex to regenerate the enzyme resulting inacetylcholine accumulation at synapses[2-4].

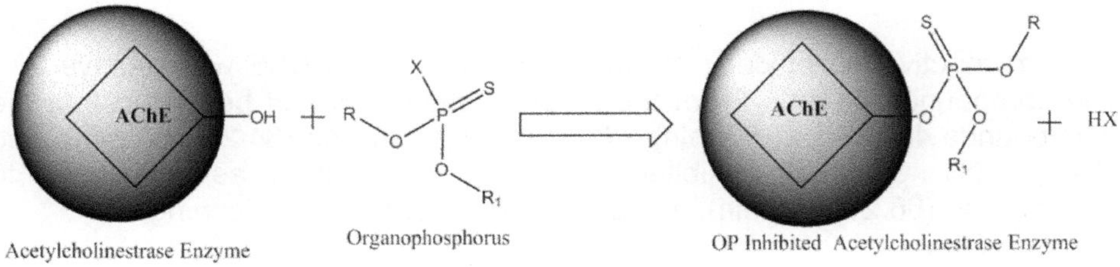

To counter the effect of OP compound, oximes catalyse the reactivation of active AChE by exerting nucleophilic attack on phosphorylgroup. Oximes theoretically removeOP compound from the complex by acting on phosphoryl

bond resultingin enzyme reactivation[5]. Reactivation of AChE inhibited by OP compounds through the above mentioned approach poses certain limitations. There is no universal antidote capable of effectivelyrestoring AChE inhibited by wide-ranging OP compounds.The oxime reactivators are efficient only when administered before the “ageing” of AChE-OP complex.Anticholinergics, like atropine, are effective only on muscarinic receptors but not on nicotinic receptors (nAChRs)[6]. Here we report the synthesize of a series of oxazoloquinolone linked to secondary heterocyclic amine and evaluation of their efficacy by *in vitro* method against chlorpyrifos inhibited AChE.

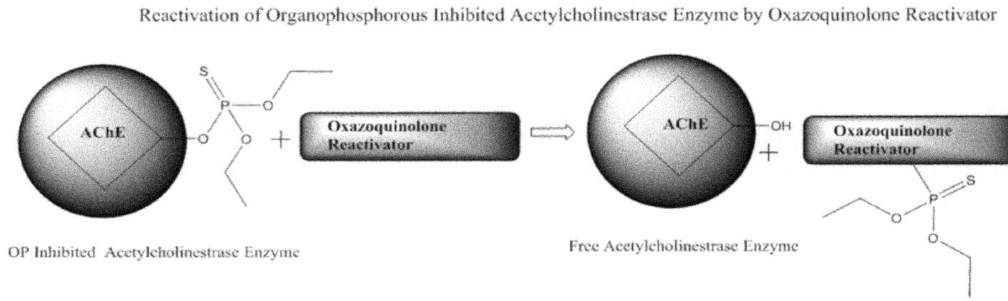

2 METHODS AND RESULTS

The present communication describes the synthesis of oxazoloquinolone linked to secondary heterocyclic amine derivatives and evaluating their efficacy by *in vitro* method against chlorpyrifos inhibited AChE. The compounds**1a-1b** were synthesized as per the method reported in our earlier paper[7].The acetyl group of compound **1a-1b** were subjected for bromination to yield compounds **2a-2b**. Further, the bromoacetyl moiety treated with secondary heterocyclic amine to achieve compounds **3a-3h**.The final compounds were obtained by refluxing the compounds **3a-3h** with hydroxylamine hydrochloride in presence of pyridine to yield the compounds **4a-4h** (Scheme 1).The structure of compounds was confirmed by physical and spectral analysis . The synthesized compounds were evaluated for their reactivation efficacy against chlorpyrifos inhibited rat brain AChE by Ellman's method[8].

In vitro reactivation efficacy of chlorpyrifos inhibited AChE was assayed for the compounds **4a-4h.** From the results obtained, it can be stated that the compounds **4b** (45.1%, 60 min), **4f** (44.8%, 60 min), **4c** (34.2%, 60 min), and **4g** (34.6%, 60 min) exhibited promising reactivation as compared to pralidoxime (68.2%, 60 min), a standard anti-dote for OP poisoning.

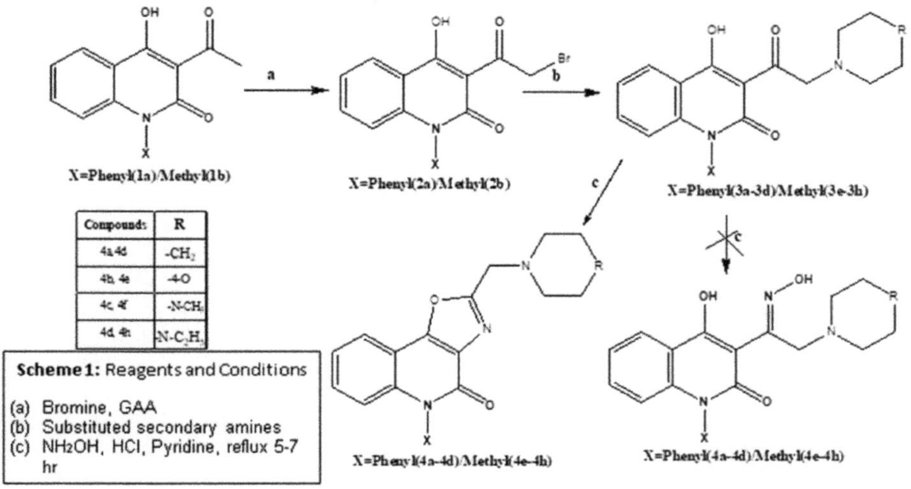

Scheme 1 *Synthesis of oxazoloquinolone derivatives*

3 CONCLUSION

Compounds 4b, 4c, 4f, and 4g having oxazoloquinolone linked substituted piperazine derivatives gave good activity against chlorpyrifos inhibited AChE. Moreover, these oxazoloquinolone derivatives appear to be potential because of their sufficient reactivation strength at lower concentration (10^{-3} M).

Acknowledgments
The work was funded by Rajiv Gandhi University of Health Sciences, Karnataka, No. P-160:2015-16 dated: 06-01-2016. The authors are thankful to Principal, Bapuji Pharmacy College, Davangere, for providing necessary facilities to carry out this research work. Thanks are extended to Indian Institute of Horticulture, Bangalore for providing chlorpyrifos.

References

1. P. Eyer, *Toxicol. Rev.*, 2003, **22**, 165.
2. C. K. Zoltani, *Handbook of Toxicology of Chemical Warfare Agents*, 2015, p. 519.
3. K. Kuca, K. Musilek, D. Jun, J. Bajgar and J. Kassa, Novel Oximes, in *Handbook of Toxicology of Chemical Warfare Agents*, 2009, p. 997.
4. K. Musilek, K. Kuca, D. Jun and M. Dolezal, *Curr. Org. Chem.*, 2007,**11**(2), 229.
5. M. Eddleston, L. Szinicz, P. Eyer, and N. Buckley, *QJM*, 2002, **95**(5), 275.
6. D. Malinak, E. Nepovimova, D. Jun, K. Musilek and K. Kuca, *Molecules*, 2018, **23**, 1.
7. M. S. Katagi, G. Bolakatti, A. Badiger, D. Satyanarayana, S. N. Mamledesai and M. L. Sujatha, *Drug Res.*, 2013, **63,** 53.
8. G. L. Ellman, D. K. Courtney, V. Andres and R. M. Feartherstone, *Biochem. Pharmacol.*, 1961, **7**, 88.

Part 4: Phytochemical Screening and Biological Evaluation

ISOLATION, CHARACTERIZATION OF RHIZOBIUM AND MESORHIZOBIUM ASSOCIATED WITH *PISUM SATIVUM* L. PEA AND *CICER AERITINUM* L. CHICKPEA AND THEIR STRESS RESPONSE TO HEAVY METAL AND HIGH SALT CONCENTRATION

Sujoy Sarkar and Tamalika Chakraborty

Department of Life Science, Guru Nanak Institute of Pharmaceutical Science and Technology, Kolkata 700114, West Bengal, India

1 INTRODUCTION

Nitrogen is essential micronutrient for plant growth and development. Leguminous plant, having rhizobia associated with the root as nodules maintains mutual symbiosis. Biological Nitrogen Fixation helps to increase agricultural productivity and food security. Pea and Brown chickpea as plant samples were studied for isolation, identification, biochemical characterization of *Rhizobium* and *Mesorhizobium* strains. Based on white mucoid growth on the specific YEMA medium and gram staining depicted rod shaped cells in microscopic view. The microorganisms were further characterized by Biochemical tests according to Bergy's Manual of Bacteriology namely Catalase test, Starch hydrolysis test, Methyl red test, Gelatin test, Indole test. The effect of NaCl and as metal (lead, Pb) stress were assessed on bacterial growth culture. Finally, mass production of the bacteria was done followed by mixing with the carrier materials for preparation of efficient Biofertilizer. Thus, the present work targets the problem of heavy metal pollution and high salt concentration in the Agricultural fields, focusing on preparation of an efficient Biofertilizer which can enhance productivity and plant yield even in the said stress conditions.

2 MATERIALS AND METHODS

2.1 Sample collection

Root nodules were collected and, samples were kept in clean sterile sample bottles sealed and stored at -20°C.[1]

2.2 Procedure for isolation of Rhizobia from root nodules

Preparation of YEMA medium, constituents are as follows: K_2HPO_4- 0.5g, $MgSo_4,7H_2O$- 0.2g, NaCl- 0.1g, $FeCl_3$- 0.1g, Mannitol- 10.0g, Yeast extract- 1.0 g, Agar- 20.0 g, Distilled water- 1000.00 ml.[2] The root system is washed in running tap water to remove adhering soil particles. Root nodules were immersed into the 0.1% $HgCl_2$, for 5 minutes to surface sterilize. Root nodules were washed 5 times with sterile distilled water to get rid of the sterilizing agent. If treated with $HgCl_2$, placed nodules in 70% ethyl alcohol for 3 minutes. Next, the nodules were washed in sterile water. The nodules were crushed in 1ml of water with a sterile glass rod. Rhizobia suspension is uniformly made up with water. Serial dilutions were made with the nodules extract. 1ml of extract is spread from each of the suspensions from various dilutions, on YEMA plates.[3] Now the plates were incubated at 35°C for 24 hours. After incubation, if proper growth comes, the serial diluent tubes were stored at 4°C in refrigerator for further characterization.[4]

2.3 Biochemical test

Table 1 *Biochemical characterization of Rhizobium and Mesorhizobium isolates.*

Host plant	Genus of Isolate	Catalase test	Starch hydrolysis test	Methyl red test	Gelatin test	Indole test
Pea (*Pisum sativum l.*)	*Rhizobium*	Positive	Positive	Negative	Negative	Negative
Brown Chickpea (*Cicer arietinum l.*)	*Mesorhizobium*	Positive	Negative	Positive	Negative	Negative

2.4 Salt (NaCl) stress

The isolates of *Rhizobium* and *Mesorhizobium* were streaked under sterile condition. Isolates growth was observed at different level of NaCl (M= 58.44 g/mol) concentration and determined by inoculating in the YEMA medium plates containing 0.5%, 1.0%, 1.5%, 2.0%, 2.5%, 3.0%, 3.5%, 4.0%, 4.5%, 5.0%, 5.5%, 6.0%, 6.5%, 7.0% and 7.5%.[5] Every experiment was done in triplicate.

2.5 Metal stress

Rhizobium and *Mesorhizobium* strains were streaked in YEMA medium on addition of lead acetate (M= 379.34 g/mol) at 10mM, 30mM, 50mM, 70mM. The experiment was done in sterile condition.[6] Each experiment was done in triplicate.

3 RESULTS

Results of Biochemical tests for different leguminous crops named pea and brown chickpea was shown in Table 1. The host plant is used for the isolation and identification of *Rhizobium* and *Mesorhizobium* strains in laboratory[7]. *Rhizobium* and *Mesorhizobium* isolates were able to grow on different salt concentrations of NaCl (Figure 1). However, Minimum Bactericidal Concentration for Rhizobium was of 7.0% and 2.0% of NaCl for *Mesorhizobium*. All isolates were found to grow on lead acetate (Figure 2).

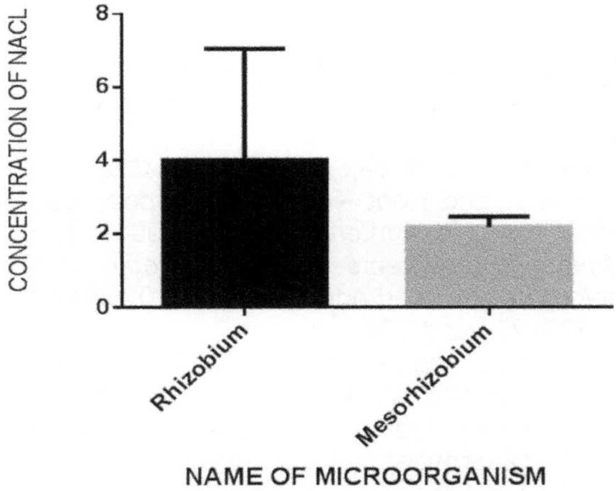

Figure 1 Survival of microorganism at high salt conc.

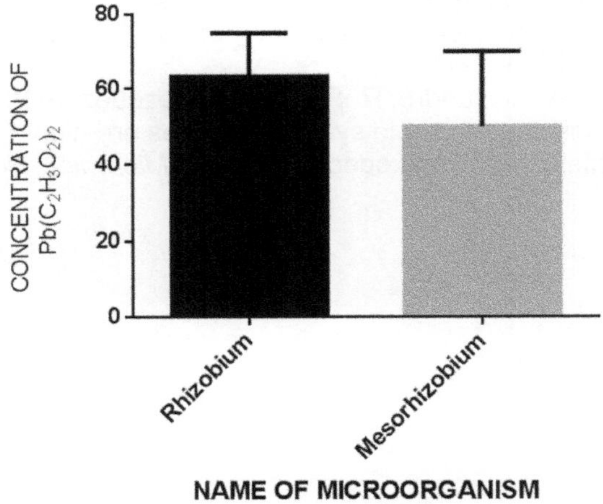

Figure 2 Survival of microorganism at high lead (Pb) conc.

4 CONCLUSION

The study showed the presence of nitrogen fixing bacteria associated with the root nodules of leguminous plant. The isolated strain was showing resistance to the heavy metal and salt stress simultaneously. Thus, the Biofertilizer prepared was efficient enough in enhancing nitrogen fixation and productivity of the plants even in stress conditions of the environment.

References

1. A. W. B Johnston and J. E. Beringer, Identification of the Rhizobium strains in pea root nodules using genetic markers, *J. Gen. Microbiol.*, 1974, 343–350.
2. T. A. Hamza and A. L. Alebejo, Isolation and Characterization of Rhizobia from Rhizospher and Root Nodule of Cowpea, Elephant and Lab Lab Plants, *Int. J. Novel Res. Interdiscip. Stud.*, 2017, **4**(4), 1–8.
3. K. R. Aneja, *Experiments in Microbiology, Plant Pathology and Biotechnology*, 4th revised edn, 2003, pp. 102–106, 178–181, 246–248, 253–255, 270–273 and 274–275.
4. A. Tyagi, V. Kumar, Purushottam and A. Tomar, Isolation, identification, biochemical and antibiotic sensitivity characterization of rhizobium strains from vigna mungo(l) Hepper, *Cicer arietinum* L and *Vigna radiata* (L) R Wilczek in Muzaffarnagar, Uttar Pradesh, India, *Int. J. Curr. Microbiol. Appl. Sci.*, 2017, **6**(12), 2024–2035.
5. S. legesse, Isolation, Identification and Authentication of Root Nodule Bacteria (Rhizobia) in Promoting Sustainable Agricultural Productivity: A Review, *Develop. Countr. Stud.*, 2016, **6**(1), **87–93**.
6. Ashok Kumar Singh and Gauri Singh, A study of multiple heavy metal tolerance in root nodulating bacteria, *Int. J. Res. Dev. Pharm. Life Sci.*, 2015, **4**(5), 1713–1721.
7. M. Laranjo, A. Alexandre, R. Rivas, E. Velázquez, J. P. W. Young and S. Oliveira, Chickpea rhizobia symbiosis genes are highly conserved across Multiple Mesorhizobium species, *FEMS Microbiol. Ecol.*, 2008, **66**(2), 391–400.

CRANBERRY AND GRAPE SEED EXTRACT AS AN INTRACANAL MEDICAMENT DURING ENDODONTIC THERAPY

Pushpalatha C.,[a] Sneha N.,[a] Gayathri. V. S.[a] and Deveswaran R.[b]

[a]Department of Pedodontics and Preventive Dentistry, Faculty of Dental Sciences, M.S. Ramaiah University of Applied Sciences, Bengaluru, 560054, India; [b]Department of Pharmaceutics, Faculty of Pharmacy, M.S. Ramaiah University of Applied Sciences, Bengaluru,560054, India

1. INTRODUCTION

Cranberry extract and grape seed extract rich in proanthocyanidin are known to exhibit anticariogenic effect and effective against oral pathogens like *Streptococcus.mutans*, *Prophyromonas.ginivalis*, *Candida albicans*, urinary tract *Escherichia coli,* and urinary tract *Enterococcus.feacalis* [1-3]. However, the effects on oral *Enterococcus. feacalis* has been not reported. Hence the aim of the present study is to evaluate and compare the antibacterial, anti-inflammatory activity, micro-hardness and fracture resistance of cranberry extract and grape seed extract on the teeth.

2. MATERIALS AND METHODOLOGY

Cranberry extract and Grape seed extract of analytical grade was formulated into intracanal medicament in the form of gel. The antibacterial and anti-inflammatory activity of the formulation was assessed. To assess the antibacterial activity of the intracanal medicament in the root canal and microhardness of the root dentine, 40 extracted single rooted teeth were randomly selected and decoronated. Canals were prepared and incubated in tryptone soya broth with *Enterococcus.feacalis* for 7-8 days for biofilm formation. Blocks were divided into 4 groups. The groups were Group 1 (Control), Group 2(Cranberry Seed extract), Group 3 (Grape seed extract) and Group 4 (Chlorhexidine). Respective medicaments were placed & sealed using sterile agarose gel. Antimicrobial activity and micro-hardness were assessed after 24hrs, 3 days and 7 days. The fracture resistance of the tooth was assessed using a universal testing machine on 7th day of the treatment procedure.

3. RESULTS

Agar well diffusion data showed both novel intracanal medicaments exhibit zone of inhibition but lesser than chlorhexidine and standard group. 99% bacterial reduction (*Enterococcus.feacalis*) was seen at 7day in chlorhexidine group followed by 95% reduction in Cranberry and grape seed intracanal medicament groups. Anti-inflammatory activity assay for grape seed intracanal medicament and cranberry intracanal medicament showed IC50 values of 67.74 µg/ml and 65.21 µg/ml, whereas for standard IC50 values were 35.7421 µg/ml. Micro-hardness of root dentine was significantly increased in grape seed and cranberry intracanal medicament group, whereas decreased significantly in the chlorhexidine group. The grape seed and cranberry intracanal medicament group showed maximum fracture resistance to load compared to chlorhexidine and control groups.

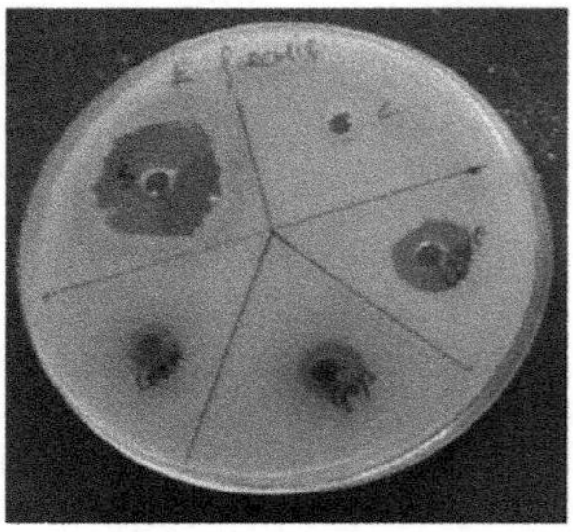

- **C**-Control
- **S**-Standard
- **Chloro**- Chlorhexidine intracanal medciament
- **Grap**-Grape seed intracanal medicament
- **Cran**-Cranberry intracanal

Figure 1 *Agar diffusion test for different groups.*

Table 1 *Anti-inflammatory activity of different groups.*

Groups	Concentration	OD at 540 nm	NO produced	% NO Inhibition	IC_{50}
Control	0	0.824	95.38	0.00	
Standard (µg/ml)	3.13	0.658	74.10	22.31	35.74
	6.25	0.601	66.79	29.97	
	12.50	0.506	54.62	42.74	
	25.00	0.385	39.10	59.01	
	50.00	0.216	17.44	81.72	
	100.00	0.102	2.82	97.04	
Grape seed intracanal medicament (µg/ml)	10	0.655	74.10	23.31	67.74
	20	0.607	62.69	39.97	
	40	0.501	51.72	44.64	
	80	0.395	45.11	54.21	
	160	0.216	37.42	61.38	
	320	0.102	26.51	78.32	
Cranberry intracanal medicament (µg/ml)	0	0.824	95.38	0.00	65.21
	10	0.651	73.21	23.25	
	20	0.554	60.77	36.29	
	40	0.498	53.59	43.82	
	80	0.427	44.49	53.36	
	160	0.358	35.64	62.63	
	320	0.261	23.21	75.67	

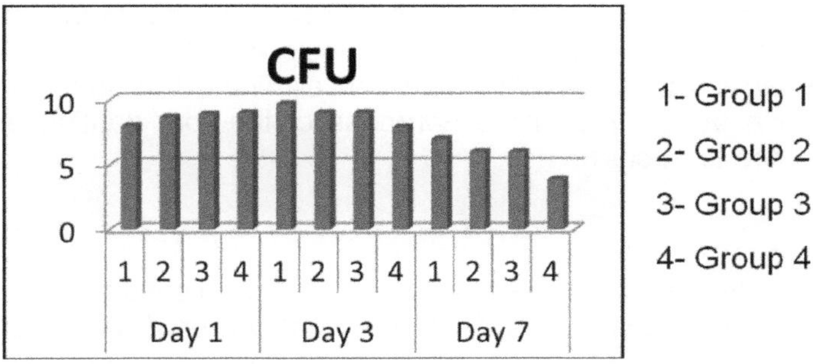

Figure 2 *Graph showing CFU counts at day 1, 3 and 7 for different groups.*

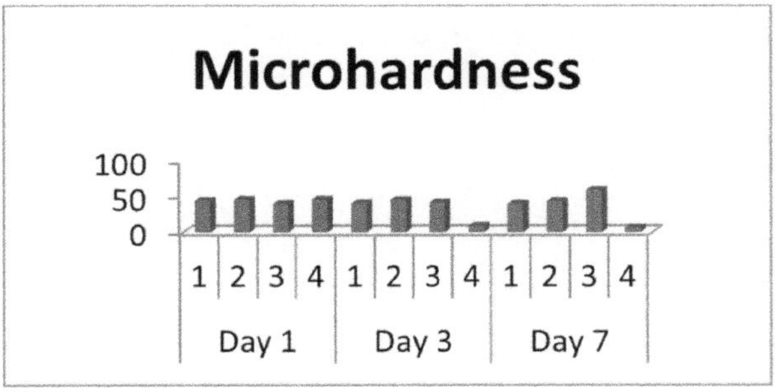

Figure 3 *Graph showing microhardness on root dentine at day 1, 3 and 7 for different groups.*

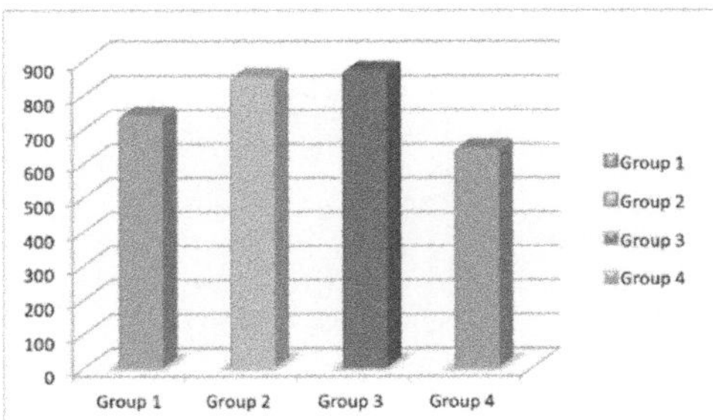

Figure 4 *Peak load at fracture for different groups.*

4. CONCLUSION

The present study shows that the developed Grape seed and cranberry intracanal medicaments have antibacterial and anti-inflammatory activities. After Grape seed and cranberry intracanal medicament application, there was an increase in both fracture resistance and microhardness of root dentine. Thus the results suggest that Grape seed and cranberry intracanal medicament can be a promising intra-canal medicament in reducing endodontics infections and increase prognosis of the endodontic treatment. The developed intracanal medicament can eliminate the adverse effects on the root dentin by enhancing the hardness of the root dentine and making tooth resistance to fracture.

References

1. K. Chatelain, S. Phippen, J. McCabe, C. A. Teeters, S. O'Malley and K. Kingsley, *Evid. Based Complement. Alternat. Med.*, 2011, **2011**, 467691.
2. W. Zhao, Q. Xie, A. K. Bedran-Russo, S. Pan, J. Ling and C. D. Wu, *Int. J. Dent.*, 2014, **42**, 1010–1018.
3. K. D. Patel, F. J. Scarano, M. Kondo, R. A. Hurta and C. C. Neto, *J Agric. Food Chem.*, 2014, ***59***, 12864–12873.

IN VIVO ANTI-ARTHRITIC ACTIVITY OF ETHANOLIC EXTRACT OF *BARLERIA MONTANA* NEES LEAVES AGAINST CHEMICAL INDUCED ARTHRITIS

J. Basini and S. Sathrapalli

Department of Pharmacology, Seven Hills College of Pharmacy, Venkataramapuram, Tirupati-517561, Andhra Pradesh, India.
Corresponding Author E-Mail: jyothi_811@yahoo.co.in

1 INTRODUCTION

Barleriamontana Nees, commonly called as mountain barleria belongs to the Acanthaceae family, widely distributed in India. In the ayurvedic system of medicine, the plant is used in treatment of urinary infections, rheumatism, jaundice, hepatic obstruction, dropsy, fever, inflammation, gastrointestinal disorders, whooping cough, *etc*. Till date there is no pharmacological report of this plant for rheumatoid arthritis. Hence, the purpose of this study was to evaluate the *Invivo* anti-arthritic activity of the leaves of Ethanolic Extract of the *Barleriamontana* Nees(EEBM) against chemical-induced arthritis.

2 METHODS AND RESULTS

2.1 Extraction of *Barleriamontana*

The leaves of plant *Barleriamontana Nees* were collected from Seshachalam hills of Tirumala in Chittoor district of Andhra Pradesh, India; washed with distilled water to remove dirt and soil and they were dried under shade at room temperature for 10 to 15 days. The dried plant leaves were powdered by using a grinding mill to obtain a coarse powder and then passed through 40 mesh sieve. The powdered material was subjected to extraction using 95% ethanol by the solvent extraction method. The percentage yield of extract was 6.75% for 100g of powder.

2.2 Estimation of preliminary phytoconstituents

Coumarins, sterols, quinones, flavonoids, terpenoids, alkaloids, proteins and tannins were present in EEBM.

2.3 Determination of acute oral toxicity studies

Acute toxicity studies were performed for EEBM according to OECD guidelines 423. 12 mice were selected for the study and oral administration of EEBM at a dose of 5, 50, 300, 2000 mg/kg. In this study, the animals were observed individually after 2 hrs administration of EEBM, periodically during the first 24 hrs and daily thereafter, for a total of 14 days for any changes in consumption of food, water, body weight, behavioral changes and mortality rate. No mortality and no physical signs of toxicity were observed for dosing up to 2000 mg/kg.

2.4 Determination of *In vivo* anti-arthritic activity

The anti-arthritic activity was evaluated *invivo* by Complete Freund's Adjuvant (CFA) method. Male Albino Wistar rats (150-200 g) were used for the study. Animals were divided into 5 groups. Group 1: Normal control-received distilled water, *p.o*, Group 2: Arthritic control-injected 0.1 ml (5 mg/ml) of CFA into the sub-plantar region of the left hind paw, Group 3: Standard control -CFA plus diclofenac sodium 10 mg/kg *p.o*, Group 4: Test 1- CFA plus a low dose of EEBM 200 mg/kg *p.o* and Group 5: Test 2- CFA plus a high dose of EEBM 400 mg/kg *p.o.*The treatment schedule was 14 days and the parameters were examined alternatively for 28days. The protocol of the study was approved by the Institutional Animal Ethical Committee (IAEC) of Krishna Teja Pharmacy College, Tirupati, Andhra Pradesh (Reg.no.1521/po/a/11/CPCSEA).

The parameters like paw volume, paw diameter and body weight on 1st, 7th, 14th, 21st and 28th day were estimated during treatment. In EEBM treatment, reduction in body weight, paw volume, paw diameter were compared to CFA induced arthritis on 21st and 28th day. On 28thday, the blood samples were collected and utilized for estimation of hematological changes (RBC, WBC, Hb% and ESR), liver marker enzymes (AST, ALT and ALP), C-reactive protein and RF. Then radiological examination was done using X-ray and histopathological studies were done for joints. In EEBM treatment, there was reduction in paw volume, polyarthritic score, paw diameter when compared to CFA induced arthritis. EEBM treatment significantly (P<0.01, P<0.001) ameliorated the hematological parameters (Table 1), liver marker enzymes, C-reactive protein and RF (Table 2) as compared to CFA induced arthritic rats with confirmation of changes by X-ray (Figure1) and histopathological examination (Figure 2).

Table 1 *Effect of EEBM on Hematological parameters in CFA induced arthritic rats.*

Groups	Dose mg/kg	Hematological parameters			
		Hb%	RBC	WBC	ESR
I-Normal control	Normal saline	12.63±0.40	7.63±0.069	5.13±0.3	4.38±0.55
II-Arthritic control	CFA- 0.1ml	6.45±0.86	4.21±0.48	9.63±0.74	9.46±0.74
III - Standard	Diclofenac sodium 10mg/kg	12.21±0.70[b]	7.48±0.28	5.68±0.48[b]	3.93±0.02[b]
IV - Test-I	EEBM 200mg/kg	11.00±0.70[a]	5.84±0.57[a]	6.46±0.51[a]	4.0±0.51[b]
V - Test-II	EEBM400mg/kg	12.08±0.50[b]	6.28±0.67[b]	5.84±0.42[b]	3.75±0.24[a]

Values are expressed as mean ± SEM (n=6). [a]$p<0.01$,[b]$p<0.001$ significant as compared with arthritic control (One –way ANOVA followed by Dunnet's test)

Table2 *Effect of EEBM on liver parameters, CRP and RF in CFA induced arthritic rats.*

Groups	Dose mg/kg	Liver parameters, CRP and RF				
		AST (IU/L)	ALT (IU/L)	ALP(IU/L)	CRP	RF
I-Normal control	Normal saline	76.29±2.76	68.75±3.80	114.25±4.45	1.85±0.7	21.14±1.71
II-Arthritic control	CFA-0.1 ml	215.51±12.50	180.35±7.35	324.80±13.50	9.35±1.15	78.76±5.65
III - Standard	Diclofenac sodium 10mg/kg	100.75±11.52[b]	80.45±2.15[b]	125.75±4.57[b]	4.60±0.76[b]	72.59±2.65[b]
IV - Test-I	EEBM 200mg/kg	115.25±3.95[a]	107.43±3.61[a]	155.52±6.50[a]	3.38±0.75[a]	21.36±1.48[b]
V - Test-II	EEBM400mg/kg	110.10±3.93[b]	89.45±3.70[b]	138.45±4.52[a]	2.75±0.91[a]	20.28±3.80[a]

Values are expressed as mean ± SEM (n=6). [a]$p<0.01$,[b]$p<0.001$ significant as compared with arthritic control (One –way ANOVA followed by Dunnet's test)

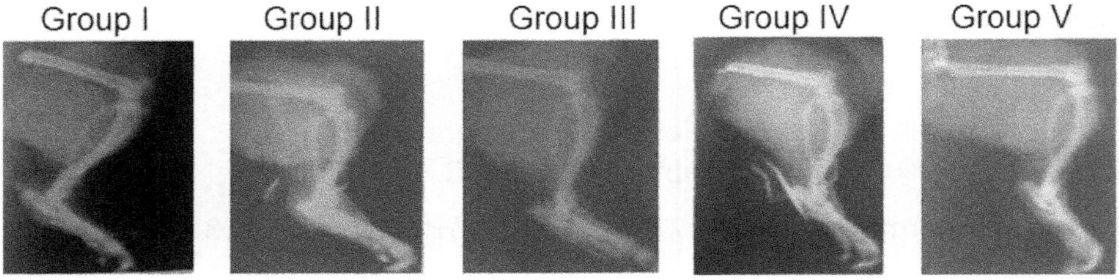

Figure 1 *Effect of EEBM on Radiographic analysis.*

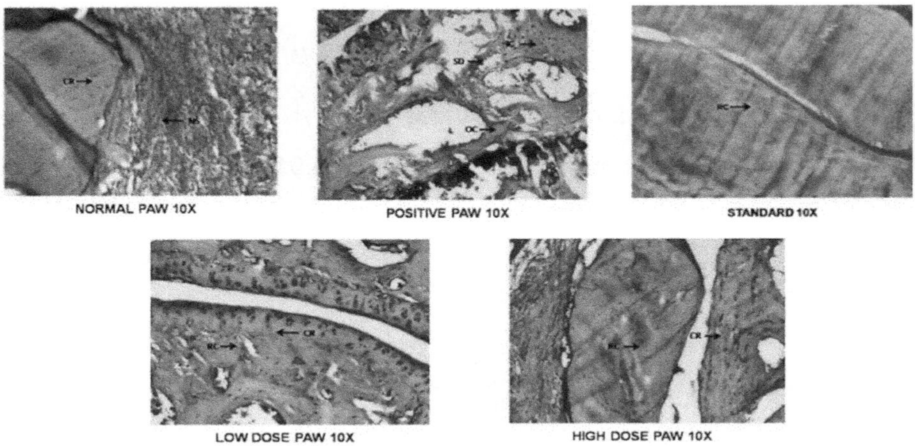

Figure 2 *Effect of EEBMon Histopathological examination.*

CONCLUSION

The results of the biochemical investigation, X-ray and histopathological studies concludes that EEBM possess a significant anti-arthritic activity against CFA induced arthritis model and justifies its therapeutic role in controlling inflammation.

References

1. D. S. Bhakuni, V. N. Sharma and K. N. Kaul, *J. Sci. Ind. Res.,* 1961, **20**, 232.
2. Y. Yang, P. Hutchinson and E.F. Morand. *Arthritis.Rheum.,*1999, **42**,538.
3. Hardiksonil, Ghanshyampatel, Megha Shah, Mayankpanchal and Krishna Murti. *IJPCS,* 2013, **2**, 14.
4. C.L.A. Ghai. Textbook of Practical Physiology. *Jaypee Brothers Medical Publishers Ltd.,* 1993, 119-202.
5. M.B. John. In: Laboratory Medicine Hematology, CV Moshy Co, St Louis, 1972,198-1209.

6. S Reitman and S Frankel. *Am.J. Clin. Pathol.,*1957, **28**, 56.
7. S.B. Nikhal,P.A. Dambe, D.B. Ghongade and D.C. Goupale. *Int. J. Pharm. Sci.,*2010, **2**, 30-32.
8. C.K. Kokate, A.P. Purohit and S.B. Gokhale. *Analytical Pharmacognosy,*Nirali Prakashan, India, 1997,119.

SANSEVIERIA ROXBURGHIANA (AGAVACEAE) AND *BAUHINIA VAHLII* (CAESALPINIACEAE) OFFER PROTECTION AGAINST EXPERIMENTALLY INDUCED CONVULSIONS

K. Sundara Saravanan and V. Madhavan

Department of Pharmacognosy, Faculty of Pharmacy, Ramaiah University of Applied Sciences, Bengaluru – 560 054, Karnataka, India

1. INTRODUCTION

In Ayurveda system of medicine, the root of *Murva* (Accepted source: *Marsdenia tenacissima* (Roxb.) Moon (Asclepiadaceae)) forms a vital component in various formulations like *Ayaskrti, Brahanmanjistadi kvatha curna,* etc. These formulations are indicated in the treatment of apasmara (epilepsy), *hrdroga* (cardiac disorders), *mutra dosa* (urinary disorders), *kustha* (skin diseases), *pandu* (anaemia), *jvara* (fever), *netra roga* (eye diseases) and *gulma* (abdominal disorders)[1]. But, there are several other plant species used in the name of *Murva* across the country and two such species are *Sansevieria roxburghiana* Schult. & Schult. f. (Agavaceae) and *Bauhinia vahlii* Wt. & Arn. (Caesalpiniaceae)[2]. Hence, in the present study, protective effect of ethanol and aqueous extract of *S. roxburghiana* rhizome and *B. vahlli* root was evaluated against Maximal Electrical Shock (MES) and Pentylenetetrazole (PTZ) induced convulsions in experimental animals.

2. METHODS AND RESULTS

2.1. Extract Preparation and Phytochemical Screening

S. roxburghiana rhizomes were collected from the Kadavur hillocks of Karur district, Tamil Nadu and the roots of *B. vahlii* from Tirumala hills of Andhra Pradesh, India during May and September 2011, respectively. The collected materials were botanically identified by taxonomist and the herbarium specimen SS46 (*S. roxburghiana*), SS48 (*B. vahlii*) were deposited in the Department of Pharmacognosy.

The collected rhizomes and roots were thoroughly cleaned to remove the extraneous materials and dried. The dried materials were coarsely powdered and used for extract preparation. Ethanol extract was prepared by continuous hot percolation method with 95% ethanol, while aqueous extract was obtained by maceration using chloroform-water. Later, the solvents were evaporated to dryness to attain the extract. The yield obtained from ethanol and aqueous extract of *S. roxburghiana rhizome* was 13.06 % w/w and 41.06 % w/w, while it was 3.2 % w/w and 9.6 % w/w for *B. vahlii* root respectively. The prepared ethanol and aqueous extract of both the plants were screened for the presence of various bioactives using standard procedures[3]. The phytochemical screening revealed the presence of fixed oils, phytosterols, gums, mucilage, alkaloids, carbohydrates, saponins, volatile oil in *S.*

roxburghiana rhizome and fixed oils, phytosterols, carbohydrates, phenolic compounds & tannins, saponins, proteins in *B. vahlii* root.

2.2 Acute Toxicity and Anticonvulsant Activity Study[4-7]

The protocol was reviewed and approved (MSRCP/P-12/2011) by the Institutional Animal Ethics Committee. Acute toxicity study was performed in Albino Wistar rats as per OECD guideline 423. Extracts reconstituted in acacia (2%) were administered orally to the animals fasted overnight in increasing dose up to 2000 mg/kg. No symptom of toxicity was observed at the tested dose of the extracts of both plants and hence 1/10th, 1/5th of this tested dose was selected for further evaluation.

MES induced convulsion

Swiss Albino mice were divided into 10 groups (n=6). Electro convulsio meter (Hicon) was used to provide electrical shock of 60 mA for 0.2 seconds to the ear pinna of each animal through electrodes. Group 1 served as control which received 2% acacia orally. Group 2 (Standard) was exposed to electrical shock 30 min after the administration of phenytoin sodium (25 mg/kg bw, i.p., Abbot Healthcare Pvt. Ltd.). Group 3 and 4 were administered with 200 and 400 mg/kg, (p.o) of *S. roxburghiana* rhizome ethanol extract; group 5 and 6 with 200 and 400 mg/kg (p.o) of *S. roxburghiana* rhizome aqueous extract; while group 7 and 8 received 200 and 400 mg/kg, (p.o) of *B. vahlii* root ethanol extract; group 9 and 10 were treated with 200 and 400 mg/kg (p.o) of *B. vahlii* root aqueous extract respectively. The animals were imposed with electric shock 60 min after the administration of extracts and vehicle. The convulsion parameters tonic flexion, hind limb extension, clonic convulsion, stupor were observed. The time duration of extensor phase, recovery in seconds (s) was recorded and the results were tabulated in Table 1.

PTZ induced convulsion

Experimental animals were divided into 10 groups (n=6). PTZ (Sigma) (60 mg/kg bw i.p) was used to induce seizures. The control group (group 1) was orally administered with 2 % acacia while group 2 was treated with PTZ 30 minutes after administration of diazepam, 4 mg/kg bw, ip (Standard, Ranbaxy Laboratories Ltd). Group 3 and 4 were administered with 200 and 400 mg/kg (p.o) of *S. roxburghiana* rhizome ethanol extract; group 5 and 6 with 200 and 400 mg/kg (p.o) of *S. roxburghiana* rhizome aqueous extract; group 7 and 8 received 200 and 400 mg/kg (p.o) of *B. vahlii* root ethanol extract; group 9 and 10 with 200 and 400 mg/kg (p.o) of *B. vahlii* root aqueous extract respectively. The convulsion inducing agent (PTZ) was administered 60 min after extract and vehicle administration. The animals were observed for the exhibition of jerks, convulsions, straub's tail. The time taken in seconds (s) for the onset of convulsion, recovery was recorded and presented in Table 2.

Table 1 *Effect of S. roxburghiana and B. vahlii on extensor phase and recovery time in MES induced convulsion.*

Group		Dose (mg/kg)	Extensor phase (s)	Recovery time (s)
Control		-	90.66 ± 3.47	306.33 ± 23.23
Standard (Phenytoin)		25	0.33 ± 0.22***	56.00 ± 14.69***
			S. roxburghiana	
Ethanol extract		200	41.33 ± 5.79***	155.63 ± 12.31***
		400	26.83 ± 1.53***	126.33 ± 18.57***
Aqueous extract		200	62.83 ± 1.37*	207.66 ± 11.68**
		400	58.66 ± 2.45**	171.33 ± 6.46***
			B. vahlii	
Ethanol extract		200	74.16 ± 4.7**	223.50 ± 11.81**
		400	52.83 ± 1.74***	166.16 ± 6.76***
Aqueous extract		200	73.33 ± 2.76**	230.50 ± 12.5*
		400	63.83 ± 1.37***	213.33 ± 7.43**

Values are expressed as Mean ± SEM.

Data were analyzed by one way ANOVA followed by Tukey Kramer Multiple Comparison test.

* $p<0.05$, ** $p<0.01$, *** $p<0.001$, in comparison with control group.

Table 2 *Effect of S. roxburghiana and B. vahlii on onset and recovery time in PTZ induced convulsion.*

Group	Dose (mg/kg)	Onset time (s)	Time taken for recovery (s)
Control	-	81.33 ± 3.91	997.16 ± 14.60
Standard (Diazepam)	4	0.00 ± 0.00***	0.00 ± 0.00***
		S. roxburghiana	
Ethanol extract	200	105.5 ± 5.13	890.83 ± 16.73**
	400	144.33 ± 11.57**	846.16 ± 13.11***
Aqueous extract	200	94.83 ± 3.29	903.66 ± 12.09
	400	114.83 ± 5.57*	882.33 ± 11.54**
		B. vahlii	
Ethanol extract	200	117.5 ± 4.47**	860.83 ± 26.62***
	400	132.5 ± 4.66***	825.00± 25.33***
Aqueous extract	200	112.16 ± 3.60*	902.5 ± 23.62**
	400	121.16 ± 2.89**	876.66 ± 22.01***

Values are expressed as Mean ± SEM.

Data were analyzed by one way ANOVA followed by Tukey Kramer Multiple Comparison test.

* $p<0.05$, ** $p<0.01$, *** $p<0.001$, in comparison with control group.

3. CONCLUSION

The ethanol and aqueous extracts *S. roxburghiana* rhizome and *B. vahlii* root were assessed for its anticonvulsant activity in MES and PTZ induced seizures. The results suggested that ethanol and aqueous extracts of both the plants possess significant effect in reducing the extensor phase and recovery phase in MES induced convulsion; and increasing the latency of convulsion and reducing the recovery time in PTZ induced seizure model. Meantime, among the extracts and doses evaluated, it was found that ethanol extract of both the plants at 400 mg/kg exhibited significant protective action against convulsion. Moreover, as per the numerical data, the protective activity offered by *S. roxburghiana* in MES induced convulsion is better than *B. vahlii*, while in PTZ induced convulsion reverse is the case; *B. vahlii* offered more protection over *S. roxburghiana*. The difference in the nature of chemical constituents of these two plants and their mechanism of action may probably be attributed to the observed variation in the activity against two screened models. The study substantiates the traditional claim and further investigations are to be taken up to identify the plant constituent/s responsible and to derive the mechanism of action.

References

1. Anonymous, *The Ayurvedic Formulary of India*, 2nd rev. edn, Part 1, Delhi, The Controller of Publications, Govt of India, 2003, pp. 6, 19, 57, 59, 68, 87, 94, 95 and 320.
2. G. S. Levekar, Kailash Cahandra, M. B. Yelne, B. P. Dhar, G. V. R. Joseph, A. K. Mangal and B. G. Chaudhari, Database on medicinal plants used in Ayurveda and Siddha, vol. 8, New Delhi, CCRAS, 2007, p. 277.
3. C. K. Kokate, Practical Pharmacognosy, New Delhi, Vallabh Prakashan, Reprint 1999, pp.107–121.
4. OECD, 2002, Acute oral toxicity. Acute oral toxic class method guideline 423 adopted 23.03.1996, in *Eleventh Addendum to the, OECD, guidelines for the testing of chemicals organisation for economical co-operation and development*, Paris, June, 2000.
5. M. Anita, C. Sudha, V. Madhavan and S. N. Yoganarasimhan, *Pharm. Biol.*, 2007, **45**, 408–410.
6. V. Madhavan, S. Arora, A. Murali and S. N. Yoganarasimhan, *J. Nat. Rem.*, 2009, **9**, 68–73.
7. A. Asma Samaunnisa, M. Riazuddin, C. H. S. Venkataramana and V. Madhavan, *Int. J. Pharm. Pharm. Sci.*, 2014, **6**, 601–605.

HEPATOPROTECTIVE ACTIVITY OF TWO SPECIES OF *SPHAERANTHUS* ON EXPERIMENTAL ANIMAL MODELS – A COMPARATIVE STUDY

Gowri R. and Madhavan V.

Department of Pharmacognosy, Faculty of Pharmacy, Ramaiah University of Applied Sciences, India

1 INTRODUCTION

Vaghbhata in *Astanga Hridaya* has mentioned that when a drug is scarce, another drug having similar therapeutic properties to the source drug may be used in the preparation.[1] Unfortunately, this tempo of replenishing additional new drugs dwindled during subsequent periods. Thus, the drugs prescribed in Ayurveda have been limited to a few hundred for the past several centuries. As such today many of these established drugs are on the verge of extinction due to indiscriminate exploitation from their natural habitats. Hence, there is an urgent need to find out alternate sources for the important drugs so as to ensure that the Ayurvedic system is accessible in the ensuring centuries. *Munditika* is an important drug in Ayurveda.[2] It is known to be used in the management of epilepsy (*apasmara*), fever (*jwara*), jaundice (*kamala*), diabetes (*meha*), cough (*ksaya*), gastric disorders (*kosthamaya*), oedema (sotha), leprosy (*kusta*), and helminthiasis (*krimi*). It is used in various Ayurvedic formulations like *Vatagajankusha rasa, Guduchi taila, Arka mundi, Navaratnaraja mriganka rasa* and *Munditika churna*. The accepted botanical source of *Munditika* is *Sphaeranthus indicus* Linn. (Asteraceae). *Sphaeranthus amaranthoides* Burm.f. occurs in South India and *Sphaeranthus africanus* Linn. which is a native of Africa but introduced and found wild in India are also used as substitutes for *Munditika*. The study was aimed at comparing the hepatoprotective activity of *S. indicus* with *S. amaranthoides*, which may provide an alternate source for the Ayurvedic drug *Munditika*.

2 METHODS AND RESULTS

2.1 Collection, identification and authentication

The whole plants of *S. indicus* and *S. amaranthoides* were collected in flowering condition from the Cheranmahadevi paddy fields of Tirunelveli District and Komaneri village of Thuthukudi region, Tamil Nadu respectively in January 2011. Identification and authentication of the plant material was done by using various floras by the taxonomist, Dr. S. N. Yoganarasimhan. Herbarium specimens were prepared following International herbarium curatorial practices as per Jain and Rao.

2.2 Preparation of plant extracts and preliminary phytochemical analysis

The plants were washed, dried separately at room temperature, powdered mechanically and passed through a 60 mesh sieve. Coarsely powdered whole plant materials (each 100 g) of both the species were extracted separately with chloroform-water by maceration technique for 7 days. The filtered extract was dried using rotary flash evaporator under reduced temperature. The percentage yield of extracts was found to be 9.43% and 11.42% w/w respectively. Preliminary phytochemical studies were carried out by using standard procedures. The result showed the presence of carbohydrates, proteins, flavonoids, tannins, saponins and volatile oil.

2.3 HPTLC Studies

HPTLC study was carried out following standard methods using Camag HPTLC system. Precoated silica gel GF_{254} was used as stationary phase; Chloroform and Methanol in the ratio 9:1 was used as mobile phase. After development the plates are sprayed with 2% $AlCl_3$ to detect flavonoids.

After derivatization both the extracts showed one yellow fluorescence spot with R_f value 0.15 at 366 nm, with Aluminium chloride indicating the presence of flavonoids. (Figures 1 and 2).

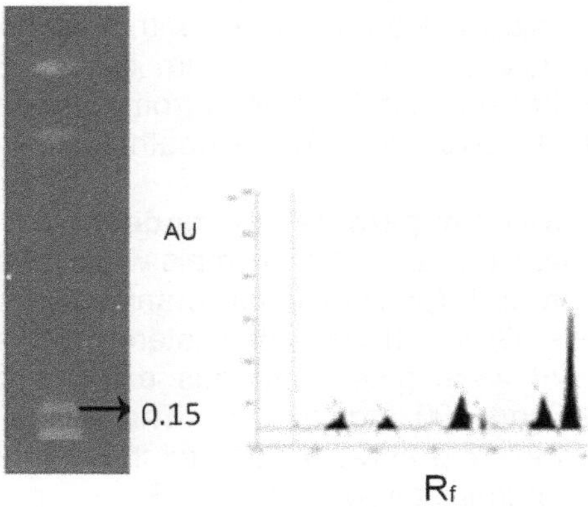

Figure 1 *HPTLC fingerprint profile of aqueous extract of S. indicus.*

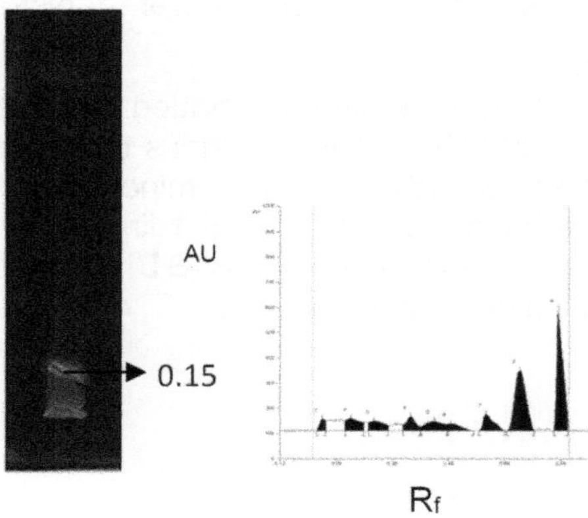

Figure 2 *HPTLC fingerprint profile of aqueous extract of S. amaranthoides.*

2.4 Pharmacological studies

Healthy Wistar albino rats (170–200 g) of either sex were used for the study.

2.5 Acute oral toxicity studies

Acute toxicity of aqueous extracts of both *Sphaeranthus* species was determined as per OECD 423 guidelines. The extracts were found to be safe up to the dose of 2000 mg/kg b.wt without any symptoms of mortality. Hence, 1/5th and 1/10th of the safe dose tested was selected for further pharmacological screening.

2.6 CCl_4 induced hepatotoxicity model[3]

Seven groups of animals (n=6) were used for the study. Group I considered as Normal Control received 10 % Tween-80 (1 ml/kg, p.o); Group II: Positive control received 10 % Tween-80 (1 ml/kg, p.o); Group III: Standard group received Silymarin (100 mg/kg, p.o); Group IV & V: Received aqueous extract of *S. indicus* at doses of 200 & 400 mg/kg, p.o; Group VI & VII: Received aqueous extract of *S. amaranthoides at doses of* 200 & 400 mg/kg, p.o. All the drug treatments were continued for 5 days. The CCl_4, standard and extracts treated animals were administered with 1 ml mixture of CCl_4 in olive oil (1:1) subcutaneously on second and third days. On the 6th day, all the animals were anaesthetized to collect blood from retro-orbital plexus (Table 1) and further sacrificed to isolate liver for histopathological examination

2.7 Paracetamol induced hepatotoxicity model[4]

Group I considered as Normal control animals were treated with 10 % Tween 80 (1 ml/kg., p.o.); group II (positive control) animals with Tween 80 (1 ml/kg b.wt., p.o.) for 5 days. Group III was administered with silymarin (100 mg/kg, p.o). Group IV to VII were given aqueous extracts of *S. indicus* and *S. amaranthoides* (200 and 400 mg/kg, p.o.). The animals were treated for a period of 5 days. On second and third day all the animals except the normal control were orally administered with 2 g/kg of paracetamol to induce hepatotoxicity. On 6th day, 18 hours after the last dose of paracetamol, blood was collected from retro-orbital plexus of all animals under anaesthesia (Table 2) and later animals were sacrificed to obtain liver for histopathology.

Separation of serum

Serum was separated by centrifuging the clotted blood at 10000 rpm for 10 min and used for estimating the levels of various biochemical parameters like aspartate aminotransferase (AST), alanine aminotransferase (ALT), Alkaline Phosphatase (ALP), total protein, and total bilirubin. All these parameters were estimated by using a Semi-automatic B4B Diagnostic Division Chemistry Analyser CA-2005, using diagnostic kits.

Table 1 *Effect of S. indicus and S. amaranthoides aqueous extract in CCl_4 induced hepatotoxicity in rats*

Groups	AST (U/L)	ALT (U/L)	ALP (U/L)	TP (g/dl)	Total Bilirubin (mg/dl)
Normal	64.85 ± 6.00	96.38 ± 0.35	175.40 ± 0.34	6.78 ± 0.35	0.908 ± 0.004
Positive control	910.33 ± 12.53[a]	465.66 ± 2.51 [a]	464.15 ± 1.41 [a]	3.53 ± 0.08 [a]	2.06 ± 0.16 [a]
Standard	225.96 ± 6.43*** (81.31)	226.40 ± 3.62*** (64.79)	209.63 ± 2.63*** (88.14)	6.21 ± 0.06***	0.79 ± 0.03*** (110.21)
Aqueous extract (SI)(200 mg/kg)	625.55 ± 3.19*** (41.60)	415.37 ± 1.70*** (13.61)	387.13 ± 2.61*** (26.67)	5.06 ± 0.08***	1.93 ± 0.008 [ns] (11.79)
Aqueous extract (SI)(400 mg/kg)	417.72 ± 3.20*** (54.41)	383.03 ± 2.46*** (22.37)	359.10 ± 2.04*** (36.38)	5.32 ± 0.13***	1.46 ± 0.02*** (52.23)
Aqueous extract (SA)(200 mg/kg)	577.97 ± 0.33*** (34.96)	395.59 ± 2.90*** (18.97)	389.07 ± 1.95*** (26.00)	4.43 ± 0.10***	1.78 ± 0.01 [ns] (24.02)
Aqueous extract (SA)(400 mg/kg)	391.72 ± 2.16*** (59.07)	351.22 ± 2.27*** (30.98)	350.34 ± 2.52*** (39.41)	5.49 ± 0.17***	1.39 ± 0.009*** (57.98)

*Values in the parenthesis indicate the % protection in individual biochemical parameters from their elevated values caused by the hepatoprotection. Values are mean ± SEM (n=6); [a] $p<0.001$ in comparison with normal control group; *** $p<0.001$, [ns] $p>0.05$ in comparison with positive control.*

Table 2 *Effect of S. indicus and S. amaranthoides aqueous extract in Paracetamol induced hepatotoxicity in rats.*

Groups	AST (U/L)	ALT (U/L)	ALP (U/L)	TP (g/dl)	Total Bilirubin (mg/dl)
Normal	71.61 ± 5.43	35.49 ± 9.96	32.56 ± 1.43	6.78 ± 0.87	0.172 ± 0.01
Positive Control	365.63 ± 29.18 [a]	125.80 ± 20.93 [a]	100.42 ± 0.09 [a]	3.08 ± 0.03 [a]	1.234 ± 0.08 [a]
Standard group	112.21 ± 4.23*** (86.19)	56.85 ± 1.56*** (76.34)	39.87 ± 0.18*** (89.22)	6.18 ± 0.19***	0.321 ± 0.03*** (85.96)
Aqueous extract (SI)(200 mg/kg)	343.37 ± 3.86*** (7.57)	112.34 ± 17.16*** (14.90)	76.26 ± 1.71*** (35.60)	3.45 ± 0.06***	1.100 ± 0.17** (12.61)
Aqueous extract (SI)(400 mg/kg)	303.44 ± 3.00*** (21.15)	80.43 ± 2.50*** (50.23)	69.80 ± 3.24*** (45.12)	4.23 ± 0.08***	0.998 ± 0.15* (22.22)
Aqueous extract (SA)(200 mg/kg)	348.36 ± 2.92*** (5.87)	95.14 ± 9.18*** (33.94)	80.32 ± 56*** (29.61)	3.64 ± 0.03***	1.101 ± 0.21** (12.52)
Aqueous extract (SA)(400 mg/kg)	308.50 ± 5.84*** (19.43)	74.06 ± 4.54*** (57.29)	65.20 ± 4.50*** (51.90)	4.45 ± 0.06***	0.932 ± 0.01* (28.23)

*Values are expressed as mean ± SEM (n=6); [a] p<0.001 in comparison with normal control; ***p < 0.001, **p< 0.01,* *p< 0.05 *in comparison with positive control. U/L – Units/litre, g/dl –grams/decilitre*

A marked elevation in alanine aminotransferase, aspartate aminotransferase, total bilirubin, ALP and decrease in total protein levels was observed due to CCl_4 intoxication which was reverted back by treatment with 400 mg/kg dose of aqueous extract of both the species. Among the extracts tested, *S. amaranthoides* aqueous extract showed highest percentage of hepatoprotection. Dense chronic inflammatory infiltration and necrosis in the histopathological section of animals treated with CCl_4 indicates severe tissue injury. However, liver sections of animals treated with 400 mg/kg dose showed regenerative changes in the liver tissue indicating the protective effect of the extract. This could be attributed to the ability of extract to scavenge the free radicals responsible for hepatotoxicity by CCl_4.

Serum enzymes like ALP, alanine aminotransferase, aspartate aminotransferase and total bilirubin levels was significantly ($P<0.001$) increased in in animals treated with paracetamol which indicates severe liver necrosis. However, pre-treatment with Silymarin 100 mg/kg and 400 mg/kg of aqueous extract of both species significantly ($P<0.001$) decreased the elevated level of serum enzymes in comparison with paracetamol intoxicated animals.

3 CONCLUSION

Several research studies have linked the hepatoprotective activity with phenolic compounds. The phytochemical screening and the HPTLC analysis confirmed the presence of phenolic compounds in aqueous extracts of both species which may be responsible for the hepatoprotective property. This hepatoprotective activity study substantiates the use of *S. amaranthoides* in Ayurveda as an alternate source for the drug, *Munditika* which is used in the treatment of *Kamala* (Jaundice).

References

1. A. D. Gupta, *Astanga Hridiya*, Varanasi, Chaukhambha Sanskrit series, 1962, p. 107.
2. P. V. Sharma, *Dravyaguna – Vijnana*, Vol-II. Varanasi, Chaukhambha Bharati Academy, 2005, 804–805.
3. T. Prakash, S. Dayalal Fadadu, U. Raj Sharma, V. Surendra, D. Goli, P. Stamina and D Kotresha, Hepatoprotective activity of leaves of *Rhododendron arboreum* in CCl_4 induced hepatotoxicity in rats, *J. Med. Plant. Res.,* 2008, **2**(11), 315–320.
4. M Nirmala, K Girija, K Lakshman and T Divya, Hepatoprotective activity of *Musa paradisica* on experimental animal models, *Asian Pac. J. Trop. Biomed.*, 2012, **2**(1), 11–15.

IN SITU EXTRACTION AND RECOVERY OF BETALAINS FROM HAIRY ROOT CULTURE OF *BETA VULGARIS*

Ashoka Babu V. L. and Madhavan V.

Department of Pharmacognosy, Faculty of Pharmacy, M.S. Ramaiah University of Applied Sciences, Bangalore, Karnataka, India-560054

1. INTRODUCTION

Today colour has become a major component of food and pharmaceutical products as it is indispensible for quick identification and eventual acceptance of the product. Colours are of two different origin such as synthetic and natural. Natural colours are known to afford natural look to the products in terms of glossy and brightness.Hairy root culture is one of the plant tissue culture techniqueswidely used for the production of root derived secondary metabolites to enhance the level of secondary metabolites[1].

Beta vulgaris belonging to Chenopodiaceae family composed of red pigment betacyanin and yellow pigment betaxanthin (betalains) has been well known for centuries as an attractive food colour. The stability of betalains is greatly influenced by light, pH, temperature, oxygen etc and become unstable under adverse conditions. Hence, in the present study, *in situ* extraction method was adopted in order to enhance the betalain content and its stability, so that the same biomass can be regrown by inoculating into the new medium. In a commercial process this leads to high profits depending on the importance of the compounds.

2. METHODS AND RESULTS

2.1 Initiation of hairy roots

Seeds of *Beta vulgaris* were washed in running tap water, sterilized with 0.1% mercuric chloride for 5 min and finally washed with sterile distilled water. Four *Agrobacterium rhizogenes* strains such as A.2/83, A.4, A.20/83 and LMG 150 were maintained in YMB medium. The sterilized seeds were transferred to MS basal media for the development of aseptic seedlings.The leaves along with the petiole portion of seedlings were injuredwith sterile needle, injured parts were treated with smear of *Agrobacterium* and incubated on MS gelled medium in dark condition. After the initiation of hairy roots, carbenicillin was added to the culture media to obtain bacterium free hairy roots and then sub cultured on MS semi solid medium, subsequently, hairy roots grown in MS liquid medium.LMG 150 was found to initiate large number of hairy roots and the morphological characters of hairy roots with different strains were observed and recorded (Figure 1 and Table 1).

2.2. *In situ* extraction of betalains[2]

Exposure of *Beet vulgaris* root cultures to acidic medium release betalains pigment. Hence, the effect of different pH and time on release of betalains into the medium was studied at pH 2, 3 and 4. Three flasks containing hairy roots grown in MS basal medium for 19 days were used. The pH of the medium in the first, second and third flask was adjusted to 3, 4 and 5 using 10% HCl respectively and incubated on a shaker. Sample contents were drawn at different time intervals of 30 min, 1 hr, 2hr, 3hr and 4hr. The betalains content of the medium and hairy root were estimated as follows.

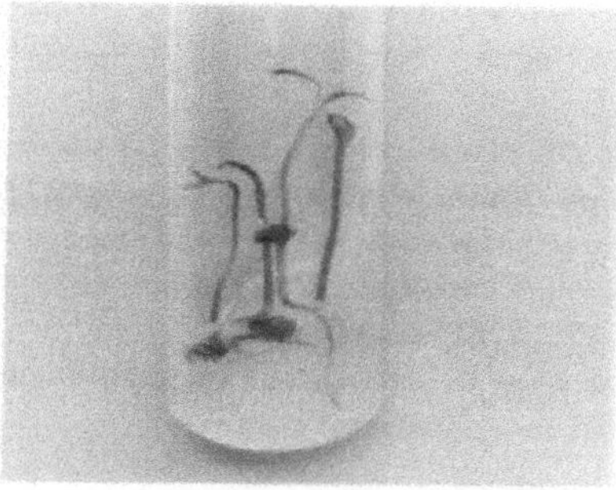

Two weeks old aseptic seedling

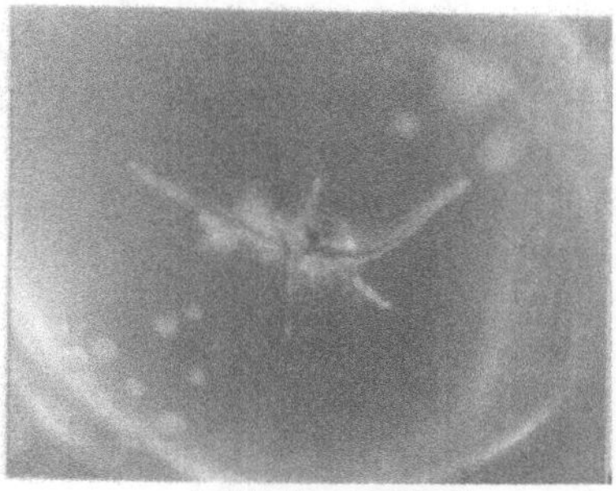

Initiation of hairy roots

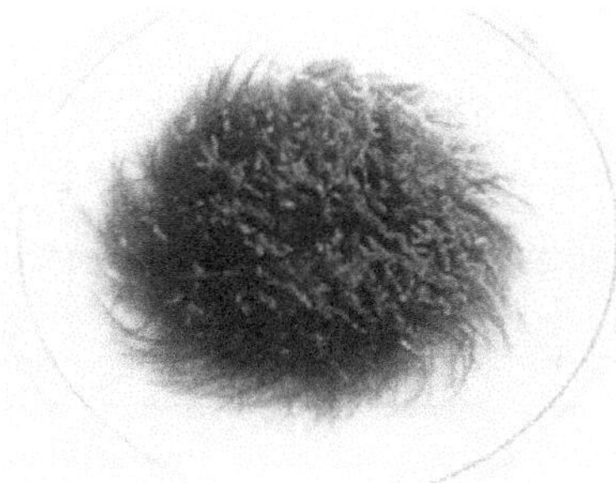

Three week old hairy roots

Figure 1 *Hairy roots induced with LMG 150 strain*

Betalains content of the medium[3]

1 ml of the sample was drawn and absorbance was measured at 480 nm (betaxanthin) and 540 nm (betacyanin) and estimated using the formula

Percentage betacyanin = a/1120 x 5 /fresh wt x 100/10 a=absorbance at 540 nm

Percentage betaxanthin = a^1/1120X5/fresh wt X 100/10 a^1= absorbance at 480 nm

Content of betalains = betacyanin + betaxanthin

Betalains content of hairy root:

100mg of hairy root was extracted with10 ml acidified water (0.1% HCl) filtered, centrifuged at 10,000g for 5 min. Supernatant was collected and estimated as above (Table 2).

2.3. Recovery of betalains by *Insitu* adsorption[4]

In situ adsorption involves direct addition of adsorbent to the medium and recovery of colour by centrifugation method. 3 gm of Silica gel, alumina and corn starchwere added directly to the three different flasks containing medium with leached pigments and left over the shaker for 10 min. The medium along with the adsorbents was centrifuged at 10,000g for 10 min, to the coloured residue 5% HCl was added with stirring for 5 min and again centrifuged at 10,000g for 10 min. Supernatant was collected and betalains content was estimated spectrophotometrically (Table 3).

Table 1 *Nature of hairy roots induced with different strains of Agrobacterium rhizogens.*

Bacterial strain	Number of explant	Number of hairy roots induced	Average hairy root per explant	Morphopological Observation
A.2/83	12	23	1.9	Colour: Slight yellowish Length: 0.25-1.25 cm
A.4	20	40	2.0	Colour: Slight yellowish Length: 0.36- 3 cm
LMG 150	15	40	2.6	Colour: Reddish yellow Length: 0.7- 3.8 cm
A.20/83		No hairy roots		

Table 2 *Effect of pH and time on release of betalains.*

pH	Time (min)	Betalains content of medium (%)			Betalains content in biomass (%)			Percent release
		Bx	Bc	Bn	Bx	Bc	Bn	
3	30	0.82	0.35	1.17				71.77
	60	0.96	0.41	1.37				84.04
	120	1.08	0.46	1.54	1.1	0.53	1.6	94.47
	180	0.82	0.39	1.21				74.23
	240	0.70	0.32	1.02				62.57
4	30	0.11	0.062	0.172				15.45
	60	0.12	0.068	0.188				16.93
	120	0.14	0.072	0.212	0.74	0.37	1.1	19.09
	180	0.15	0.080	0.230				20.72
	240	0.18	0.092	0.272				24.50
5	30	0.03	0.022	0.052				5.97
	60	0.04	0.037	0.077				8.80
	120	0.06	0.052	0.112	0.58	0.29	0.8	12.6
	180	0.07	0.061	0.131				15.05
	240	0.08	0.071	0.151				17.35

Bx= Betaxanthin Bc= Betacyanin Bn=Betalains

Table 3 *Effect of different adsorbents on recovery of betalains.*

Sl. No	Adsorbents	Total betalains content (%)	Betalains content in Eluent (%)	Percent elution
1	Silica gel		1.34	87.01
2	Alumina	1.54	0.85	55.19
3	Corn starch		0.67	35.06

Bx = Betaxanthin Bc = Betacyanin Bn =Betalains

3. CONCLUSION

The findings of the present study demonstrate that *in situ* extraction at pH 3 for 2 hours was found to be more effective in releasing betalains and silica gel was found to be better adsorbent for recovery of betalains.

References

1. K. Shimomura, H. Sudo, H. Saga and H. Kamada, *Plant Cell Rep.*, 1991, **10**, 282–285.
2. U. Mukundan, V. Bhide, G. Singh and W. R. Curtis, *Micro. Biotech.*, 1998, **50**, 241–245.
3. T. Nilson, *Lantbrushoegskolan*, 1970, **36**,179.
4. T. Rudrappa, B. Neelawarne and R. G. Aswathanarayana, *Biotechnol. Prog.*, 2004, **20**(3), 777–785.

Part 5: Formulation Design and Characterization

COCOA BASED DENTAL VARNISH AS A DELIVERY SYSTEM FOR EARLY CARIOUS LESION MANAGEMENT

Pushpalatha C.,[a] Swaroop Hegde[b] and Deveswaran R.[c]

[a]Department of Pedodontics and Preventive Dentistry, Faculty of Dental Sciences, M.S. Ramaiah University of Applied Sciences, Bengaluru, 560054, India; [b]Department of Conservative and Endodontic Dentistry, Faculty of Dental Sciences, M.S. Ramaiah University of Applied Sciences, Bengaluru, 560054, India; [c]Department of Pharmaceutics, Faculty of Pharmacy, M.S. Ramaiah University of Applied Sciences, Bengaluru,560054, India

1 INTRODUCTION

Recently, the polyphenols and theobromine present in cocoa beans have shown to exhibit anti-cariogenic and remineralization effect. The studies have been demonstrated that cocoa polyphenols inhibit the enzyme dextransucrase, cellular growth and acid production by cariogenic microorganisms which are responsible for caries formation.[1, 2] The reports suggest that theobromine readily available in cocoa displays enhanced surface recrystallization, increased resistance to acid dissolution by enamel crystals and increased surface enamel microhardness in comparison to commercially available fluoride agent.[3] However, literature search on isolated components of cocoa bean in a drug delivery system is not been reported. Hence the isolated compounds of cocoa beans such as theobromine, catechin, and epicatechin were coupled with benzoin powder and agarose in the dental varnish formulation. The novelty of this new dental varnish lies in its ability to possess sustained drug release potential which is suitable for oral conditions, and it can eliminate the adverse effect of dental fluorosis associated with commercially available fluoride varnish since it is composed of natural compounds. This novel varnish possesses both antibacterial and remineralizing effect with increased patient compliance.

2 METHODS AND RESULTS

2.1 Preparation and characterization of developed cocoa varnish

The 25ml of novel varnish was prepared by adding 1.8% theobromine, 0.005% catechin, 0.005% epicatechin, 0.3% agarose, 2% benzoin powder, and 2% ethanol. The developed dental varnish was characterized for physical, mechanical, rheological properties, biocompatibility, and drug release.

2.2 Assessment or remineralization potential of developed varnish

45 healthy human first maxillary premolars extracted and enamel samples were prepared. The enamel samples were randomly divided into 3 groups: Group A was control group(C), Group B treated Fluoride varnish (S) and Group C was treated with developed varnish (F1). The baseline data for mineral content mainly calcium and phosphorus level using DIAGNOdent™ pen and SEM-EDS

of the all groups were measured. The surface microhardness of the groups measured using Vickers microindenter and depth sensor nanoindenter. All the groups were subjected for demineralization for 96 hrs and later for the pH cycling model. After 28 days of pH cycling, the samples were recorded for changes in mineral content and surface microhardness of the enamel.

2.3 Statistical analysis

The data of mineral content and enamel surface microhardness were analyzed using SPSS statistical software version 22 at 0.05 level significance and 95% confidence interval. One way analysis of variance followed by post hoc multistep comparisons (Tukey HSD) was used for analysis.

2.4 Results

The developed varnish showed satisfactory physical, mechanical and rheological properties. The in-vitro drug release showed a sustained drug till 4hrs using both UV-Spectroscopy and HPLC methods. The biocompatibility assay revealed that the developed varnish at dilutions of 2.5% showed 96.32% cells viability whereas at 20% dilutions showed 82.22% cells viability. Assessment of mineral content by DIAGNOdent^{TM} pen after remineralization showed a gain in mineral content in Group B and Group C. Group B and Group C also showed increased mean values for both calcium and phosphorus level after remineralization treatment by SEM-EDX analysis. The depth sensor nanoindenter showed equal elastic modulus of the tooth for Group B and Group C after remineralization procedure. Group B and Group C showed similar surface microhardness using Vickers microindenter and depth sensor nanoindenter. Whereas,Group A showed no significant changes in mineral content and surface microhardness values after remineralization. SEM images of Group B and Group C showed a decrease in the porosity of the demineralized surface lesion, whereas Group A showed no change in the size of the porosity.

Table 1 *Mean value, Standard deviation and P-value between the groups at baseline, after demineralization and after demineralization using DIAGNOdent pen.*

Comparison of mean Diagnodent values between groups using One-way ANOVA Test							
Time	C		S		F1		P-Value
	Mean	SD	Mean	SD	Mean	SD	
BL	2.53	0.52	2.53	0.52	2.67	0.49	0.87
DM	8.53	0.52	8.40	0.51	8.40	0.51	0.87
RM	8.07	0.46	5.07	0.26	5.53	0.52	<0.001*

Table 2 *P-value of all the groups at baseline, after demineralization and after remineralization using Vicker's Hardness Test.*

Comparison of mean Vicker's Hardness Test values between groups using One-way ANOVA Test							
Time	C		S		F1		P-Value
	Mean	SD	Mean	SD	Mean	SD	
BL	323.76	2.60	325.14	3.76	324.98	3.27	0.59
DM	260.91	7.40	260.64	11.16	260.64	11.16	1.00
RM	261.04	7.41	284.42	3.66	282.49	3.68	<0.001*

Table 3 *P-value of all the groups at baseline, after demineralization and after remineralization using DSI for surface microhardness.*

Comparison of mean Surface Hardness values by DSI between groups using One-way ANOVA Test							
Time	C		S		F1		P-Value
	Mean	SD	Mean	SD	Mean	SD	
BL	283.40	3.08	283.44	3.08	283.41	3.09	0.83
DM	153.10	2.15	153.92	2.11	154.05	2.06	0.29
RM	154.83	1.92	198.14	5.68	209.78	4.24	<0.001*

Table 4 *Elastic modulus of the tooth at baseline, after demineralization and after remineralization using depth sensor nanoindenter.*

Comparison of mean Elastic Modulus of Tooth values between groups using One-way ANOVA Test							
Time	C		S		F1		P-Value
	Mean	SD	Mean	SD	Mean	SD	
BL	73.45	1.35	73.45	2.06	73.33	2.24	0.94
DM	49.20	0.87	48.99	0.91	49.11	1.73	0.98
RM	49.01	1.28	62.30	1.81	65.21	2.15	<0.001*

Table 5 *Calcium level of all the groups at baseline, after demineralization and after remineralization using SEM-EDS.*

Comparison of mean Calcium levels in tooth between groups using One-way ANOVA Test							
Time	C		S		F1		P-Value
	Mean	SD	Mean	SD	Mean	SD	
BL	47.59	0.87	47.83	1.18	48.76	2.20	0.14
DM	22.28	1.86	22.11	1.05	22.96	0.88	0.23
RM	22.28	0.91	41.05	1.95	41.49	1.49	<0.001*

Table 6 *Phosphorus level of all the groups at baseline, after demineralization and after remineralization using SEM-EDS.*

Comparison of mean Calcium levels in tooth between groups using One-way ANOVA Test

Time	C		S		F1		P-Value
	Mean	SD	Mean	SD	Mean	SD	
BL	47.59	0.87	47.83	1.18	48.76	2.20	0.14
DM	22.28	1.86	22.11	1.05	22.96	0.88	0.23
RM	22.28	0.91	41.05	1.95	41.49	1.49	<0.001*

Note: * - Statistically Significant, BL - Baseline, DM - Demineralization, RM - Remineralization

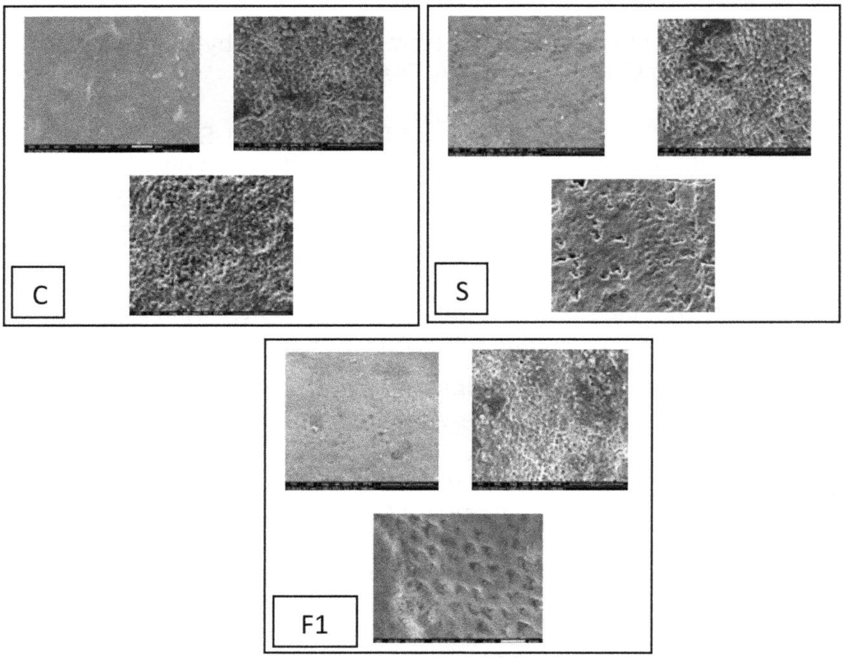

Figure 1 *SEM images of different groups at baseline (1), after demineralization (2) and after remineralization (3).*

3. CONCLUSION

The results of the present study showed both novel and commercially available varnish are equally effective in remineralization of enamel lesion. The amount of theobromine in the present study to produce the remineralization effect was 1.8% compared to commercial varnish which contained 5% sodium fluoride. Hence the frequency of application of the novel varnish can be increased because of lesser concentration of drug and excellent biocompatibility compared to commercial fluoride varnish. The present study demonstrates that the novel varnish containing isolated natural compounds like catechin, epicatechin and theobromine possessing can be a promising alternative for remineralization of early enamel lesions with additional antibacterial activity.

References

1. M. N. Naylor, Nutrition and dental decay, *Proc. Nutr. Soc.*, 1984, **43**(3), 257–263.
2. N. V. Babu, D. K. Vivek and G. Ambika, *Euro. Arch. Paed. Dentistry*, 2011, **12**(5), 245–249.
3. B. T. Amaechi, N Porteous, K Ramalingam, P. K. Mensinkai, R. C. Vasquez, A. Sadeghpour and T. Nakamoto, *Caries Res.*, 2013, **47**(5), 399–405.

SOLID STATE MANIPULATION OF PALIPERIDONE FOR ITS BIOAVAILABILITY ENHANCEMENT USING PRECIPITATION INHIBITORS

J. Thimmasetty,[a] T. Ghosh,[b] Shashank N. N.[a] and A. Raheem[a]

[a] Department of Pharmaceutics, Bapuji Pharmacy College, Davangere and Rajiv Gandhi University of Health Sciences, Karnataka, India
[b] Department of Pharmaceutics, Faculty of Pharmacy, MS Ramaiah University of Applied Sciences, Bangalore, Karnataka, India
Corresponding Author E-mail: tanmoysakti@gmail.com

1 INTRODUCTION

The discovery of new drug molecules, which belongs to BCS class II and class IV, are suffering from poor aqueous solubility related issues. Pharmaceutical scientists are seeking various approaches such as polymorphism, cocrystals, salt formation, etc. to overcome these issues. Among the various methods, modifications of drugs into pharmaceutically useful cocrystals using coformers have been emerged in recent years. This improves the physicochemical properties such as solubility, dissolution, stability, and bioavailability. In addition to this, precipitation inhibitor (PPI) such as Hydroxy propyl methyl cellulose (HPMC) can provide a 'parachute' effect in biological medium, which further improves the bioavailability of the drug by preventing its precipitation upon the dissociation of cocrystals. Selection of coformers, from the Generally Recognized As Safe (GRAS) list by FDA, can be done by predicting the cocrystal formation using Hansen solubility parameter (HSP).[1]

2 METHODS AND RESULTS

2.1 Prediction and preparation of cocrystals

Preparation of cocrystals is one of the methods of solid state manipulation techniques to achieve desired physicochemical properties. An attempt has been made to prepare paliperidone cocrystals using coformers. Cocrystal formation can be forecasted if the difference in the calculated HSP of drug and coformer is less than 7 $MPa^{1/2}$.[2] The solubility parameter for the drug and several coformers were calculated using group contribution methods such as Fedors, Hoys, and Van Krevelen methods[3]. Three coformers namely benzamide, nicotinamide, and PHBA were selected based on the criteria. On the other hand, boric acid was also selected to verify its suitability as a coformer though it did not fit into the criteria. Calculated solubility parameter of the drug and coformers are shown in the Table 1. Paliperidone cocrystals were prepared using the coformers with 1:1 molar ratio employing solvent evaporation technique.

Table 1 *Calculated HSP for drug and coformers.*

Drug/Coformer	Solubility parameter ($MPa^{1/2}$)		
	Fedors	Hoys	Van Krevelen
Paliperidone	25.39	20.06	25.29
Benzamide	28.85	25.52	23.56
Nicotinamide	31.80	24.65	23.52
PHBA	29.49	28.89	27.69
Boric acid	46.31	44.46	59.62

2.2 Characterization of paliperidone cocrystals

The prepared cocrystals were characterized for the properties such as melting point, particle size, saturation aqueous solubility, and *in vitro* drug release. Instrumental techniques such as FTIR, DSC, PXRD, and SEM were used to confirm cocrystal formation. The results of these instrumental techniques are shown in the Figure 1 for PAL-PHBA as sample representation.

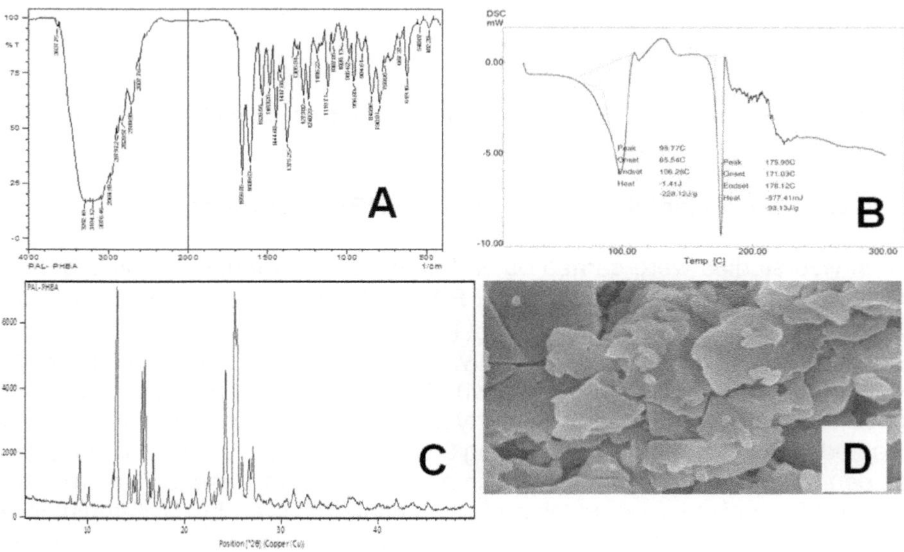

Figure 1 *A: FTIR spectrum, B: DSC thermogram, C: PXRD Diffractogram, D: SEM of PAL-PHBA cocrystals.*

The instrumental analysis revealed the formation of paliperidone-benzamide (PAL-BED), paliperidone-nicotinamide (PAL-NID), and paliperidone-PHBA (PAL-PHBA) cocrystals as predicted. On the other hand, cocrystals with boric acid (PAL-BOA) were not confirmed. The order of the cocrystals for mean saturation solubility of three trials with standard deviation is paliperidone pure drug (0.115±1.74) < PAL-BOA (0.17±1.33) < PAL-NID (2.13± 0.54) < PAL-BED (7.21± 0.62) < PAL-PHBA (16.29±0.69). Similarly, the dissolution of the paliperidone, as depicted in Figure 2, was increased in case of all cocrystals except PAL-BOA when compared to the pure drug. The highly significant observation was made with PAL-PHBA cocrystals, which had shown 141.6 fold increase in the solubility when compared to pure drug and 100 % release of the drug within 30 min. Hence these cocrystals were further selected for animal studies.

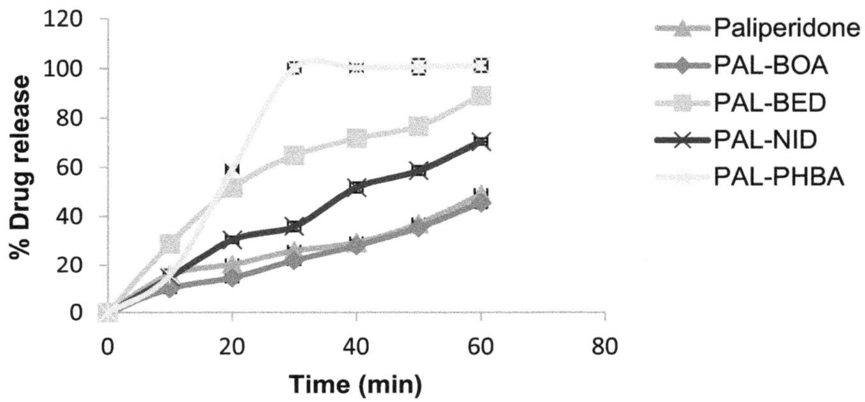

Figure 2 *Dissolution profiles of paliperidone and its cocrystals (Mean ± SD, n=3).*

2.3 *In vivo* bioavailability studies of PAL-PHBA cocrystals

The *in vivo* studies were carried out in healthy albino rabbits by dividing them into 3 groups. Pure drug (0.6 mg/kg), PAL-PHBA cocrystals (0.8 mg/kg), and PAL-PHBA cocrystals with 5% HPMC (1.0 mg/kg) were administrated to Group 1, Group 2 and Group 3, respectively, through the oral route using an oral feeding tube. The blood was withdrawn at a time intervals half, one, two, four, eight and twelve twenty four, and fourty eight hours. The withdrawn blood was collected in Eppendorf tubes containing 0.5% sodium citrate solution. The blood was centrifuged at 3500 rpm for 10 min at 4 °C to obtain plasma. The drug was extracted from plasma on addition of acetonitrile followed by centrifugation at 10,000 rpm, 15 min at 10 °C. The supernatant was used for HPLC analysis. The results showed an increase in the bioavailability of paliperidone administered in the form of cocrystals compared to the pure drug. Moreover, addition of PPI (HPMC) delayed the precipitation of paliperidone, which further improved the bioavailability. The plasma concentration-time profile and pharmacokinetic parameters are shown in Figure 3 and Table 2, respectively.

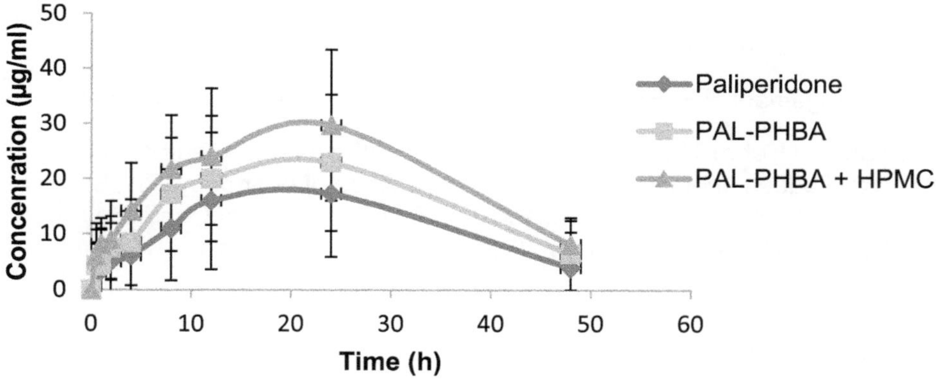

Figure 3 *Mean paliperidone plasma concentration-time profiles of paliperidone, PAL-PHBA cocrystals, PAL-PHBA cocrystals + HPMC (PPI) after oral route administration to rabbits (Mean ± SD, n=6).*

Table 2 *Pharmacokinetic parameters of paliperidone in plasma (Mean ± SD, n=6).*

Sample	T_{max} (h)	C_{max} (µg/ ml)	AUC_{0-48} (µg/ ml)	Relative bioavailability
Paliperidone	24	17.23 ± 11.36	560.27±235.23	-
PAL-PHBA cocrystals	24	22.89±12.35	744.82±321.65	1.62 ± 0.87
PAL-PHBA cocrystals + HPMC	24	29.69± 13.69	962.52±298.57	2.05± 0.68

3 CONCLUSION

The work presented herein exemplifies how cocrystals can generate novel crystalline forms of existing drug with different physicochemical properties. The coformers can be successfully selected for the preparation of paliperidone cocrystals using the solubility parameter concept. It was evidenced by the characterization of cocrystals using instrumental techniques. The bioavailability of paliperidone was improved using PAL-PHBA cocrystals, which was further enhanced using PPI (HPMC) in the formulation. Enhanced bioavailability of paliperidone in the method explained increases the validity of the work.

References

1. *Pharmaceutical Salts and Co-crystals*, ed. J. Wouters and L. Quéré, Royal Society of Chemistry, Cambridge, 2012.
2. A. Salem, S. Nagy, S. Pal and A. Szechenyi, *Int. J. Pharm.*, 2019, **558**, 319–327.
3. F. M. B. Allan, *Handbook of Solubility Parameters and Other Cohesion Parameters*, 2nd edn, CRC Press LLC, 1991.

OPTIMIZATION OF EZETIMIBE NANO FORMULATION BY TWO-LEVEL FULL FACTORIAL DESIGN

D. Balanagamani,*[a] K. Bhaskar Reddy,[b] and K. Sesha Maheswaramma[c]

[a] Jawaharlal Nehru Technology University Ananthapuram, Ananthapuram. Andhra Pradesh, India; [b] Sri Venkateswara college of pharmacy, R.V.S Nagar, Chittoor (Dt), Andhra Pradesh, India; [c] Department of Chemistry, JNTUACEP, Pulivendula, Kadapa (Dt), Andhra Pradesh, India

1 INTRODUCTION

Ezetimibe (Ez) is a lipid lowering compound which inhibits intestinal cholesterol and phytosterol absorption, hence it is used in primary hyperlipidemia treatment. It selectively inhibits the intestinal absorption of dietary and biliary cholesterol without affecting the absorption of fat-soluble nutrients. Ez is used as antihyperlipidemic agent in conventional dosage but there is a need to control release of dosage. The present research deals the formulation and optimization of nano Ezetimibe by design of experiment regarding interactions of each factor and to get optimized conditions for preparation of nano formulation by two level full factorial design of quadratic model with Response surface method.

2 METHODS AND RESULTS

The formulation includes Ezetimibe with Hydroxy Propyl Methyl Cellulose (HPMC) as polymer as it is a controlled release polymer, good suspending agent and Poloxomer as surfactant for its good drug release charecteristics, compatibility with other chemicals and its suitable mucoadhesive property by high phacoemulsification probe method. As the HPMC and Poloxomer both were less toxic and compatible to other chemicals there were less chances of toxicity. By the nature of controlled release of drug will enhance the bioavailability than the conventional dosage form. Physical and chemical compatibilities were checked with drug before the selection of polymer and surfactant, shown in Scheme 1.

2.1 Design of experiment

The prepared formulation was optimized by Design of experiment (DoE) by Design-Expert 11 software. DoE is a structured and organized method for determining the relationships between input factors (xi – independent variables) affecting one or more output responses (y – dependent variables), through the establishment of mathematical models[2] (y = f(xi))[2]. Present experiment was designed by two level factorial design of Quadratic model with Response Surface Method (RSM).

2.2 Preparation of Ez nano formulation

The different ratios of polymer, surfactant with completely dissolved drug subjected to high phacoemulsification probe method as per the literature.[3] The prepared nano formulation was optimized by DoE. Here polymer ratio, surfactant ratio, homogenization time, homogenization speed, sonication time were taken as independent variables[4] and particle size, Zeta Potential (ZP), poly dispersion Index (PI), Z – Average were dependent variables i.e. responses.

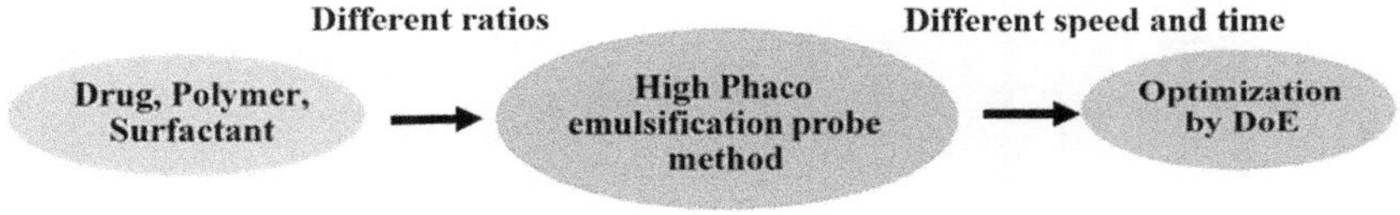

Scheme 1 *Preparation of Ez nano formulation.*

2.3 Analysis of variance

The experimental design reports have shown the effect of coded factors on the responses and their interactions with each other. Response surfaces were shown in terms of counter plots, 3D - Graphs. Analysis of variance (ANOVA) table (Table 1) indicates the significant levels of the designed experimental model, fit summary and statistics indicates that the model is good or not. The statistical report including correlation coefficient values has shown the optimized values of each factor to proceed the formulation process to get effective results.

Table 1 *ANOVA for selected factorial model.*

Source	Sum of Squares	Degree of freedom	Mean Square	F-value	p-value	Significance
Model	0.0000	0	-	-	-	-
Curvature	0.3378	1	0.3378	0.445	0.05101	Significant
Residual	22.80	30	0.7601	-	-	-
Lack of Fit	20.73	27	0.7679	1.11	0.5454	Not significant
Pure Error	2.07	3	0.6902	-	-	-
Correlation Total	23.14	31	-	-	-	-

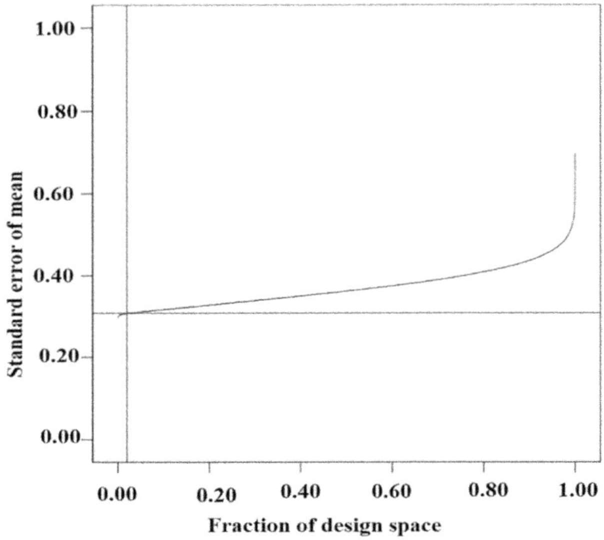

Figure 1 *FDS Graph.*

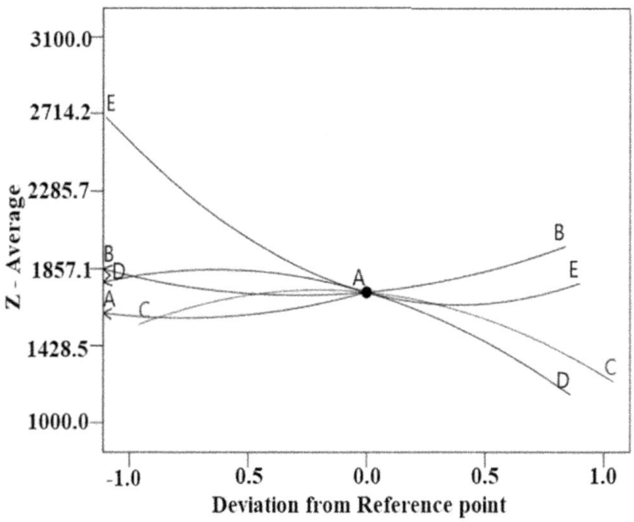

Figure 2 *Perturbation graph.*

The Standard Error of Mean (SEM) between 0.37 and 0.697, indicates the changes of mean present with in the design space, shown in Figure 1. The deviation from the reference point indicates that the interactions of factors, shown in Figure 2. DoE optimized conditions for each factor were specified as in Figure 3 which were used in the further formulation process.

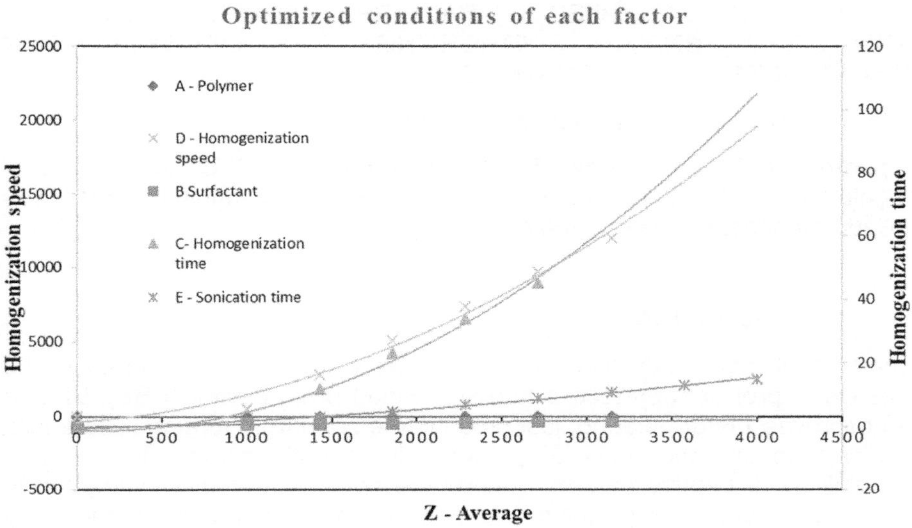

Figure 3 *Consolidated data for different factor interactions showing optimized conditions.*

3 CONCLUSION

The DoE results shown better response for the optimized conditions A,B,C,D,E as 2.0,1.580, 22.120, 7055.0, 8.700 respectively to prepare Ez nanoformulation. ANOVA results confirms that the designed experimental method has significant model and insignificant lack of fit which ensures that the model is good enough to be used. High phacoemulsification probe method was used to prepare Ez nano formulation, which shown all the charecteristics features with in the tolerance limits.

References

1. M. H. Shukra, S. Ismaila and S. M. Ahmedb, *J. Drug Delivery Sci. Technol.*, 2019, **49**, 383–395.
2. I. M. Fukuda, C. F. F. Pinto, C. D. S. Moreira, A. M. Saviano, and F. R. Lourenço, *Braz. J. Pharm. Sci.*, 2018, **54**(Special), 1–16.
3. T. Neelima Rani and Y. Indira Muzib, *Asian J. Pharm.*, 2016, **10**(4), 518–526.
4. J. T. Lalwani, V. T. Thakkar and H. V. Patel, *Int. J. Pharm. Pharm. Sci.*, 2013, **5**(3), 513–522.

A NOVEL NANO-ENABLED NASAL PHYTOFORMULATION – A POTENTIAL BRAIN TARGETING STRATEGY FOR NEURODEGENERATIVE DISORDERS

Hajira Banu H.[a*] and Dhrubojyoti Mukherjee[b]

[a] Department of Pharmacology, M S Ramaiah University of Applied Sciences, Bengaluru, India; [b] Department of Pharmaceutics, M S Ramaiah University of Applied Sciences, Bengaluru, India

1. INTRODUCTION

A major challenge associated with the delivery of drug to the brain is the existence of proficient protective barrier – Blood Brain Barrier (BBB). BBB, a CNS microvasculature that not just protects the brain from foreign substances but also impedes the entry of potential therapeutic agents due to their hydrophilicity or peptide nature. Among the several strategies attempted in order to overcome this problem and achieve effective treatment for many neurological disorders, properly tailored nanoparticles have a greater potential. Additionally, the administration route also becomes a point of interest in order to attain higher drug delivery to the brain. Intranasal route is one such non-invasive method to bypass the BBB and attain therapeutic concentration in CNS. Moreover, the development of new technologies that would help in delivery of herbal drug to a target site is an interesting approach to consider. Thus, this study was designed with the aim of improving herbal drug delivery to brain via intranasally delivered nanocarriers.

Use of herbal drugs for the management of neurodegenerative diseases have become prevalent in the recent years. A number of drugs from herbal origin have been investigated for the management of Alzheimer's and dementia. One such drug is *Centella asiatica* that is reported to improve memory and also has neuroprotective activity against amyloid beta peptide[1]. This neuroprotection is attributed to the presence of asiaticoside and asiatic acid in the extract. However, the extract with immense pharmacological activity is hygroscopic[2] in nature and has poor bioavailability. Hence, development of nanoparticles that could improve the stability of this herbal drug and also deliver it to brain in sufficient quantity would be beneficial. Thus, the present study was designed to prepare a nano-formulation containing thiolated chitosan complex of standardized extract of *C. asiatica* to enhance its stability and also achieve better therapeutic activity in brain by intranasal administration.

2. METHOD AND RESULTS

2.1 Selection of Polymer

Many studies have reported the beneficial effect of using thiolated chitosan wherein thiolation of chitosan gives rise to a chitosan derivative with higher mucoadhesive properties[3,4] compared to unmodified chitosan. Thiolated chitosan is described to have enhanced the bioavailability of insulin peptide when administered via nasal route. Additionally, a preliminary docking studies using Autodock vina also showed the affinity of thiolated chitosan to receptors like Sphingosine 1- phosphate receptor (SIP-1) and Liporotein receptor related protein-1 (LRP-1) with a binding energy of -8.5 and -6.4 respectively. SIP-1 promotes barrier function in peripheral vessels and inhibition of SIP-1 may lead to opening of BBB. Similarly, LRP-1 exhibit transcytosis capacity and acts as a receptor for ligand mediated delivery of drug to brain. Thiolated chitosan was selected for the study as it demonstrates an ability to bind to these receptors and hence it might help in the delivery of drug across BBB.

2.2 Synthesis of thiol functionalized chitosan polymer

For the present study, thiolated chitosan was prepared by a method with slight modifications reported by Kast and Bernkop-Schnurch. Briefly, chitosan (obtained from HiMedia laboratories, India) was dissolved in 1M HCl to obtain chitosan hydrochloride. To this, an intermediate 1-(3-dimethyl amino propyl) 3-ethyl carbodiimide hydrochloride (EDC-HCL) and thioglycollic acid was added and allowed to stir for 30 minutes. NaOH was added to maintain the pH between 10-12, further stirred and incubated for 12 hours at room temperature. The reaction mixture was dialyzed using dialysis tubing's for 3 d in order to isolate polymer conjugates and unbound thioglycolic acid. They were first dialyzed using 5mM HCl and then with 1% NaCl and 5mM HCl for two times to quench ionic interactions between the cationic polymer and anionic sulfhydryl compounds. Then the samples were lyophilized at -85°C and 0.063 mbar and stored at 50°C until further use.

2.3 Preparation of *Centella* - thiolated chitosan (CtC) complex and its spectral analysis

The standard ethanol extract of *Centella* procured from The Himalaya Drug Company, Bangalore was used for the study. 1:1 ratio of *Centella* and thiolated chitosan was used wherein 1% solution of *Centella* was mixed with EDC hydrochloride. Thiolated chitosan dissolved in 1M HCl was added and mixed for 1-2 hours. To this resultant solution, 5% NaOH solution was added to obtain the precipitate, further mixed for an hour and incubated overnight to obtain the complex that was lyophillised and stored. The prepared CtC complex was subjected to FTIR and X-Ray Diffraction studies. FTIR spectral data of the CtC complex was obtained to ascertain the possible interaction between the extract and the polymer. CtC showed peaks from 800-666 cm^{-1} indicating -NH wagging, bands between 1488-1576 cm^{-1} indicating the amide bond that arises from the interaction between, -CN stretching, -C-N-H group and -NH bending. An absorption at 1630-1680 cm^{-1} showed amide bonds with stretching vibrations as seen in tertiary amides (Figure 1 c). These bands indicate the interaction of amine group of thiolated chitosan with the carbonyl group of asiaticoside present in *Centella* leading to the formation of CtC complex.

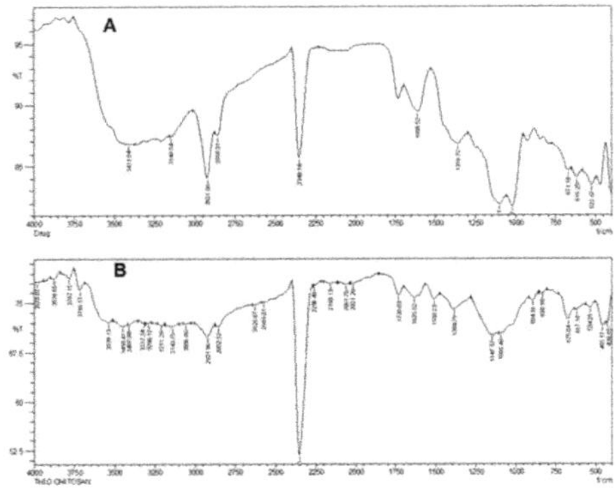

Figure 1 *(A) FTIR Spectrum of Centella. (B) FTIR Spectrum of Thiolated Chitosan. (C) FTIR Spectrum of CtC Complex .*

The XRD patterns of the *Centella*, thiolated chitosan and CtC complex reveled characteristic peak. *Centella* showed the prominent sharp peaks at 2θ values 28.209˚, 40.415˚, 50.063˚ which revealed the crystalline property of the drug. Thiolated chitosan showed sharp peaks at 31.736˚, 45.478˚, 56.496˚. Interpretation of the XRD patterns of CtC complex showed a shift of the peak heights and d-spacing. It demonstrated peaks at 20.075˚ (d=4.41 Å), 31.67˚ (d=2.82 Å) and 45.45˚ (d=1.99 Å) indicating its amorphous nature and a change in physical nature of the complex.

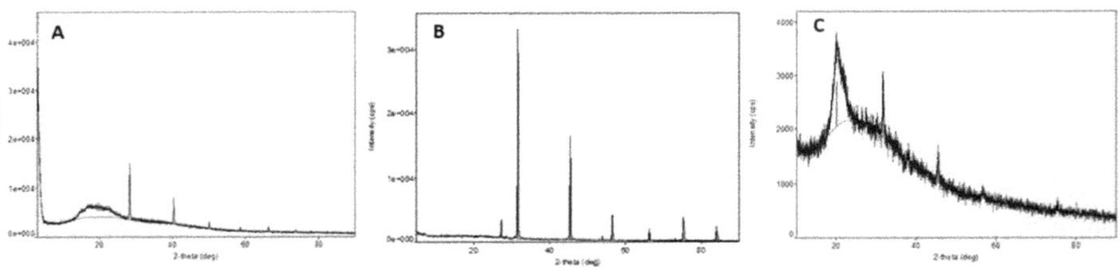

Figure 2 *(A) XRD pattern of Centella. (B) XRD pattern of Thiolated Chitosan. (C) XRD pattern of CtC Complex.*

2.4 Preparation and Characterisation of nanoformulation of CtC Complex

A nanoformulation of CtC complex was prepared using acid-base reaction method. Briefly, 1% solution of CtC complex in 2% glacial acetic acid was added dropwise into a mixture of phosphate buffer (pH – 7.4) containing di sodium hydrogen phosphate and sodium dihydrogen orthophosphate (1:1) using homogenizer. The solution was stirred continuously for a period of 4-5 h followed by probe sonication. Resultant solution was subjected to centrifugation at 8000 RPM for 25-30 minutes. The sediment was washed with saline phosphate buffer and redispersed into demineralized water to obtain end product. The formulation was subjected to Scanning Electron

Microscopy (SEM) to study the surface morphology, Atomic Force Microscopy (AFM) to determine the particle size and shape, Energy Dispersive X-ray Spectroscopy (EDX) to check for impurities and Dynamic light scattering (DLS) to obtain zeta potential of the formulation that indicates stability of the nanosystem.

2.4.1 SEM and EDX Analysis

CtC complex nanoparticles showed a particle diameter of 125-200 nm. The particles were spherical, non- aggregated and were found to be distributed uniformly. The EDX characteristics showed the presence of sulphur group indicating the strong sulphur bonding in the conjugation. Formation of these bonds may be involved in the altered density, particle compressibility eventually leading to nanosize of the particle. (Fig 3A and 3B).

2.4.2 AFM and DLS Characterization

AFM analysis revealed the height of CtC Complex nanoparticles and it was found to be in the range of 154 nm - 211 nm. The AFM analysis showed the nano size of particle in all the dimensions. The data correlates with the results shown in SEM and are also in good agreement with those indicated in the literature. The minor variation may be due the difference in method of treatment and sample preparation during analysis.

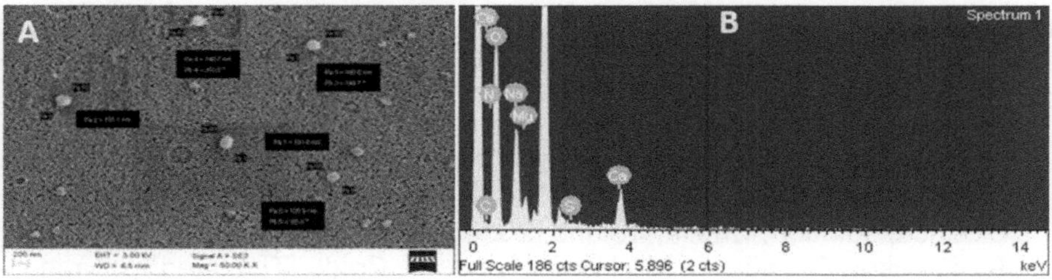

Figure 3 *(A) SEM of CtC Complex. (B) EDX of CtC Complex.*

Prepared nanoformulation was subjected to measurements like mean particle size, particle size distribution and stability by DLS. The data showed a well dispersed colloidal system of nanoparticles with an average particle size of 210.5 d.nm. Zeta potential was measured to determine the stability of the formulation, the result showed the value of -14.5 mV. This potential value is well within the range and signifies that the prepared formulation is stable. Poly dispersity index (Pdi) was determined to verify the distribution of nanoparticles within the formulation. The formulation showed a Pdi of 0.260 and as per literature, the pdi values within 0.6 are acceptable indicating homogeneity of particles.

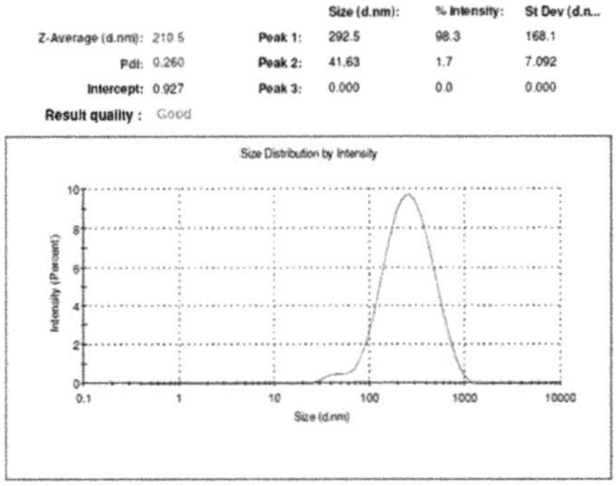

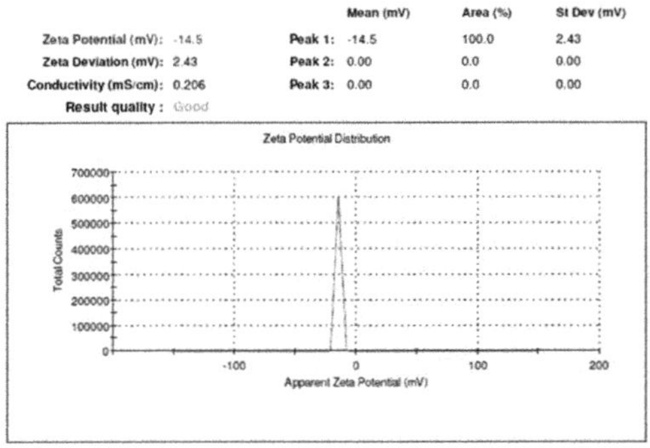

3. CONCLUSION

Centella has been established as an efficient memory enhancer and a potential candidate for the management of neurodegenerative diseases by many researchers. In the present study, an attempt has been made to enhance this activity through a novel phytoformulation that could have an ability to increase the concentration of *Centella* in the brain. The results showed the successful preparation of a brain targeted formulation. However, *in vitro* and *in vivo* studies should be performed to substantiate the claim.

References

1. S. S. Jamil, Q. Nizami, M. Salam and L. Urban, Centella asiatica A Review, *Nat. Pro. Rad.*, 2007, **6**, 158–170.
2. S. Okonogi, J. Sirithunyalug and Y. Chen, Nanoencapsulation of Centella Asiatica Bioactive Extract, XVIth International Conference on Bioencapsulation, Dublin, Ireland, 2008.
3. A. Bernkop-Schnürch, M. Hornof and T. Zoidl, Thiolated polymers—thiomers: modification of chitosan with 2-iminothiolane, *Int. J. Pharm.*, 2003, **260**, 229–237.
4. A. Bernkop-Schnürch, D. Guggi and Y. Pinter, Thiolated chitosans: development and in vivo evaluation of a mucoadhesive permeation enhancing oral drug delivery system, *J. Control. Release*, 2004, **94**, 177–186.

STABILITY ENHANCEMENT AND FORMULATION DEVELOPMENT OF L-CARNITINE FAST DISSOLVING PELLETS THROUGH PRO-DRUG APPROACH

N. Jawahar[a] and S. Jubie[b]

[a]Department of Pharmaceutics, JSS College of Pharmacy, Ooty-643001, Tamilnadu, India
[b]Department of Pharmaceutical Chemistry, JSS College of Pharmacy, Ooty-643001, Tamilnadu, India

1 INTRODUCTION

L-Carnitine is an imperative compound in human digestion. It is considered as "restrictively fundamental" since few people, including newborn children, can't combine the compound endogenously at satisfactory dimensions. Since L-Carnitine is exceptionally hygroscopic in nature, it is exceptionally hard to define it in pharmaceutical dose shapes with ideal security and bioavailability. Oral drug administration has been a standout amongst the most advantageous and generally acknowledged courses of conveyance for most remedial operators. The most ordinarily utilized pharmaceutical strong measurement frames today incorporate granules, pellets[1], tablets and cases, out of which tablets being the most mainstream dose structure, representing 70% of all moral pharmaceutical arrangements delivered.[2] Due to the hygroscopic nature of L-Carnitine, it is difficult to stabilize the drug in various dosage forms. In order to increase the stability and to mask the hygroscopic nature, the current research work was aimed to synthesize L-Carnitine pro-drug and formulated as pellets. The carboxylic group of L-Carnitine has been converted into the corresponding hydrazine via esterification followed by hydrazinolysis approach.

2 METHOD AND RESULTS

2.1 Preparation of L-Carnitine Hydrazide

A gauged measure of L-Carnitine (1M), in ethanol (2M) and concentrated sulphuric corrosive (1ml) was refluxed on steam bath at 70-80oC for 6-8 h. To a mixture of ethyl L-Carnitine (1M), in ethanol, hydrazine hydrate (2M) was included and refluxed steam bath at 70-80^0 C for 6-8 h. The strong isolate was gathered by filtration, dried and recrystallized from ethanol (Scheme1)

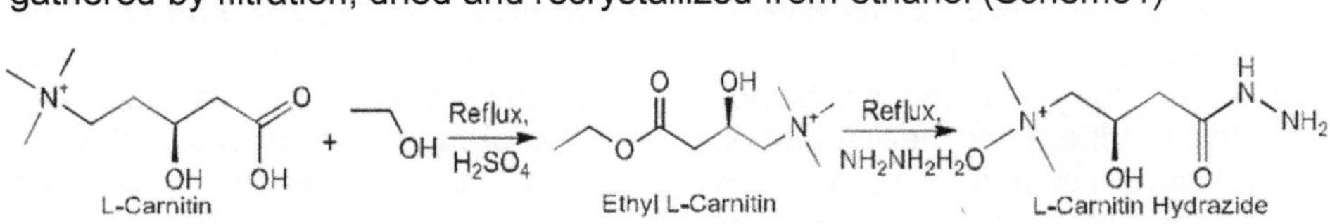

Figure 1 *Synthesis of L-Cartinin hydrazide.*

2.2 Determination of Hygroscopicity (Moisture uptake study)

Sample material (100–300 mg) was weighed into bare and weighed plastic petri-plates and placed in open environment. At regular intervals, final mass of the sample was determined until 24 hours. The percentage weight gain of the sample was calculated using the equation given below. Hygroscopicity of the sample was found using this method.

Table 1 *Moisture uptake study.*

Time (h)	% Moisture Pure drug	Absorbed Prodrug
0	0	0
0.5	10	0
1	22.5	0
2	28	0
4	32	0
8	33	0.2
16	39.5	0.2
24	31.70	0.3

2.4 Pro-Drug Metabolism

Equivalent weight of L-Carnitine was taken and dissolved in simulated gastric fluid. This was subjected for UV-Spectroscopic determination. The process was conceded at wavelength of 222.5nm. The amount of L-Carnitine formed from L-Carnitine hydrazide was determined. This results shows that L-Carnitine hydrazide is hydrolyzed in to the parent drug L-Carnitine at pH of 1.3.

Table 2 *Determination of pro-drug metabolism.*

Absorbance	Conc (µg/ml)	DF	CONC(mg/ml)	Total Amount	% Drug
0.03209	9.996	99.96	0.0996	9.996	99.96

2.5 Preparation of Pellets

Different formulations of pellets containing MCC, PVP k-30, lactose, ethanol were set up with various grouping of croscarmellose of 2%, 4%, and 5%. The dry elements of the plans were mixed for couple of minutes in a mortar and pestle until to get a uniform blend. The suitable amount of cover fluid was included gradually for steady blending, and the procedure was proceeded for further couple of min to acquire a wet mass of wanted consistency. The wet mass was expelled. The results of the expulsion procedure were handled utilizing a spheronizer turning at 1000 rpm for times of up to 8-10 min. The pellets shaped were gathered and dried in a sight-seeing oven at 60-80°C.

Table 3 *Composition of formulations.*

Ingredient	F-1	F-2	F-3
MCC	4.3g	4.3g	4.3g
PVP k30	0.3g	0.3g	0.3g
Croscarmellose	0.2g	0.4g	0.5g
Lactose	5.2g	5.0g	4.9g
Isopropylalcohol	12ml	12ml	12ml

2.6 *In vitro* Dissolution

Relating the drug releasing profile of F-1, F-2 & F-3 formulations, it is found that F-3 has speedy disintegration and dissolution than other two formulations (F-1 & F-2). This is due to the advanced concentration of CROSMELLOSE in the formulation. About 92% of drug is released within 5 minutes of the release. From the above results it is found that F-3 has fast dissolving property when equated to other formulations.

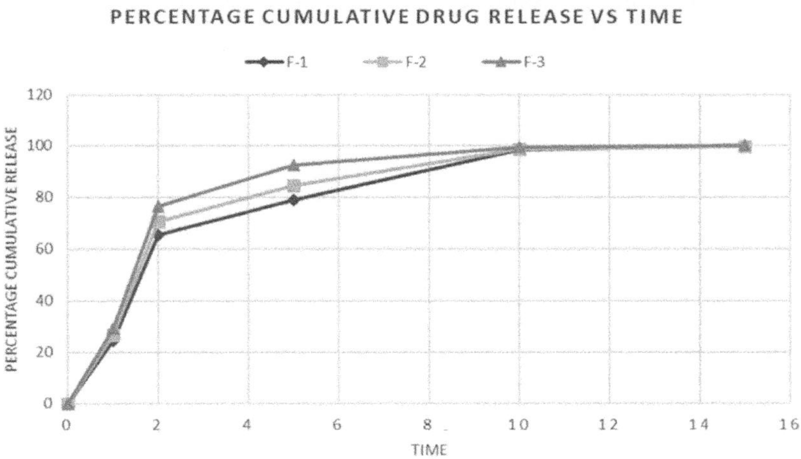

Figure 2 *In Vitro Dissolution Profile.*

2.7 *Stability Studies*

Stability studies, can be used to assess chemical effects at non-accelerated conditions and to evaluate the effect of short term excursions outside the label storage conditions. The drug powder and drug loaded pellets were stored at room temperature for a period of 6 months. After the storage time, the drug was dissolved in a suitable solvent, diluted and examined under UV-Spectrometer for the final concentration.

Table 4 *Stability studies for powders and pellets.*

Sl. No	Sample	Absorbance	Amount Present	%Drug Estimation
1	Powder	0.390	9.994	99.94
2	Pellet	0.398	9.982	99.82

3 CONCLUSION

A favorable formulation of fast dissolving pellets of stability enhanced L-Carnitine was effectively prepared by extrusion/spheronization. The drug loaded pellets exhibited very less disintegration time and tremendous drug release in 5 minutes. The study recommends that the pro-drug developed has a better stability in terms of moisture absorption than precursor (L-Carnitine).

References

1. P. S. Shankar, P. D. Ganesh and S. R. Bhanudas, *World. J. Pharm. Res.*, 2014, **3**(6), 1972.
2. K. lavanya, *Int. J. Pharm. Sci. Res.*, 2011, **2**(6), 1337.

DEVELOPMENT OF TASTE MASKED ORAL RECONSTITUTABLE SUSPENSION OF CEFIXIME TRIHYDRATE

Abhishek Gurung,[a] Sindhu Abraham,[a]* Neelamma,[a] Anbu Jayaraman[b] and Srinivasan Bharath[a]

[a]Department of Pharmaceutics, M. S. Ramaiah University of Applied Sciences, Gnanagangothri Campus, New BEL Road, MSR Nagar, MSRIT Post, Bangalore- 560054, Karnataka, India; [b]Department of Pharmacology, Faculty of Pharmacy, M. S. Ramaiah University of Applied Sciences, Gnanagangothri Campus, New BEL Road, MSR Nagar, MSRIT Post, Bangalore- 560054, Karnataka, India

1 INTRODUCTION

The palatability of paediatric medicines is one of the most important formulation factors with potential to influence patient compliance and a drug's taste has to be masked without compromising on its safety and efficacy.[1,2] One of the popular approaches in the taste masking of bitter drugs is the use of Ion Exchange Resins (IER). IERs are solid and suitably insoluble high molecular weight polyelectrolytes that can exchange their ions of equal charge with the surrounding medium. Cefixime Trihydrate, an oral third generation cephalosporin antibiotic is effective in the treatment of gastroenteritis in young children.[3] The amine and nitrogen group of the drug is responsible for its bitter taste and these groups can be blocked by forming a complex with the ion exchange resin Kyron T-104, thereby restricting the release of drug into the saliva.[4] Thus the resin effectively reduces the interaction between the drug and the taste buds.[5]

2 METHODOLOGY AND RESULTS

2.1 Preparation of drug resin complexes

The drug-resin (Cefixime Trihydrate- Kyron T-104) complexes were prepared in ratios 1:1, 1:2, 1:3, 1:4, and 1:5.

2.2 Characterization of drug resin complexes

The drug- resin complexes were characterized by Fourier Transform Infrared Spectroscopy (FT-IR), Differential Scanning Calorimetry (DSC) and X- Ray Diffraction (XRD).

The FT-IR spectra of pure drug showed characteristic peaks at 1666 cm^{-1},1728 cm^{-1}, (C=O stretch), together with C-O stretch at 1064 cm^{-1}, C=N stretch at 1529 cm^{-1} and NH stretch at 3967 cm^{-1}. Similar peaks were observed in the drug resin complex spectra with slight shifts in the data, confirming the compatibility of the selected ingredients. (Figure 1).

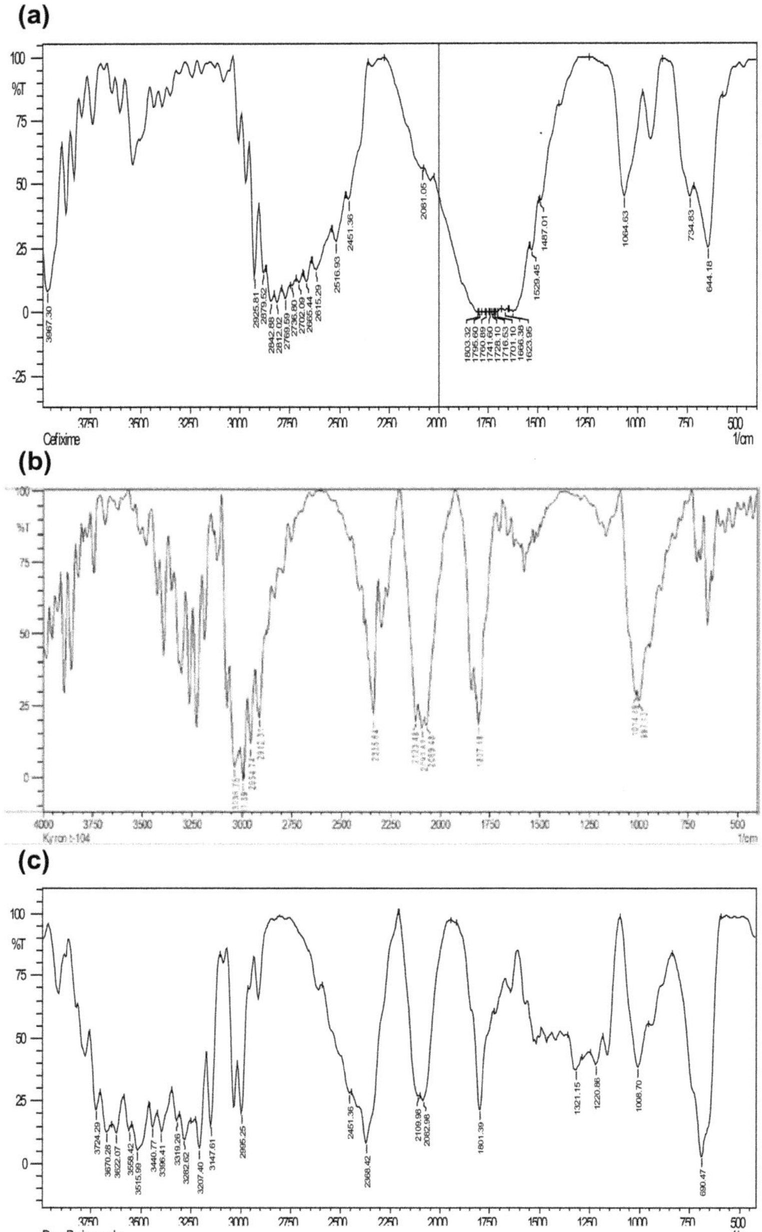

Figure 1 *FT-IR spectra of (a). Cefixime Trihydrate (b) Kyron T-104 (c) Drug-resin complex.*

The DSC thermogram of Cefixime Trihydrate showed one endothermic peak at 113.92°C and two exothermic peaks at 189.27°C and 251 °C. The thermogram of drug-resin complex showed two endothermic peaks at 74.93°C and 102.07°C and two exothermic peaks at 189.13°C and 259 °C. Absence of peaks at 251°C in drug-resin complex indicated complexation of drug with resin (Figure 2).

(a)

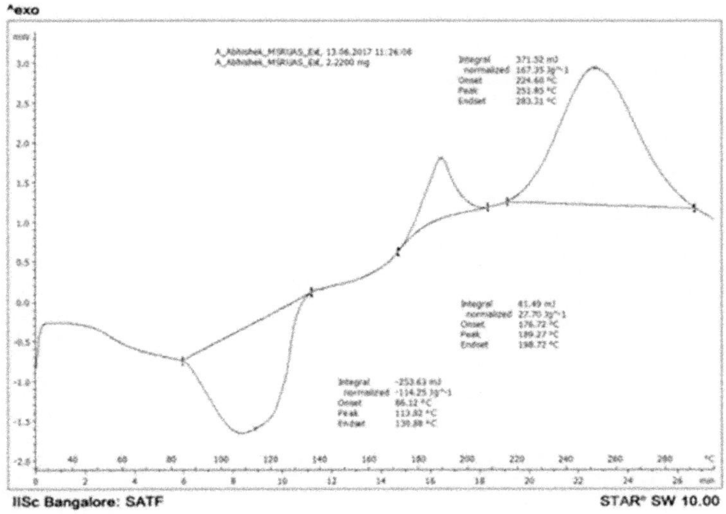

(b)

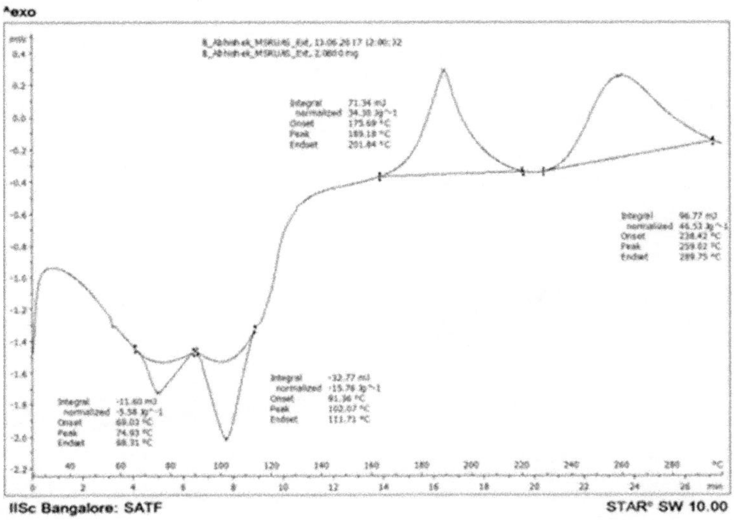

Figure 2 *DSC thermograms of (a) Cefixime Trihydrate (b) Drug-resin complex.*

The XRD spectra of pure Cefixime Trihydrate showed sharp distinct peaks at 2θ diffraction angles 8°, 15°, 19°, 22°, 23° and 27°. A number of sharp peaks were observed indicating the crystalline nature of Cefixime Trihydrate. The spectra of Kyron T-104 showed some diffused peaks, revealing the amorphous nature of resin. The spectra of the drug-resin complex showed diffused peaks suggesting the conversion of drug from crystalline to amorphous form due to formation of a complex (Figure 3).

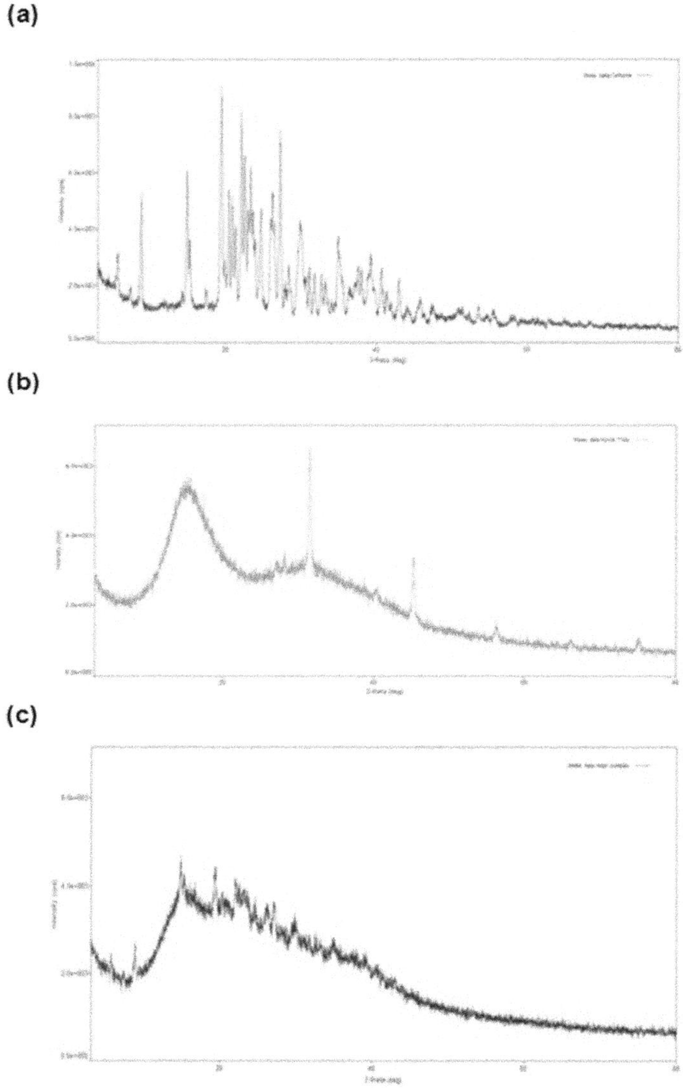

Figure 3 *XRD spectra of (a) Cefixime Trihydrate (b) KyronT-104 (c) Drug-resin complex.*

Drug content

Drug-resin complex equivalent to 100mg of the drug was sonicated with 0.1N HCl in a 100ml volumetric flask. The solution was then further suitably diluted and analyzed using UV spectrophotometer (UV-1700, Shimadzu) at 285.5 nm. The drug content of all the drug- resin complexes were in the range 75±1.20 to 92±0.87.

2.3 Preparation of dry suspension

Dry suspension powder containing Cefixime Trihydrate – Kyron T-104 complex (ratio 1:3) equivalent to 100mg of Cefixime Trihydrate was prepared using Xanthan gum as suspending agent at various concentrations of 0.2%, 0.4%, 0.6%, 0.8% and 1.0% w/v. The preservatives used were methyl and propyl paraben. Citric acid served as a buffer and Aspartame as flavouring agent (Table 1).

2.4 Evaluation of taste masked suspension

The dry powder mixtures were reconstituted with purified water to form suspensions and evaluated for viscosity, sedimentation volume and redispersibility. The results are shown in Table 2.

Table 1 *Composition of reconstitutable dry suspension formulation.*

S. No.	Ingredients	F_1	F_2	F_3	F_4	F_5
1.	Drug-resin complex equivalent to 100mg CefiximeTrihydrate (g)	2.4	2.4	2.4	2.4	2.4
2.	Xanthan Gum (g)	0.06	0.12	0.18	0.24	0.30
3.	Sucrose (g)	0.980	0.980	0.980	0.980	0.980
4.	Citric Acid (g)	0.05	0.05	0.05	0.05	0.05
5.	Aspartame (g)	0.025	0.025	0.025	0.025	0.025
6.	Methyl Paraben (g)	0.005	0.005	0.005	0.005	0.005
7.	Propyl Paraben (g)	0.005	0.005	0.005	0.005	0.005
8.	Tartarzine	q.s	q.s	q.s	q.s	q.s
9.	Pineapple Flavour	q.s	q.s	q.s	q.s	q.s

Table 2 *Physical properties of reconstituted suspensions.*

Parameters	Formulations					
	F1	F2	F3	F4	F5	ZIPRAX
Viscosity (cps)	3.78	3.79	3.80	3.81	3.80	3.80
pH	8.10±0.98	8.10±0.95	8.15±1.4	8.10±0.86	8.20±1.40	8.10±1.10
Sedimentation value	0.70	0.80	0.80	0.85	0.95	0.96
Redispersibility	Easy	Easy	Easy	Easy	Easy	Easy

2.5 In Vitro Drug Release

In vitro drug release of formulations F1-F5 and marketed formulation (ZIPRAX, Cipla) were carried out using USP XXIII type II (Paddle type) dissolution apparatus.Paddle speed and bath temperature were set at 50 rpm and 37°C ± 0.5°C respectively. Suspension samples equivalent to 100mg of Cefixime Trihydrate resinate were suspended in 900ml 0.1N HCl and 5 ml of the dissolution media was withdrawn at predetermined time intervals and equivalent amount of fresh dissolution media was replaced.The study was conducted for 45 minutes to ensure complete drug release from all the formulations. At the end of 45 min, F-3 and F-4 showed 90% and 85% drug release respectively. Drug release was better than the marketed oral suspension (M) which showed only 25% drug release at the end of 45 min.

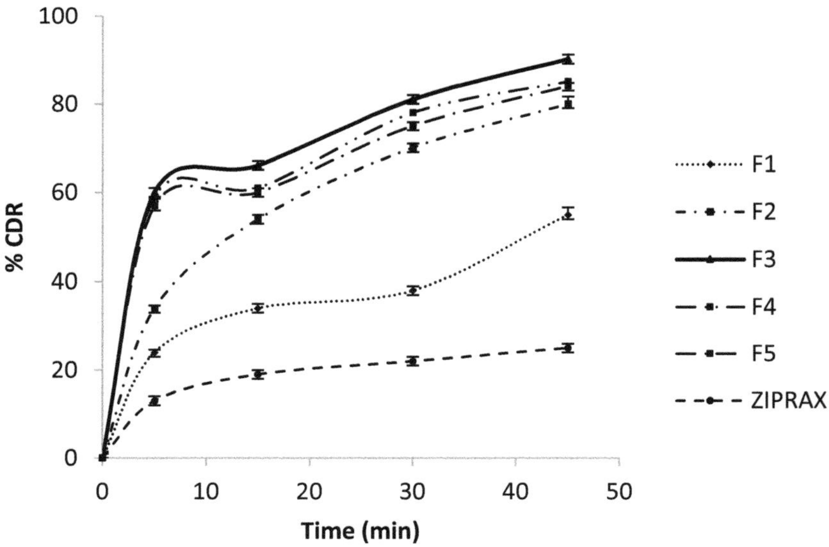

Figure 4 *Comparative in vitro drug release profile of formulations F1 – F5 and marketed formulation (ZIPRAX).*

2.6 *In vivo* taste masking study – Rat Behavioural Avoidance model

The model is based on the principle that administration of a bitter solution to water deprived rats reduces the drinking frequency.[6] The rats were deprived of water for a period of 22 hrs, made to lick bottles containing water for 5 min and the licking activity obtained. Water was taken as the baseline for the study. The rats were divided into 5 groups namely control, plain Cefixime Trihydrate suspension, marketed formulation (ZIPRAX,Cipla) and taste masked suspensions F3, F4. Each test solution was presented for 8 sec, twice during the experiment. Test trials were alternated with 2 sec water rinse trials to minimize the carry over effects of the previous test solution. The number of times the rat licked the sample was counted. If the rat finds the sample presented to be aversive it will lick less often in comparison to the water sample. The average number of licks for test solution was then divided by average number of licks for water to generate the % inhibition of licking.[7] The licking frequency for plain Cefixime Trihydrate suspension was 64 ± 0.85% whereas marketed preparation (ZIPRAX) showed 90 ± 1.31%. A higher licking frequency of 99.3± 0.21% was obtained for taste masked suspension F-3 and 98± 0.36% for taste masked suspension F-4 indicating that the bitter taste of the drug was efficiently masked. Hence it could be concluded that KyronT-104 efficiently masked the bitter taste of Cefixime Trihydrate.

CONCLUSION

Complexation using ion exchange resin is asimple and cost effective technique for taste masking. Taste masked suspensions of Cefixime Trihydrate were successfully prepared by complexation method using Kyron T-104 as ion exchange resin. *In vitro* studies revealed that the drug was released from the complex significantly. *In vivo* study conducted on rats proved that the bitter taste of Cefixime Trihydrate was efficiently masked. The results obtained for this work shows that the complexation of Cefixime Trihydrate with Kyron T-104 effectively masked the bitter taste of the drug, thus making it suitable for formulation as a pediatric oral suspension.

References

1. K. P. S. Kumar, D. Bhowmik, L. Deb, A. Yadav and A. S. Dutta, Recent trends in taste masking of bitter drugs, *J. Drug Delivery Res.*, 2012, **1**(1), 2319–1074.
2. B. V. Akbari, B. P. Patel, R. B. Dholakiya, B. G. Shiyani and D. J. Lodhiya, Development and evaluation of taste masked suspension of prokinetic agent by using ion exchange resin, *Int. J. PharmTech Res.*, 2010, **2**(1), 240–245.
3. A. M. Rajesh, S. A. Bhatt, H. Brahmabhatt, P. S. Anand and K. M. Popat, Taste masking of ciprofloxacin by ion-exchange resin and sustain release at gastric-intestinal through interpenetrating polymer network, Asian J. Pharm. Sci., 2015, **10**(4), 331–340.

4. Cefixime use in children: When and why, *Can. J. Infect. Dis.*, 1995, **6**(4), 204–205.
5. S. P. Kela, B. S. Kuchekar, S. C. Jagdale and S. A. Patil, Formulation, Characterization and Evaluation of Taste-Masked Rapid Disintegrating Tablet of Cefixime by Ion Exchange Resin Technique, *Res. J. Pharm. Tech.*, 2012, **5**(8), 1320–1325.
6. S. Sana, A. Rajani, N. Sumedha and B. Mahesh, Formulation and Evaluation of taste masked oral suspension of Dextromethorphan Hydrobromide, *Int. J. Drug Dev. Res.,* 2012, **4**(2), 159–172.
7. M. G. Bhat, R. M. Jordt, M. A. Khan, C. E. Foley and T. A. Gilbertson, Validation of a rat behavioral avoidance model from a drug delivery perspective, *Int. J. Pharm.*, 2005, **303**(1–2), 31–36.

NANOCOMPOSITE HYBRID POLYMERIC SCAFFOLD FOR SKIN TISSUE ENGINEERING

Devanand Kamnoore,[a] Dhrubojyoti Mukherjee,[a]* Bhagyashree K.,[a] Shovik Kumar Debnath,[a] Damodar Nayak[b] and Parsuraman Pavadai[c]

[a] Department of Pharmaceutics, M S Ramaiah University of Applied Sciences, Bengaluru, India; [b]Department of Pharmacology, M S Ramaiah University of Applied Sciences, Bengaluru, India; [c]Department of Pharmaceutical Chemistry, M S Ramaiah University of Applied Sciences, Bengaluru, India

1. INTRODUCTION

Biodegradable polymeric scaffolds have received much attention in tissue engineering as they provide a spatial and temporal environment for cell regeneration. The scaffold can be defined as "any material, natural or artificial that comprises of entire or portion of a living structure or biomedical device which performs, extends or substitutes a natural function". Biomaterials based scaffolds are projected to interact with living systems and to treat, augment, assess and substitute any function, tissue, and organ of the body.

Tissue engineering aims to facilitate the growth and replace the diseased or damaged tissue by using the combination of bioactive molecules, cells, and biomaterials. Many clinical studies were carried out to repair or replace tissue in the human body such as damaged or diseased tissue using autografts, xenografts or allografts (i.e. donor graft tissue). But a major drawback of these treatments is a shortage of donor site or donor rejection of grafts, morbidity, donor site pain, the transmission of disease, the possibility of harmful immune responses and the volume of donor tissue. In the last two decades, rapid progress has been seen in the emerging field of tissue regeneration and tissue engineering.[1]

2. METHOD AND RESULTS

2.1 Preparation and characterization of nanocomposite functionalized hydrogel scaffold

The freeze-thawing technique was followed to prepare the hydrogel scaffolds. Thiolated chitosan (1.5 g) was dissolved in a solution of 1% v/v acetic acid. Thiolated eudragit (1.5 g) was separately solubilized in 1% w/v of sodium hydroxide solution. A solution of polyethylene glycol (PEG) 1.5% w/v was added to the thiolated chitosan solution followed by the addition of thiolated eudragit solution into the mixture. The mixture was kept under mechanical stirring at room temperature for a period of 12 h at 1000 rpm. The resultant mixture was then poured into a fabricated mould and subjected to freeze-thawing technique with 18 h of freezing and 6 h of thawing for 6-7 days. During the freeze-thawing process, the scaffolds were functionalized with previously prepared hydroxyapatite nanoparticles. The hydroxyapatite nanocomposite was characterized by FESEM, AFM, and DLS and the prepared nanocomposite embedded scaffold was characterized by FTIR, XRD, gel fraction, swelling, water vapour transmission, SEM, optical Profilometry and Micro-CT analysis.[2]

XRD profile of the scaffold (Figure 1) exhibited sharp nature of the peaks signifying stability of the scaffold. When compared to the peaks of the individual polymers (not shown) there has been a shift in the theta and d-spacing value. This could be due to the polymeric interlinking in the scaffold matrix during the freeze-thaw technique. The shift in peak height and d-spacing signified changes in the configuration of crystallinity in the cross-linked polymers and may be a result of changes in the atomic density in the crystal lattice of a specific plane subjecting to fractional miscibility between the polymers during the formation of the scaffold. The swelling degree of the scaffold was exhibited in Figure 1. A1 is scaffold without nanocomposite and A2 with nanocomposite. The overall swelling degree was observed to be low, which could be due to the high intensity of the polymeric crosslinking leading to reduced intramolecular and intermolecular spaces leading to less water channeling space (and capacity) of the scaffold. The size and morphology of the nanocomposite were characterized by SEM and AFM (Figure 2). The combination of SEM and AFM indicated the particle size of the nanoparticle was in the range of 100-300 nm. SEM of the cross-section of the scaffold was exhibited in Figure 3, which exhibited the nanocomposite functionalization in the scaffold matrix and also signify the efficiency of nanodispersion technique during the freeze-thawing process in developing nanocomposite functionalized scaffold. Optical Profilometry (VSI) and micro-CT of the scaffold system was represented in Figure 4. The surface of the scaffold was observed to be smooth with less tortuousness and micro roughness. The presence of micropores could have led to micro-roughness onto the surface of the scaffolds. Density related changes during the polymeric interlinking process in the freeze-thawing technique could have led to micro roughness, which is also evident from the results of micro-CT. The controlled porous structure of the scaffold was confirmed by micro-CT analysis. The polymeric cross-linking during freeze-thawing technique in addition to the concentration of the polymers could have led a controlled pore formation in the scaffold matrix.

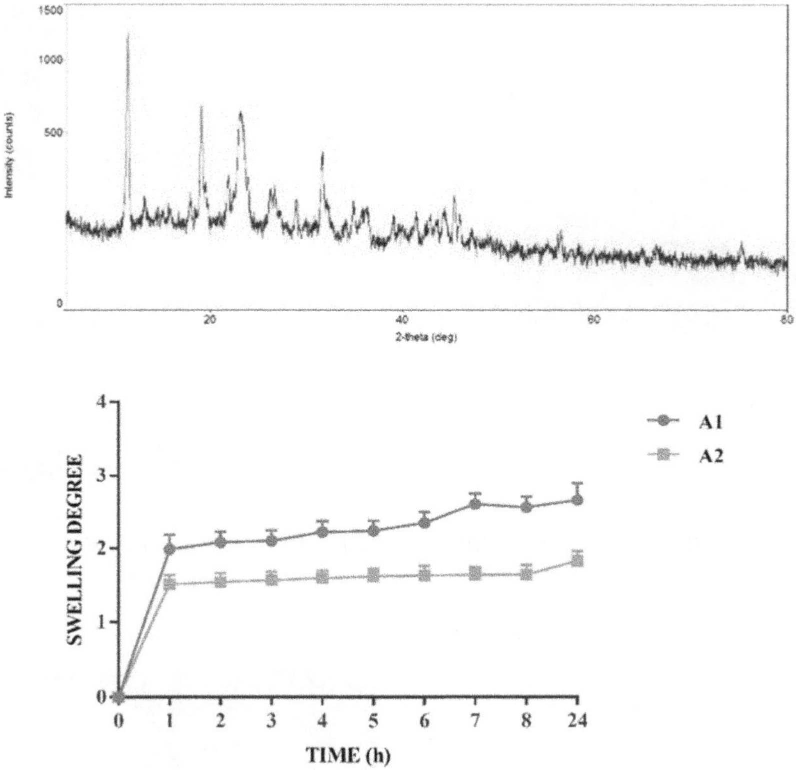

Figure 1 *XRD and swelling degree profile of the developed scaffold.*

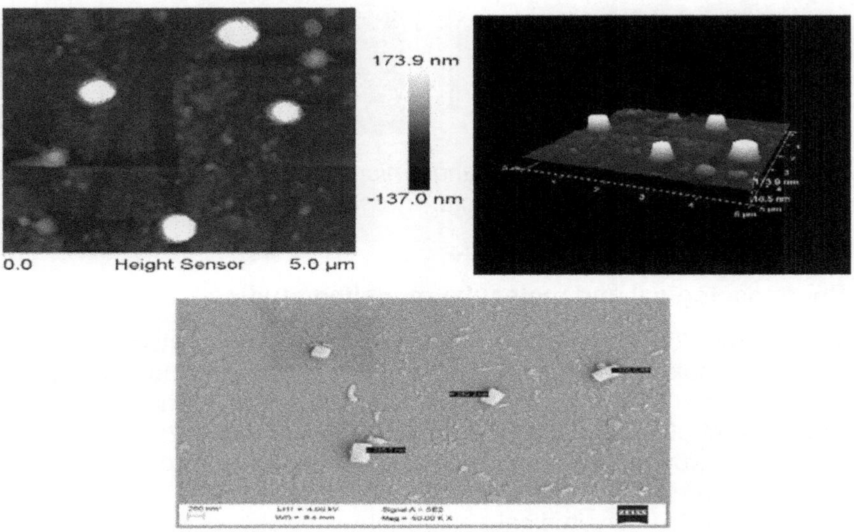

Figure 2 *AFM and SEM of HAP nanocomposite.*

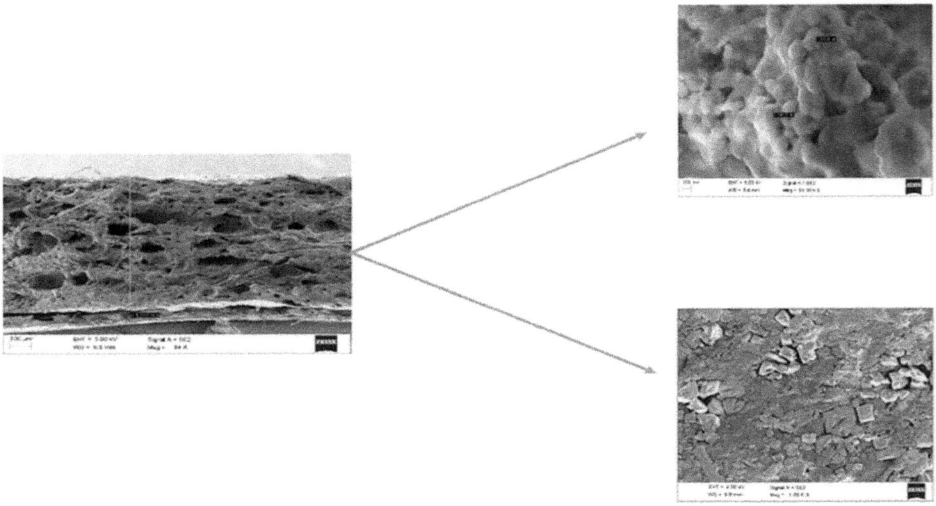

Figure 3 *SEM of HAP nanocomposite functionalized scaffold.*

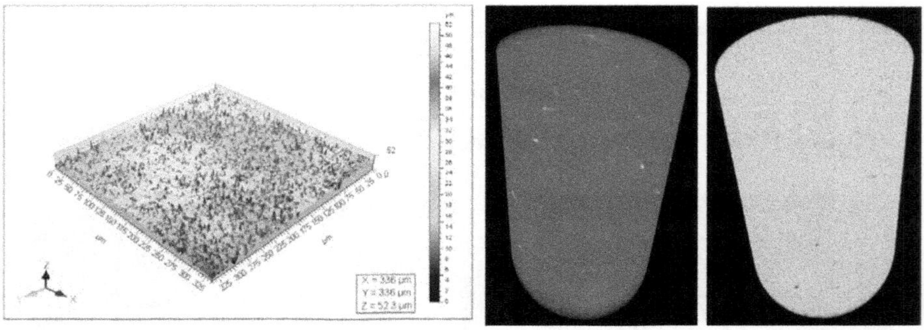

Figure 4 *Optical Profilometry and micro-CT characterization of HAP nanocomposite functionalized scaffold.*

2.2 Pharmacological and molecular modeling studies

Albino rats (200-250 g) were taken for the pharmacological studies where four groups of 6 animals each were used. The pharmacological study was approved by the Institutional Animal Ethics Committee of Faculty of Pharmacy (Ref No: XXI/MSRFPH/M-09/10.09.2018).The rats were anesthetized prior to the creation of the wounds, with a combination of ketamine and lignocaine (1:4) injection. The dorsal fur of the animals was shaved with an electrical hair remover (trimmer). A longitudinal paravertebral incision of 8mm diameter was made through the skin and cutaneous tissue on the back surgically. After creating two full-thickness incisions the wounded areas were disinfected with a suitable disinfectant. Sterile gauze was used to cover the wounds (Positive control), the standard marketed product, and hydrogel scaffold A1 (without nanocomposite) and A2 (with nanocomposite) respectively. On 3rd, 6th, 12th and 15th day after the surgery, the size of each wound was measured using Vernier caliper.[2] Simulation studies were performed to explore the binding interactions and conformations of the polymers present in the scaffold with the proteins responsible for skin tissue regeneration and re-epithelialization.[3]

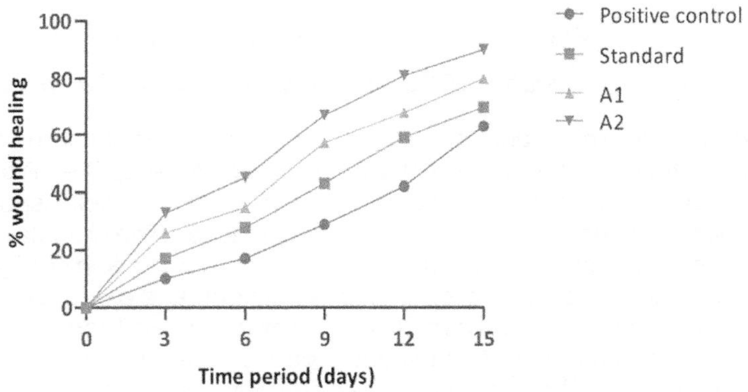

Figure 5 *Comparaitive graphical representation of wound healing.*

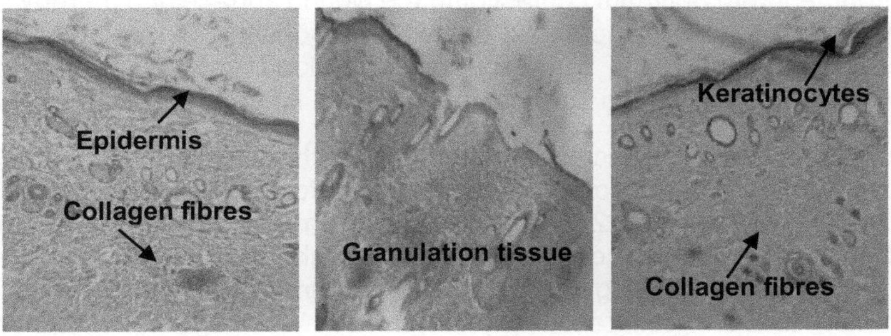

Figure 6 *Histopathological observations of normal control, positive control, and scaffold respectively.*

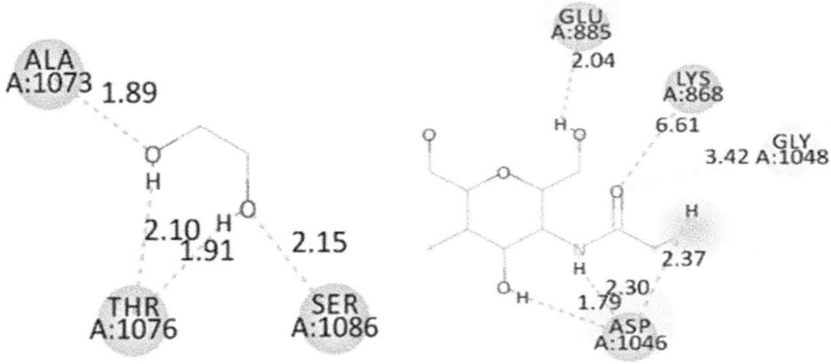

PEG Thiolated chitosan

Figure 7 *Interaction of Ligands (PEG and Thiolated chitosan) with VEGFR2.*

The *in vivo* wound healing and the histopathological study was carried out post 15 days of treatment. In the treatment groups, complete re-epithelialization was observed with the newly formed dermis (Figure 6). The results showed that epidermis with keratinocyte formed the keratinized layer. The granulation tissue was clearly seen with specifically arranged active fibroblasts and collagen fibers which plays an active role in the synthesis of collagen and extracellular matrix. Wounds treated with scaffold A2 (hydrogel) showed the well-formed epidermis and could be due to the alteration in controlled moisture presence surrounding the wounded tissues. The scaffold treated group showed almost complete re-epithelialization as compared to other groups. Neovascularization and collagen proliferation was observed in the treatment group as compared to the control group. The results of micro-CT, SEM, and VSI has indicated the presence of microroughness and micropores onto the scaffold. This surface properties of scaffold could have significantly led to controlled moisture content in the wound environment with optimum contact onto the wound surface and thereby helping in altering the protein metabolic pathway responsible for reepithelialization, tissue regeneration and formation of new blood vessels by altering the environment in the wound surface, thus, making it more feasible for healing.

Molecular docking of vascular endothelial growth factor receptor 2 (VEGFR2) kinase with the primary components of the scaffold was represented in Figure 7. Thiolated chitosan (-5.27), PEG (-3.22) and hydroxyapatite nanocomposite (-3.8) exhibited optimum docking score with high intensity. Thiolated eudragit did not show any docking score with the protein. The correlation between *in vivo* skin tissue regeneration experiment and docking score supports the hypothesis that the nanocomposite functionalized interpolymeric scaffold was significantly superior in skin tissue regeneration and re-epithelialization confirming its therapeutic potential.[4]

3. CONCLUSION

The developed nanocomposite functionalized scaffold was evaluated for physicochemical parameters and characterized by scanning electron microscopy, optical profilometry and micro-CT. Tissue regeneration efficiency of the scaffold was evaluated by reepithelialization ability on an excision wound model. The hydroxyapatite nanocomposite was having a particle size within 300 nm and was successfully functionalized (embedded) in the polymeric scaffold system. The scaffold was observed to have optimum swelling and significant surface morphological properties. The porous nature of the scaffold was successfully controlled and exhibited a significant and superior level of skin tissue regeneration and reepithelialization property.

References

1. P. Turon, L. J. Del Valle, C. Alemán, J. Puiggalí, *Appl. Sci.*, 2017, **7**(1), 60.
2. D. Mukherjee, M. Azamthulla, S. Santhosh, G. Dath, A. Ghosh, R. Natholia, *et al.*, *J. Drug Deliv. Sci. Technol.*, 2018, **46**, 498–510.
3. S. Santhosh, D. Mukherjee, J. Anbu, M. Murahari, B. V. Teja, *J. Microencapsul.*, 2019, **36**(4), 338–355.
4. S. Thönes, S. Rother, T. Wippold, J. Blaszkiewicz, K. Balamurugan, S. Moeller, *et al.*, *Acta Biomater.*, 2019, **86**, 135–147.

IN SILICO PREDICTION AND *IN VIVO* WOUND HEALING INDICATORS OF NANO CHITIN INCORPORATED LYOPHILLISED ALGINATE MATRICES

Sharon Caroline Furtado,[a] Bharath Srinivasan[a] and Judy Jays[b]

[a]Department of Pharmaceutics, Faculty of Pharmacy, M S Ramaiah University of Applied Sciences, Bengaluru-560 054, India; [b]Department of Pharmaceutical Chemistry Faculty of Pharmacy, M S Ramaiah University of Applied Sciences, Bengaluru-560 054, India

1. INTRODUCTION

Chitin (β-(1-4)-poly-N-acetyl-D-glucosamine) is considered the second most abundant polysaccharide in nature. Chitin can be degraded into shorter oligosaccharides known as chito-oligosaccharides by either chemical or enzymatic methods. Chitin and its oligosaccharides, being biocompatible and biodegradable, have been reported to have several biomedical applications. The wound healing properties of this polymer have been determined primarily using *in vivo* studies on various animal models. However little has been studied regarding the mechanism of action of this polymer with regard to wound healing.

Wound healing is a highly co-ordinated process involving inflammation, cell proliferation, matrix deposition, tissue remodeling, collagenation and epithelialization. Matrix metalloproteinases (MMPs) are a class enzymes that that are structurally similar and contribute in degradation of extracellular matrix. They have a role in both normal and pathological matrix remodelling. Over expression of metalloproteinases is generally related to delayed or impairment in wound healing.[1] Studies carried out on human dermal fibroblasts have reported, chitooligosaccharides ranging from 3–5 kDa to have a strong inhibitory effect on gene expressing MMP-2.[2] However, any inhibitory activity of chitin on the protein itself has not been reported.

Molecular docking is an established computational technique used to predict the interaction energy between two molecules and to find the best orientation of ligand which would form a complex with overall minimum energy. In the present study, molecular docking studies of chitin on the enzyme collagenase has been performed. The work may provide a better understanding of the potential of chitin as an antagonist for collagenase. *In vivo* would healing property of nanochitin incorporated dressings was screened in Wistar Albino rats employing excision wound models. An attempt has been made to correlate the docking studies with *in vivo* collagen synthesis in wounds treated with chitin containing alginate matrices.

2. METHOD AND RESULTS

Docking studies of chitin on the enzyme collagenase was performed using Schrodinger Drug Design Suite of 2016–3. The glide parameters and nature of interactions was compared with reported collagenase inhibitor OED.

2.1. Preparation of chitin containing alginate matrices

The chitin incorporated sodium alginate matrices (NSA) were prepared using 3% w/w of sodium alginate, 1.5% w/w of chitin and glycerol (40% of the polymer weight) as the plasticizer. Acid hydrolysed chitin was dispersed in deionized water and subjected to probe sonication (Qsonica- Q700) at the following conditions: Amplitude: 40%; Pulse: 5 sec. ON; 5 sec. OFF with duration 15 min. The weighed amount of sodium alginate and glycerol was then added to the colloidal chitin using a digital propeller operated at 100 rpm for 1h. Similarly sodium alginate dressings containing 3% sodium alginate and glycerol (40% of the polymer weight) without nanochitin (SA) was also prepared. Mixture (20g) was casted on petri plates (9cm diameter). The casted mixture was frozen overnight at -40°C and subjected to lyophilization at a vacuum of 0.11mbar (Martin Christ, ALPHA 1-2 LD plus) for a period of 8 hours.

2.2. Ligand preparation

The molecular structure of chitin was retrieved from the PubChem database (Compound ID: 21252321) (Fig.1). LigPrep module was used for geometrical refining of chemical structures and to generate energy minimised 3D structures using OPLS2005 (Optimised Potential for Liquid Simulations) force field.

2.3. Protein preparation

The protein preparation wizard of Maestro 10.7 was used. The crystal structure of fibroblast collagenase complexed with an inhibitor (N-[(1s)-3-{[(benzyloxy) carbonyl] amino}-1-carboxypropyl]-l-leucyl-N-(2-morpholin-4-ylethyl)-l-phenylalaninamide) (OED) was downloaded from Protein Data Bank (PDB ID: 1CGL).[3] The catalytic domain of the enzyme was solved at 2.4 Å resolution and refined to a crystallographic R factor of 0.18. The PDB structure had two identical chains. Missing hydrogen atoms were added. Co-crystallized water molecules and ions were removed from the protein. The protein structure was minimized to the default Root Mean Square Deviation (RMSD) value of 0.30. The co-crystallized ligand was separated from the protein structure and a receptor grid was automatically generated using the glide settings.

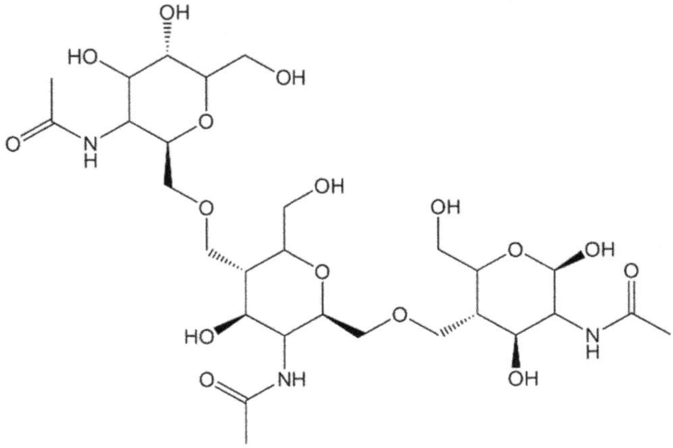

Figure 1 The molecular structure of chitin.

2.4. Ligand Docking

The prepared ligand (chitin) and the native ligand (OED) were docked onto the active site of the protein using the extra precision (XP) feature of Glide module. The docking poses were ranked according to their docking scores (GScore), their binding energy with the enzyme (Emodel), nature of interactions.

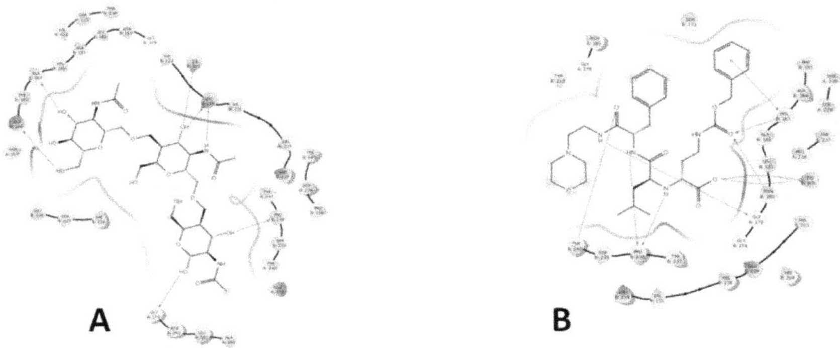

Figure 2 *Binding interactions of Chitin (A) and OED (2) with human collagenase.*

The molecular docking studies of chitin in comparison with the known inhibitor (OED) showed similar interactions (Fig.2). Chitin exhibited hydrogen bonding interactions with a Docking score of -8.606. OED had hydrogen bonding and Pi-Pi stacking and a Glide score of -8.570. Glide Emodel observed for chitin and OED was -141.032 and -130.013kcal/mol respectively.

2.5. *In vivo* wound healing studies in excision rat wound models

The polysaccharide chitin was broken down using a combination of acid hydrolysis and probe sonication techniques to form nano chitin. Sodium alginate, 3%w/w dressings were prepared with and without the incorporation of 1.5% w/w nanochitin by lyophillisation. Wistar rats of either sex, weighing 250-300g were used in this study with prior IAEC approval. The animals were anaesthetised using ketamine and xylazine. Four full thickness excision wounds were created on the dorsal thoracic region on the depilated part of the skin using a sterile biopsy punch of 10 mm diameter. The area of the wounds were measured by tracing the wounds onto graph paper on the day of wounding and subsequently on 3rd, 5th, 7th, 9th, 11th and 14th day post-wounding. The percentage of wound size reduction was calculated using the following equation.

$$\text{Wound size reduction (\%)} = \frac{A_0 - A_t}{A_0} \times 100$$

A_0 and A_t are initial wound area and wound area after time interval 't'.

The healing potential was assessed based on percentage wound contraction, histopathlogical analysis and hydroxyproline concentration in the healed tissue.

2.6. Estimation of hydroxyproline

Collagen is responsible for providing structural integrity to the extracellular matrix and is the major component of the granulation tissue. Hydroxyproline is a biochemical marker for collagen deposition since the amino acid proline is a component of collagen fibres. Hence increase in hydroxy proline content in the healed skin tissue is a positive indication of wound healing progression. The animal subjected to wound healing study was sacrificed using excess ether anaesthesia, a portion of the healed skin was excised and taken for estimation of hydroxyproline content using HPLC method.[4]

The percentage wound contraction for wounds treated with alginate dressings was 82.34±5.10 and the same for nanochitin incorporated alginate dressings was 90.26±3.98. Hydroxyproline concentration in the sodium alginate dressing without chitin was 61mg/g sample and for chitin incorporated sodium alginate dressing it was 66.87mg/g sample.

3. CONCLUSION

The similarities in the docking score and Emodel for chitin and the known reported inhibitor, OED, support the hypothesis of chitin being a collagenase inhibitor which is further evidenced in improved collagen deposition in wounds treated with chitin incorporated dressings as seen in *in vivo* studies.

References

1. S. Gill and W. Parks, *Int. J. Biochem. Cell Biol.*, 2008, **40**, 1334–1347.
2. N. Thomas and S. Kim, *Biochem. Res. Int.*, 2010, **2010**, 1–10.
3. B. Lovejoy, A. Cleasby, A. Hassell, K. Longley, M. Luther, D. Weigl, G. McGeehan, A. McElroy, D. Drewry, M. Lambert *et al.*, *Science*, 1994, **263**, 375–377.
4. F. Vázquez-Ortíz, O. Morón-Fuenmayor and N. González-Méndez, *J. Liq. Chromatogr. Rel. Technol.*, 2004, **27**, 2771–2780.

ENHANCEMENT OF ATAZANAVIR SOLUBILITY USING β-CYCLODEXTRIN BASED NANOSPONGES

D. N. Venkatesh

Department of Pharmaceutics, JSS College of Pharmacy, JSS Academy of Higher Education & Research, Ooty – 643 001. Tamil Nadu, India
Email: nagasamyvenkatesh@jssuni.edu.in

1 INTRODUCTION

Oral route is considered as the major route for the delivery of drug, especially in the treatment of chronic diseases. A drug with a high lipophilicity more than 50%, however regarded as deterrent for oral delivery.[1] Especially drugs under Biopharmaceutical Classification Systems (BCS) II and IV causes challenges to the formulators as their poor bioavailability caused due to poor aqueous solubility resulting in reduced drug absorption. To overcome such issues various strategies have been developed such as inclusion complexation.[2] Recently nanosponges have gained considerable attention, owing to their biocompatibility, nanoporous in nature, capable of forming supra molecular inclusion complexes with both hydrophilic and lipophilic drugs. The toxicological aspects of nanosponges have been proven to be safe for use in pharmaceutical products. Also the drug loading can be enhanced by modifying the cyclodextrin to cross linker ratio to obtain the desired release profile of the drug. Nanosponges are nanostructured carriers prepared by reacting cyclodextrin with suitable cross linkers such as carbonyl diimidazole, hexamethylene and diphenyl carbonate.[3] Nanosponges were employed as carriers for solubility enhancement and stability of many drugs. In the present study, Atazanavir (ATZ) was selected as it is BCS II drug having low solubility (0.11 mg/l) with a log P value of 5.6.[4] It is used in the treatment of Human Immunodeficiency Virus (HIV) infection with better potency, but low oral bioavailability (35%) imparts poor aqueous solubility, slow dissolution rate and delays onset of action. In the present work, binary complexes of ATZ was prepared using β-cyclodextrin with varying concentration of 1,1 carbonyl diimidazole (CDI) as cross linker were prepared. The prepared nanosponges were characterized for FTIR spectroscopy study, scanning electron microscopy and *in vitro* dissolution study.

2 METHOD AND RESULTS

2.1 Formulation of β-Cyclodextrin Nanosponges

A 250 ml of round bottom flask contained 17.5 m mol of anhydrous β-cyclodextrin and 9.96 m mol of 1, 1' carbonyl diimidazole (CDI) were solubilized using 100 ml of anhydrous N,N dimethyl formamide (DMF). The resulting solution was heated in an oil bath at 90° C for 24 hours until a gel is formed. After 24 hours the solid is ground in a mortar and washed with deionized water to remove the solvent. Finally, the polymer was rinsed and extracted with ethanol for 45 hours using a soxhlet extractor. The white powder product (nanosponges) so obtained was dried at 60° C in an oven and subsequently ground in mortar for studies. The composition of the nanosponges was shown in Table 1.

Table 1 *Composition of different formulation of nanosponges*

Code of formulation	β-cyclodextrin (mg)	1, 1' carbonyl diimidazole (mg)	Vehicle (ml)
F1	1	2	10
F2	1	4	10
F3	1	6	10
F4	1	8	10
F5	1	10	10

2.2 Formulation of ATZ loaded β-Cyclodextrin Nanosponges

ATZ was dispersed in aqueous suspension of nanosponges in different ratio's were centrifuged at 2000 rpm for 10 minutes to separate the uncomplexed drug. The resultant product was freeze dried to obtain drug loaded nanosponges.

2.3 Characterization of ATZ nanosponges

2.3.1 Compatibility study using Fourier Transform Infrared Spectroscopy (FTIR)
FTIR analysis was conducted to verify the possibility of chemical interaction between the drug and polymer. Samples were scanned in the range of 4000-400 cm^{-1}(Perkin Elmer, Japan). The appearance of the functional groups in the drug such as -C-H stretching at 3427 cm^{-1}, -OH stretching at 2955 cm^{-1}, C-C stretching at 1648 cm^{-1} and –C=C stretching at 1166cm^{-1} were present in both drug and drug and polymer mixture. These results indicating that there was no chemical interaction between the drug and the polymer (Figure 1 and 2).

2.3.2 Determination of particle size
The nanosponges of ATZ were chracterized for particle size (Malvern Instruments, UK), the particle sizes of the different formulations were shown in Table 2.

2.3.3 Determination of drug content
The ATZ content in the nanosponges were estimated by UV spectrophotometer. A known quantity of ATZ nanosponges from different formulations were added with phosphate buffer saline (PH 7.4) in a 100 ml volumetric flask. Then the solution was filtered and suitably diluted to measure the concentration of ATZ at 301 nm. The drug content was in the range of 95.11 to 98.98% for different nanosponges formulations as shown in Table 2.

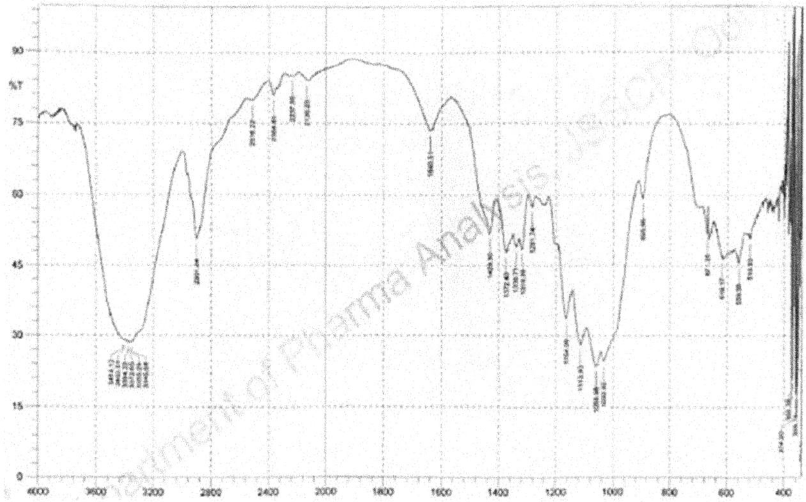

Figure 1 *FTIR Spectra of ATZ.*

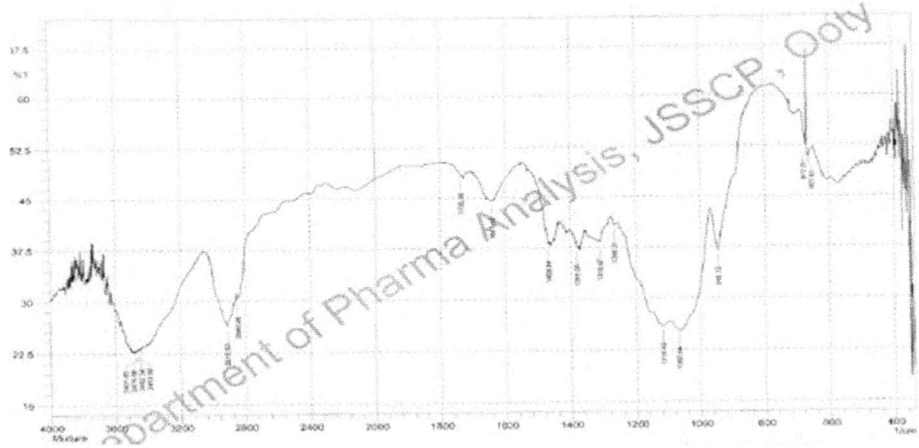

Figure 2 *FTIR Spectra of ATZ nanosponges.*

2.3.4 Determination of surface morphology by scanning electron microscopy (SEM)

Scanning electron microscopy (JSM-5200, Tokyo, Japan) was used to analyze the particle topography was operated at 15Kv acceleration voltage. A concentrated aqueous suspension was spread over a slab and dried under vaccum. The samples was given a gold layer coat of 20 nm thick and images were measured the surface topography. The representative SEM analysis revealed that the nanosponges were uniform in shape with spongy and porous nature (Figure 3).

Figure 3 *Scanning electron micrography of ATZ nanosponges (F3)*

2.3.5 Evaluation of entrapment efficiency

To evaluate entrapment efficiency, accurately weighed quantity of nanosponges (10 mg) from different batches were mixed with 10 ml of phosphate buffer saline (PBS 7.4) and shaken for 5 minutes using vortex mixer. Then suitable dilutions were made with same buffer, filtered, and the concentration was determined by UV spectrophotometer at 301 nm. The entrapment efficiency of the different formulations was shown in Table 2.

Table 2 *Evaluation of ATZ-βCD nanosponges*

Code of formulation	Drug content (%)	Particle size (nm)	Entrapment efficiency (%)
F1	95.87	610.5	60.2
F2	96.41	410.2	70.1
F3	98.82	250.5	86.5
F4	95.11	micron size	54.1
F5	96.25	micron size	52.8

2.3.5 *In vitro* drug release studies and determination of release kinetics

The *in vitro* release study was performed in phosphate buffer saline (pH 7.4) for 1 hour. Among the different batches batch F3 was selected for the *in vitro* dissolution study based on the higher drug entrapment efficiency (86.5%). The release of the drug from the nanosponge's formulation was found to be 94.12% at the end 1 hour. (Figure 3). Various kinetic models such as zero order, first order, Higuchi and Peppa's were employed to understand the order and mechanism of drug release from the nanosponges. The release of drug from the nanosponges was found to follow zero order kinetics obeying non-fickian diffusion as their prominent mechanism (Table 3).

Table 3 *Release kinetics of ATZ-βCD nanosponges*

Code of formulation	Zero order	First order	Higuchi	Peppa's	
	r^2	r^2	r^2	r^2	n
F1	0.9662	0.8461	0.9732	0.9811	0.5611
F2	0.9781	0.8692	0.9722	0.9837	0.5508
F3	0.9848	0.8770	0.9840	0.9876	0.5774
F4	0.9719	0.8505	0.9745	0.9745	0.5622
F5	0.9685	0.8622	0.9673	0.9765	0.5519

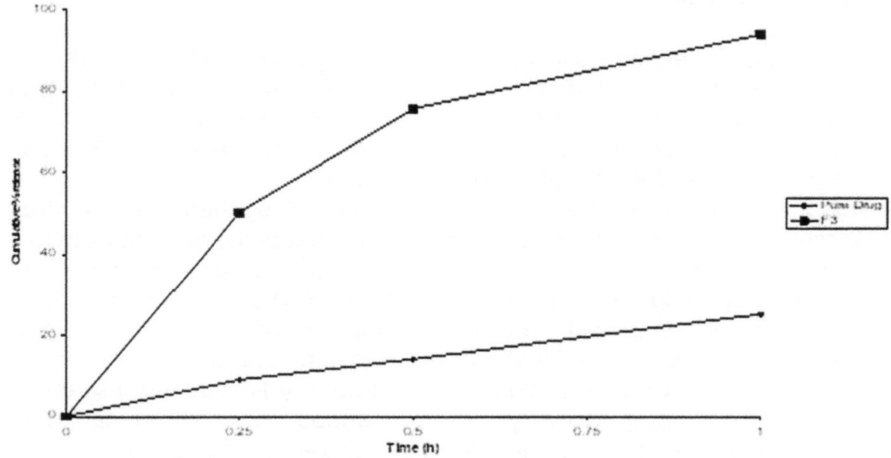

Figure 3 *In vitro dissolution profile of ATZ from pure drug and nanosponges (F3).*

3 CONCLUSION

Cyclodextrins are capable in forming inclusion complexation with various molecules resulting in increased solubility of drugs. Nanosponges are biocompatible cross linked cyclodextrin polymer with the ability to encapsulate diverse molecules. In the present study, atazanavir - β-cyclodextrin based nanosponges were prepared. A remarkable increase in the solubility and dissolution rate was observed. However, *in vivo* studies are required to establish the potential effect of this formulation is necessary.

References

1. G. Neslihan and S. Benita, *Biomed. Pharmacother.*, 2004, **58**, 173.
2. D. Peri, R. Cleary, M. Wyandt and A. Jones, *Drug Dev. Ind. Pharm.*, 2008, **20**(8), 1401.
3. R. Cavalli, F. Trotta and W. Tumiatti, *J. Incluson. Phenom. Macrocyclid. Chem.*, 2006, **56**, 209.
4. J. W. Erickson and S. K. Burt, *Annu. Rev. Pharmacol. Toxicol.*, 1996, **36**(11), 545.

EVALUATION OF CARDIOMYOCYTE TARGETED NOVEL NANO-FORMULATION OF PIRFENIDONE ON DOXORUBICIN INDUCED CONGESTIVE HEART FAILURE IN RATS

M. Azamthulla,[a] Dhrubojyoti Mukherjee,[b] J. Roy[a] and C. Ravishankar[a],

[a] Department of Pharmacology, Faculty of Pharmacy, M.S. Ramaiah University of Applied Sciences, Bangalore, Karnataka, India – 560054; [b] Department of Pharmaceutics, Faculty of Pharmacy, M.S. Ramaiah University of Applied Sciences, Bangalore, Karnataka, India - 560054.

1. INTRODUCTION

Cardiovascular diseases have become a major public health concern and represent one of the most important causes of mortality and morbidity with an escalating prevalence of 12.5% every year since the past decade. It is estimated that 31% of all-cause of mortality is due to cardiovascular diseases (according to 2015 World Health Organization data) [1].
Pirfenidone (PFD) was selected in the present study because the drug possesses multifaceted pharmacological activity, but delivering the drug to a specific target was a challenging approach. Biomedical application of nanotechnology could help deliver the drug to a specific target at a minimum concentration of drug with maximum efficacy and reduced toxicity. Nanotechnology, being the functional system that operates at molecular level, yields the opportunity to address and treat cardiovascular diseases[2]. Over conventional drug delivery system, nanoparticles (NP's) can offer significant advantages in terms of high drug-carrying capacity, high specificity, high stability, versatility to use in different routes of administrations, the ability for controlled release and capacity to deliver both hydrophobic and hydrophilic drugs. Hence, the objective of current investigation was to develop a suitable nanoformulation with pirfenidone (PFDNF) and to evaluate its cardioprotective and cardiotonic efficacy.

2. METHODS AND RESULTS

2.1. Development of PFD Nano-formulation

PFD, Eudragit L100, polyethylene glycol, acetone, distilled water, citric acid buffer were used for the preparation of nanoparticles and characterized by using Atomic Force Microscopy (AFM), Scanning Electron Microscopy (SEM) and X-Ray Diffraction studies (XRD).

AFM studies:

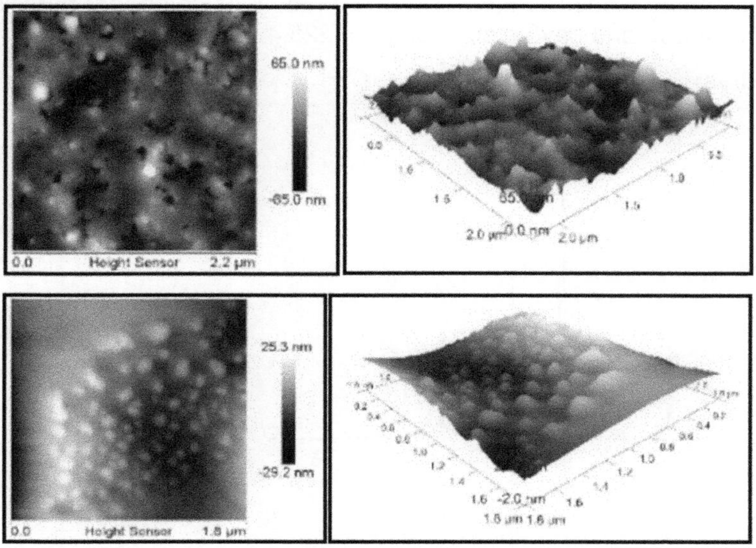

Figure 1 *Top: AFM of PFDNF. Bottom: AFM of Blank Formulation.*

Atomic force microscopy indicated (Figure 1) that the size of the nanoparticles was within 65 nm.

SEM:

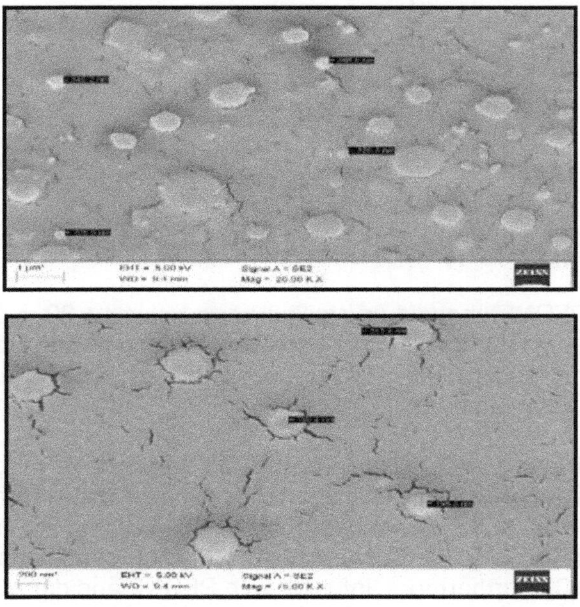

Figure 2 *Top: SEM of PFD. Bottom: SEM of PFDNP.*

The surface morphological characterization of developed nanoformulation was done by scanning electron microscopy. The result indicated that particles are spherical in shape and have a size range of 225.4 – 414.3 nm.

XRD studies:

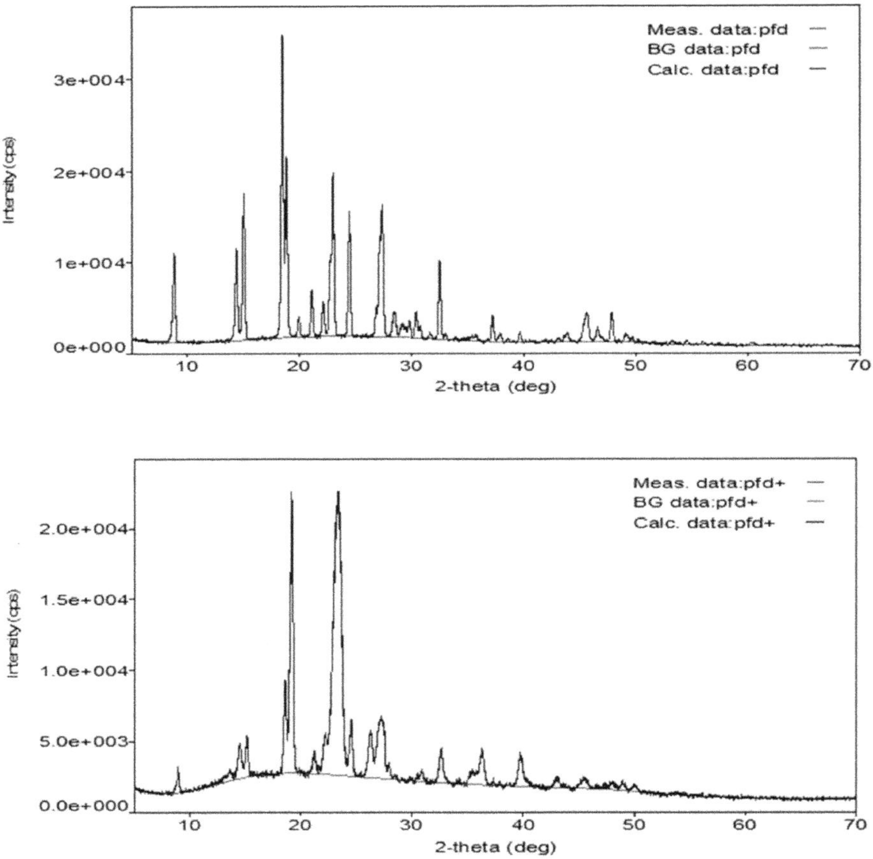

Figure 3 *Top: XRD of Blank. Left: XRD of PFDN*

XRD profile of PFDNF showed crystalline zone between 2θ = 18.59^0, 23.27^0 and 28.06^0. The highest peaks (cps) were compared between blank and drug-loaded nanoparticles and the results exhibited that drug and polymer are compatible with each other.

***In vitro* cardio-tonic activity using modified Langendroff's apparatus:**
Perfusion with Normal Ringer Lock Solution showed normal Heart Rate (HR), Cardiac Output (CO) and Force of Contractions (FOC) and the heart perfused with Hypodermic Ringer Lock Solution showed failing conditions and reduced HR, CO, and FOC. Furthermore, infusion with PFDNF and digoxin of 5, 10 and 20 mcg showed dose dependant / monotonic responses and increase in HR, FOR and CO (Figure 4).

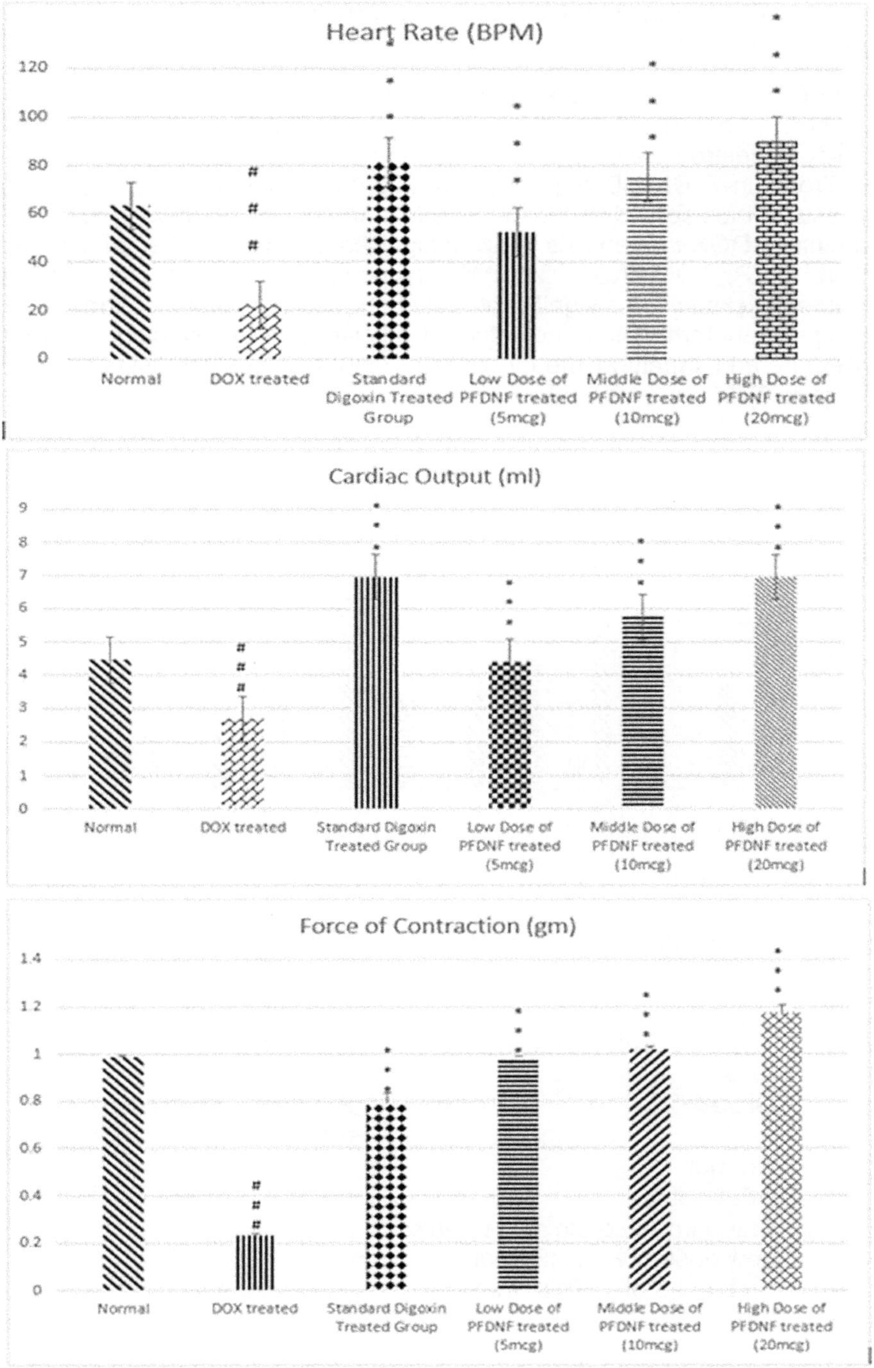

Figure 4 *Effect of PFDNF on HR, FOC, and CO.*

In vivo cardioprotective activity in Doxorubicin (DOX) induced cardiotoxic model[3]:

Effect of PFDNF on Serum enzyme levels

DOX chronic treatment is a well know model for inducing cardiac toxicity due to its ability to destroy the myocardial cell. As a result of their toxicity, ALP, ALT, LDH, Troponin-T, CK-MB were released into the bloodstream by their storage facilities and thus served as the diagnostic markers for determining myocardial tissue injury. DOX treated rats showed significant elevation of such diagnostic markers (ALP, LDH, ALT, CK- MB, Troponin-T). In pre-treated animals, biomarker level showed a significant reduction. This reduction confirms that the developed nanoformulation indicated its efficiency in maintaining the normal architecture and structural integrity of the cardiomyocytes and tissues (Table 1 and Figure 5).

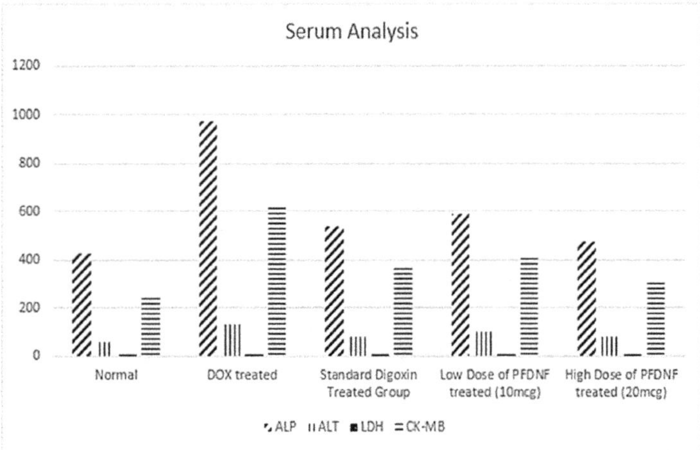

Figure 5 *Cardiac enzyme biomarker level of different experimental groups.*

Table 1 *Analysis of cardiac troponin-T.*

Group	Presence of cardiac troponin-T
Normal	-
DOX treated	+
Standard Ascorbic acid treated	-
Low dose of PFDNF treated	-
High dose of PFDNF treated	-

Effect of PFDNF on ECG parameters:

DOX treated groups showed significant alteration in ECG pattern in comparison to the normal group. The characteristic finding includes elevation of ST-segment, depression of T wave, QRS complex and increased or prolongation of QT and R-R intervals. PFD treated groups showed a significant protective effect against the challenges of DOX cardiotoxicity, where ECG patterns or parameters were restored to normal and fatal complications were successfully eliminated thus protecting the cellular damage. Furthermore, treatment concerning all the groups resulted in significant changes which are portrayed in Figure 6 and 7.

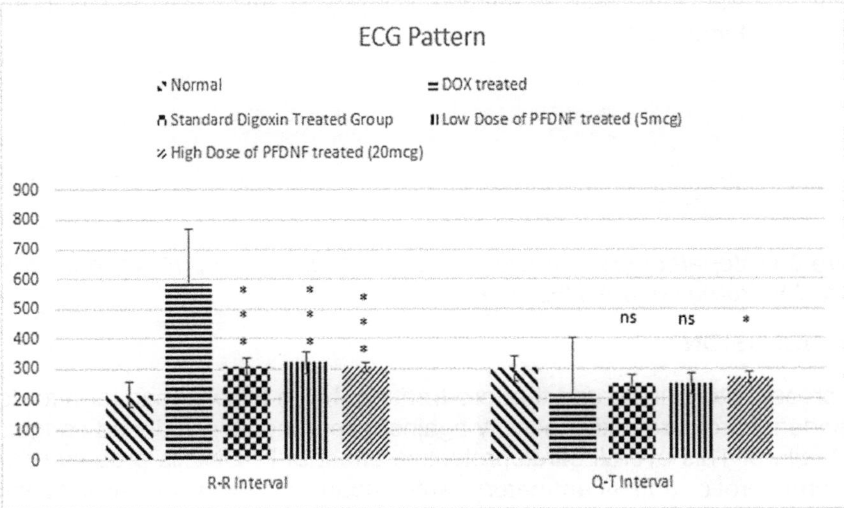

Figure 6 *R-R and Q-T interval changes in different experimental groups.*

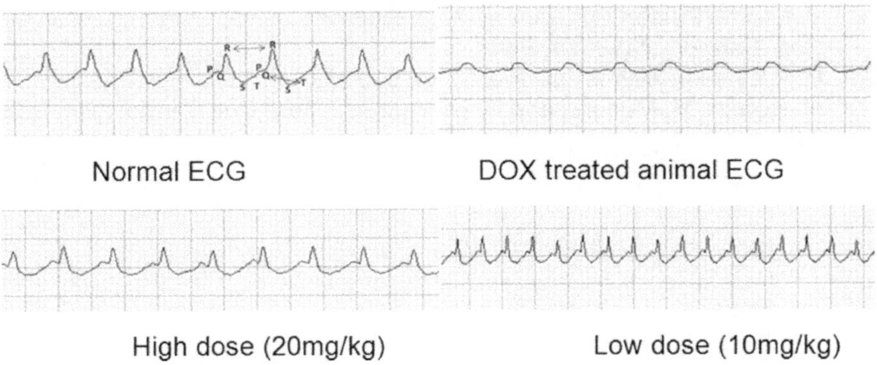

Figure 7 *ECG pattern of different experimental groups.*

Histopathological observations:

Histologically, heart muscle consists of muscle fibres but it differs from both smooth and skeletal muscles. The normal group showed clear integrity of the myocardial membrane without the loss of myofibrils and cellular infiltration of the cardiac muscle and muscle fibres showed intact myocardial cell membrane with striations. The interstitial space also appeared to be intact (Figure 8.1). The positive group showed necrosis of the cardiac muscle, vascular dilation and fibrosis in the focal areas with cellular infiltration (Figure 8.2). PFDNP groups showed improved integrity of myocardial cell membrane and myofibrils and better arrangement of cardiac muscle fibres (Figure 8.3 and 8.4). High dose-treated animal showed significant improvement in the arrangement of cardiac muscle fibres with enhanced integrity of myofibrils and myocardial membrane (Figure 8.4)

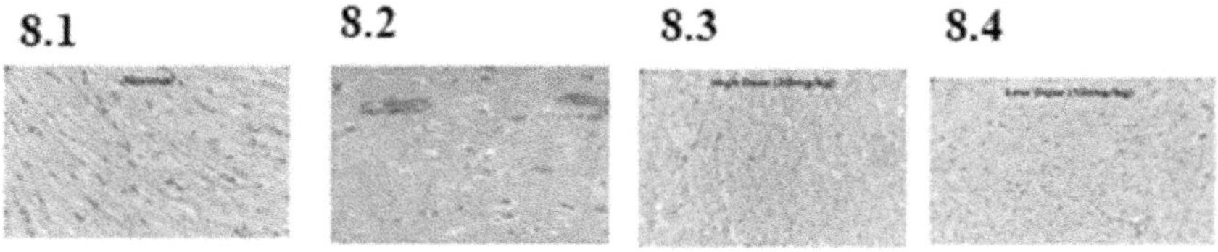

Figure 8 *Histopathological studies of heart* ***8.1*** *Normal,* ***8.2*** *Positive control,* ***8.3*** *PFDNP20mg/kg,* ***8.4.*** *PFDNP10mg/kg.*

3. CONCLUSION

The present study indicated that pre-treatment with PFDNF can lead to dose-dependent cardioprotective activity against experimentally induced congestive heart failure. This overall cardioprotective effect of PFDNP is probably due to their anti-fibrotic, anti-inflammatory, anti-oxidant and free radical scavenging activity.

References

1. S. Kishore, E. Blank, D. J. Heller, A. Patel, A. Peters, M. Price, M. Vidula, V. Fuster, O. Onuma, M. D. Huffman and R. Vedanthan, *J. Am. Coll. Cardiol.*, 2018, **71**, 564.
2. F. A. Shiekh, *Nanomedicine*, 2014, **9**, 2259.
3. M. Robert, M. Mocanu and D. M. Yellon, *J. Mol. Cell Cardiol.*, 2011, **50**, 940.

FABRICATION OF DONEPEZIL ENCUMBERED MANNOSE ASSISTED POLYCYANOACRYLATE NANOCARRIERS

Mayank Handa,[a] Rewati Raman Ujjwal[a] and Rahul Shukla[a]

[a] Department of Pharmaceutics, National Institute of Pharmaceutical Education and Research-Raebareli, India

1 INTRODUCTION

Blood-brain barrier (BBB) is the preventive membranous sheath which act as barrier between blood and brain. Barrier observed at junction between brain and blood act as sealant along with tight junctions which inhibits any engulfing activity and restricting the molecular exchange of any cellular type transport. Passive transport is limited to drug with unionized, lipophilic and low molecular weight properties which enables the free diffusion or exchange of molecules from blood to brain. Carrier mediated transport system is best suited for brain transport across BBB [1,2]. Amino acids, sugar based moieties, neuropeptides, proteins use carrier mediated transport system for permeation into brain. Disadvantage with BBB permeation is that most of therapeutic potential of moiety diminishes or the concentration delivered beyond BBB is below effective levels, thus compromising the activity and act as main cause in decreased activity to brain related disorders.

Nanocarriers based on polymeric nanoparticles is the promising approach for brain delivery. These formulation approaches are employed for enhancing the therapeutic efficacy and reduction in dose of various drugs. Alone polymeric nanoparticles suffer from rapid clearance from blood due to rapid uptake from mononuclear phagocyte system (MPS) which is main cause for delivery to the targeted organ [3]. For overcoming MPS uptake, polymeric nanoparticles are coated with hydrophilic moieties which will not only prevent the aggregation of polymeric nanoparticles among themselves but also inhibit MPS uptake. Most commonly used hydrophilic moieties are polysorbate 80, poloxamers, PEGylation. They act as stealth and their hydrophilic ends prevent macrophageal uptake. Hydrophilic coatings prevent the opsonisation process and reduces macrophage identification. Targeting of polymeric nanoparticles can be done by coating or fabricating with mannose which act on the receptors present on the dendritic cells. Mannose selection is due to high rate of binding with receptors present in brain comparison to fucose and galactose [4]. Further glial cells have more potential for binding with mannose in comparison to neuronal cells, which is an appropriate approach for targeting in Alzheimer's disease. Cyanoacrylates have property of adhesiveness and being a hydrophilic polymer it also minimizes the crisis for macrophageal uptake. Cyanoacrylate entraps the drug inside its core and deliver the drug to targeted site [5].

Donepezil which acts as reversible and non-competitive acetylcholinesterase (AChE) inhibitor. Donepezil is FDA approved moiety for management of

Alzheimer's disease. Donepezil is first choice of drug for both severe and mild type of Alzheimer's. It has high selectivity for tissues and targeted organs. In marketable dosage form it is available as solid oral tablet in dose of 5 and 10 mg, while sustained release as 23 mg. But issue with Donepezil is that it has high first pass metabolism and it is the reason for its low brain bioavailability. Another reason for low brain bioavailability is low permeation into brain from circulating blood. Approximately 25% of the administered dose reaches the cerebrospinal fluid after 24 h and 1/4th of the dose encountered with hepatic first pass metabolism [6–8].

Formulation scientist are exploring the alternative approaches for targeting the brain for overcoming the issues related to brain drug delivery. Traditional approaches include intracerebral injection directly into brain or another method is disruption of BBB by using hypertonic solutions. These approaches are invasive techniques which involves painful surgery and risking life of patient. Another approach is chemical modification of active moieties by using conjugation techniques for enhancing the lipophilicity but this concept is not appropriate because of increase in the molecular weight of newly conjugated moiety. Various workers across the globe have fabricated liposomes, transferrin coated moieties for brain delivery with an aim to enhance the bioavailability but they end up in unconvinced manner.

The present communication aims to formulate stable donepezil loaded polymeric nanoparticles by entrapping them in poly cyanoacrylate and followed by coating with mannose. Primarily the synthesis of poly cyanoacrylate and formulating donepezil loaded synthesised polymeric nanoparticles and encumbering of mannose over formulated polymeric nanoparticles and study its release behaviour followed by various characterization techniques.

2 METHODS AND RESULTS

2.1. Method for synthesis of Poly (ethylcyanoacrylate) (PECA)

Synthesis of Poly (cyanoacrylate) was done by using Ethyl-2-cyanoacrylate (2g, 0.04 mol) in 20 mL of triple distilled water in 250 mL flask, followed by addition of 2 mL of potassium bromate solution (0.025 g, 0.12 mmol), with further addition of sodium hydrogen sulphite solution (0.06 g, 0.6 mmol). The mixture was stirred continuously for 15 minutes. Now this thoroughly mixed mixture was poured into concentrated sodium chloride solution and product was obtained as precipitant. Followed by washing with water for removal of unreacted monomer and salts.

2.2. Method for formulation of Donepezil loaded poly cyanoacrylate polymeric nanoparticles

Solvent evaporation method was employed for preparation of Donepezil loaded Poly cyanoacrylate polymeric nanoparticles. Fixed amount of Donepezil (10 mg) was added to varying amount of polymer as described in Table 1. Both drug and polymer were solubilised in mixture of organic solvent i.e. acetone:

methanol (8:2). Now this organic phase was added dropwise with the help of syringe-needle to 10 mL of aqueous solution containing varying concentration of Poloxamer 407 kept under continuous stirring of high speed homogenizer (HSH). After complete addition under HSH solution was kept under probe sonication. After completion of probe sonication, solution was kept over magnetic stirrer for overnight to evaporate the organic phase [9].

Table 1 *Various batches with variation in processing parameters and polymer and stabilizer concentration.*

Batch Code	R1[a]	R2[b]	R3[c]	Particle size (in nm) ± S.D.	PDI ± S.D.
PCA 1	100	50	10	158.75±7.57	0.179±0.032
PCA 2	75	50	10	182.48±9.68	0.222±0.034
PCA 3	50	50	10	267.56±10.01	0.444±0.029
PCA 4	25	50	10	498.53±17.38	0.680±0.040
PCA 5	10	50	10	757.31±26.90	0.954±0.034
PCA 6	100	25	7.5	224.70±16.37	0.374±0.056
PCA 7	75	25	7.5	386.50±29.53	0.556±0.043
PCA 8	50	25	7.5	718.01±19.38	0.746±0.038
PCA 9	100	75	5	356.79±18.91	0.172±0.025
PCA 10	75	75	5	560.27±20.40	0.381±0.057
PCA 11	50	75	5	640.40±27.27	0.646±0.030

[a]Amount of Polymer incorporated (in mg) - R1, [b]Amount of stabilizer (in mg) - R2, [c]Probe sonication time (in min) - R3

Particle size was measured on Zeta sizer by taking small aliquots from each batch and filtration with 0.45μm syringe filter. Batch with particle size less than 200 nm and PDI below 0.35 was preferred for further coating with mannose. PCA 2 batch was preferred for coating with mannose and various concentration (0.5%, 1.0%, 1.5%, 2.0%, 3.0% *w/v*) was added to solution. Solution was kept for overnight on rotatory shaker and characterised for zeta potential, particle size and PDI.

Table 2 *Batch with variation in concentration of mannose.*

Batch Code	Amount of Mannose (in mg)	Particle size (in nm) ± S.D.	PDI ± S.D.	Zeta potential (in mV) ± S.D.
PMA 1	5	352.29±24.93	0.352±0.025	-23.02±1.64
PMA 2	10	430.0±26.40	0.445±0.053	-44.28±3.18
PMA 3	15	534.0±21.02	0.573±0.05	-
PMA 4	20	847.32±19.53	0.830±0.048	-
PMA 5	30	940.66±43.09	0.832±0.16	-

2.3. Results

Ethylcyanoacrylate (monomer) was polymerized by using free radical polymerization. Peak for carbonyl group (C=O) at 1733 cm^{-1} suggests the possible formation of polymer. Formulation trials were carried out. Particle size evaluation on Zeta Sizer predicted that batches PCA 1 and PCA 2 were showing desirable results. Particle size of PCA 1 predicted to be 158.75±7.57 nm and polydispersity of 0.179±0.03, while PCA 2 predicted 182.48±9.68 nm and polydispersity of 0.222±0.03 respectively. Particle size of these batches were below 200 nm, which favours to inhibit macrophageal uptake and prevent the opsonisation. PCA 2 was selected for further coating with mannose moiety. Zeta potential was observed for PCA 2 batch and predicted to be -13.26±1.60 mV. After overnight shaking with mannose, particle size and zeta potential vary with concentration of mannose. Zeta potential of optimized batch PMA 2 was -44.28±3.18 mV, which confirmed that effective coating take place. Drug loading and entrapment efficiency calculated for PCA 2 was 7.94±0.74% and 69.94±4.31% respectively. Characterization studies includes FTIR analysis, DSC studies, microscopic studies and *In-vitro* release studies. FTIR studies were performed for lyophilized mannose coated polymeric nanoparticles which predicted the characteristic peaks of mannose and Poloxamer 407. DSC studies also predicted the endothermic peaks of mannose and Poloxamer 407 in lyophilized nanoparticles. SEM image predicted spherical shaped nanoparticles. *In-vitro* release studies were performed in phosphate buffer pH 6.8 for 36 h. Free drug was rapidly released in comparison to coated nanoparticles and uncoated nanoparticles. Uncoated nanoparticles showed sustained release for 36 h and in first 12 h, 51.93±1.85% of drug was released. In case of mannose coated nanoparticles in first 12 h 42.94±2.26% of drug was released. Sustained release of drug was might due to polymeric property which follow matrix or diffusion release model. **Figure 1** provides the release profile of Donepezil free, uncoated Donepezil and coated Donepezil nanoparticles.

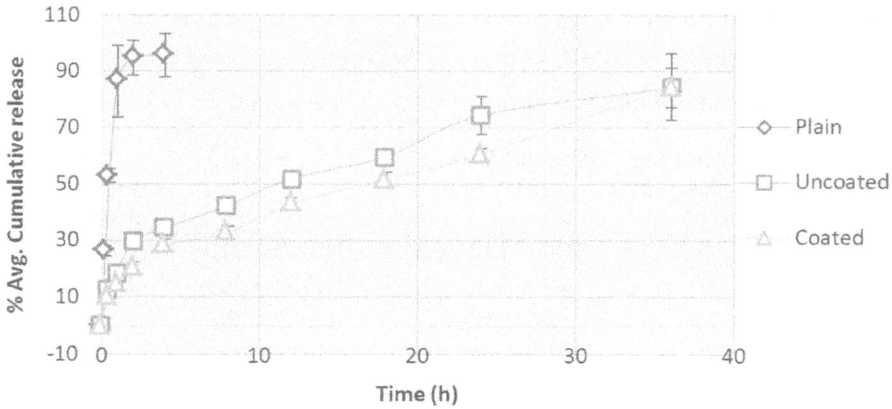

Figure 1 *The release profile of Donepezil free, uncoated Donepezil and coated Donepezil nanoparticles.*

3. CONCLUSION

The present study shows the Donepezil-PECA nanocarriers have been coated with mannose and evaluated for its *In-vitro* release studies in phosphate buffer pH 6.8. Coating with mannose will provide a novel approach for targeting the moieties directly into brain via GLUT transporters (at periphery of BBB) and lectin receptors (in glial cells). The cyano moiety in PECA provides sustained mechanism of drug release due to its bio-adhesive nature of polymer and decreases the dose of moiety thus preventing side effects from excess dose. Futuristic studies involve *In-vivo* evaluation by measuring organ/tissue biodistribution. Targeted approach might be successful for management of Alzheimer's disease where people from across the globe are affected with this disorder. This study will represent novel carrier for brain drug delivery in near future.

References

1 W. M. Pardridge, *J. Cereb. Blood Flow Metab.*, 2012, **32**(11), 1959–1972.
2 Y. Chen and L. Liu, *Adv. Drug Delivery Rev.*, 2012, **64**(7), 640–665.
3 R. Shukla, M. Handa, S. B. Lokesh, M. Ruwali, K. Kohli and P. Kesharwani, in *Polymeric Nanoparticles as a Promising Tool for Anti-cancer Therapeutics*, 2019.
4 F. Umezawa and Y. Eto, *Biochem. Biophys. Res. Commun.*, 1988, **153**(3), 1038–1044.
5 P. Calvo, B. Gouritin, H. Chacun, D. Desmaile, J. D'Angelo, J. P. Noel, D. Georgin, E. Fattal, J. P. Andreux and P. Couvreur, *Pharm. Res.* 2001, **18**(8), 1157–1166.
6 M. Shigeta and A. Homma, *CNS Drug Reviews*, 2001, **7**, 353-368.
7 A. Singh, K. Vaishagya, R. K. Verma and R. Shukla, *AAPS PharmSciTech*, 2019, **20**(5), 213.

FABRICATION, CHARACTERIZATION AND *IN VITRO* EVALUATION OF MONOISOAMYL DIMERCAPTOSUCCINIC ACID (MiADMSA) ENCAPSULATED CHITOSAN MICROPARTICLES ON NEURONAL CELL LINES

Vishwas P. Pardhi,[a] Kshirod B. Sathua[b] and S. J. S. Flora[a,b*]

[a]Department of Pharmaceutics, National Institute of Pharmaceutical Education and Research (NIPER), Raebareli, Transit Campus Lucknow. UP, India; [b]Department of Pharmacology, National Institute of Pharmaceutical Education and Research (NIPER), Raebareli, Transit Campus Lucknow. UP, India

* Corresponding author

1. INTRODUCTION

Chitosan is a well-known linear polymer with antioxidant activity and is being used extensively in drug delivery systems. Its use has been widely explored due to its biocompatibility, biodegradability, mucoadhesive property and regulatory acceptance(1)(2). Although chitosan microparticulate systems have been extensively studied to deliver molecules with common diseases and disorders, very few reports have been published to explore its applicability towards metal toxicities such as arsenic, copper and lead poisoning. For such certainties' chelation therapy (use of chelating agents in metal poisoning conditions) is a foremost preference and considered as a standard treatment(3). Such chelating agents can be delivered via formulating them into microparticulate system made of chitosan with a notion to use the antioxidant property of chitosan to reduce the oxidative load in metal poisoning(1)(4)(5). Additionally, chitosan had been reported to exhibit the chelation properties which could add extra advantages in metal poisoning therapy (6). MiADMSA(under preclinical investigations)is an lipophilicMonoisoamyl derivative of Dimercaptosuccinic acid (DMSA) having promising results in the treatment of metal poisoning (especially Arsenic) and has been reported to possess better pharmacokinetic properties than parent drug(7). MiADMSA forms chelatewith metals through the thiols groups present in its structure (Figure 1.)(8). Regardless of better pharmacokinetic performance, MiADMSA is associated with certain dose dependent cytotoxicity due to its lipophilic nature (3). This cytotoxicity demonstrated by MiADMSA is thought to be overcome by concomitant administration of antioxidants and controlling the release profile. Therefore, formulating MiADMSA with the excipient possessing antioxidant property might be beneficial in reducing the side effects and may act synergistically with chelation therapy in heavy metal poisoning.In the proposed work, we have fabricated chitosan microparticles encapsulating MiADMSA, characterize it and evaluate the optimized formulation for drug loading capacity, *in vitro* drug release profile and cell line studies for investigating the cellular viability.

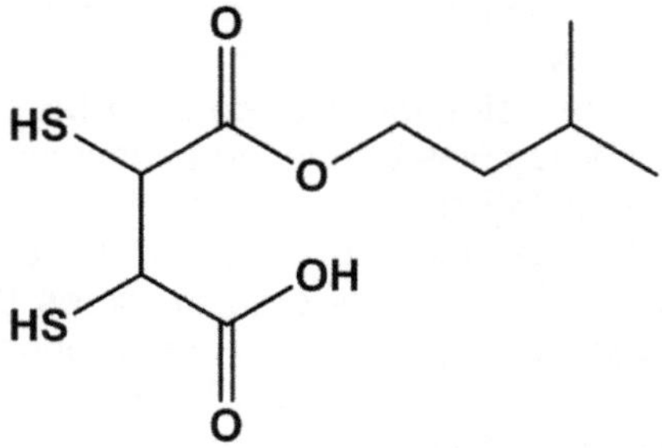

Figure 1 *Chemical structure of MiADMSA.*

2. METHODS AND RESULTS

2.1. Methods

MiADMSA loaded chitosan microparticles (MCMP) were prepared by employing ion gelation method using tween 80 as surfactant and sodium tripolyphosphate (TPP) as crosslinking agent. Method include, dissolution of 0.5 % w/v chitosan (low molecular weight) in 1 % v/v acetic acid solution followed by addition of surfactant (2 % v/v). To this, methanolic solution of MiADMSA (20 mg/ml) was added along with homogenization using high shear homogenizer followed by dropwise addition of crosslinking agent. The formulation was kept on stirring for overnight to evaporate the organic solvent which might be present in trace amount. The prepared microparticles were subjected to characterization for sizeand zeta potential using Laser diffraction analysis (LD) and Photon Correlation Spectroscopy (PCS), respectively. The drug-excipient interaction and thermal behaviour of prepared microparticleswere investigated usingfourier transform infrared spectroscopy (FTIR) and differential scanning calorimetry (DSC), respectively. The *invitro* drug release study was performed in phosphate buffer (pH 7.4) to evaluate the release profile of MCMP and compared with that of plain MiADMSA (MiADMSA drug powder used as such without any further processing). Finally, efficacy of the optimized batch of MCMP was evaluated for toxicological properties by exposing it to neuronal cell lines to assess the cell viability by MTT assay and compared with cell viability of plain MiADMSA exposed cells. The % viability of control cells, toxicant exposed cells (Arsenic (10μM)), plain MiADMSA and MiADMSA loaded microparticle treated cells was determined by MTT assay. Additionally, the cellular recovery in MiADMSA treated cells was investigated by microscopic observations.

2.2. Result and discussion

The prepared chitosan microparticles were analysed for size and zeta potential as shown in figure 1using laser diffraction and PCS. The particle size distribution was found to have d (0.9) value of 3.00 µm with uniformity of 0.30 and span of 0.93. The zeta potential was found to be positive with potential distribution value of 8.29 which was attributing to cationic surface on chitosan. The stability of prepared MCMP for particle size distribution was investigated at room temperature (RT), freeze temperature (2-8 °C) and 45 °C temperature for the period of 1 month. The stability data indicates that prepared MCMP was stable, as no aggregation of particles has been observed that would lead to increase in particle size. FTIR and DSC analysis were carried out to evaluate the molecular interaction and thermal behaviour of optimized chitosan microparticle, respectively. FTIR data reveal the formation of hydrogen bonds which is indicated by broad peak at the region of 3253.90 cm^{-1} along with the reduction in intensities of other characteristic peaks of MiADMSA. The DSC thermogram shows characteristic melting endothermic peak of MiADMSA at 158.87 °C followed by initiation of degradation. The degradation of MiADMSA after melting was indicated by downward shift of thermogram immediately after the melting endotherm.The chitosan had shown broad endothermic peak at 115.37 °C (midpoint) which has been started from 40 °Cand continues upto 180 °C. This is corresponding to typical polysaccharide behaviour of chitosan representing the dehydration associated with the first degradation stage. The MiADMSA loaded microparticles (MCMP) demonstrate the shift in endotherm of chitosan towards higher melting point which was found at 143.33 °C. However, no melting endotherm for MiADMSA was found which represent encapsulation of MiADMSA inside the chitosan matrix. The additional peak at 165.77 °C representsmelting endotherm of mannitol which was incorporated in formulation during lyophilization at the concentration of 1 % w/v.

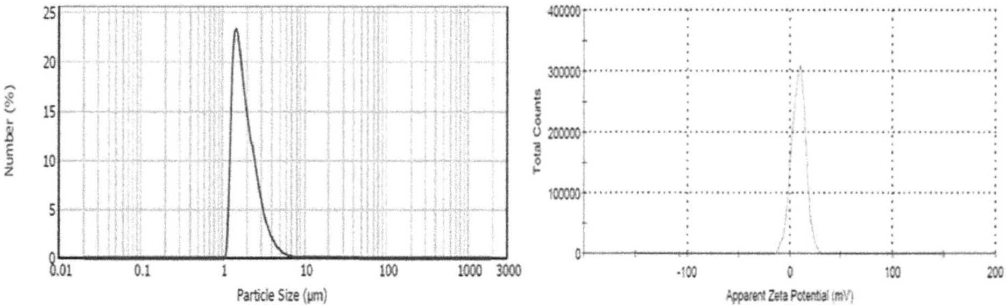

Figure 2 *Size and zeta potential of prepared microparticles.*

The MiADMSA loading capacity in chitosan microparticle was found to be 23.59 %. The drug dissolution study has been performed to assess the release pattern of MiADMSA from MCMP and compared with MiADMSA plain. Controlled release of MiADMSA was demonstrated by MCMP as compared to MiADMSA plain which had shown only 55.13 % of total drug release within 480 minutes, whereas, MiADMSA plain drug had been completely dissolved at this time point with the total drug release of 99.68 %. This controlled release behaviour of MCMP attributes to the entrapment of drug within the polymeric matrix where the dissolution medium needs to penetrate into the matrix and diffuse out the drug. The release of the drug from such polymeric matrix usually controlled by means of three mechanisms, i.e., erosion, swelling and diffusion(9).

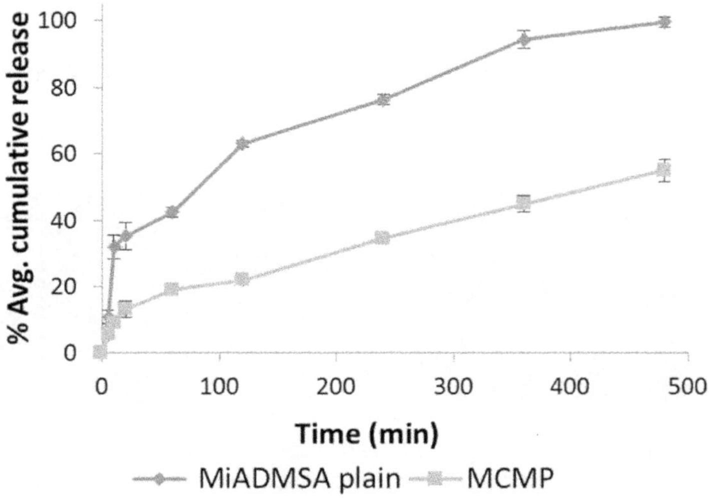

Figure 3 *Comparative release profile of MiADMSA plain drug and MCMP.*

In vitro efficacy of MiADMSA encapsulated microparticles was evaluated in neuronal cell lines (Neuro-2A). Cell viability was evaluated by performing MTT assay (Figure 4.). The cell viability in MiADMSA treatment groups was assessed against approximately 50 % cell viability in arsenic exposed cells. The MTT results showed average % cell viability of 86.04 % in MCMP group which was found to be significantly higher than MiADMSA plain drug (75.15 % cell viability). Morphological investigation using inverted microscopy reveals that arsenic exposure leads to cell death possibly by increasing oxidative stress as compared with normal cells. However, treatment with MiADMSA plain drug (10 μM) and MCMP were shown its recovery with respect to arsenic exposed cells. The recovery was found to be significantly better with MCMP when compared with plain MiADMSA. The improved efficacy of microparticles may be due to alteration in release profile of MiADMSA as well as antioxidant effect of chitosan. This suggests that chitosan microparticles loaded with MiADMSA have beneficial effect fortreatment for arsenic poisoning.

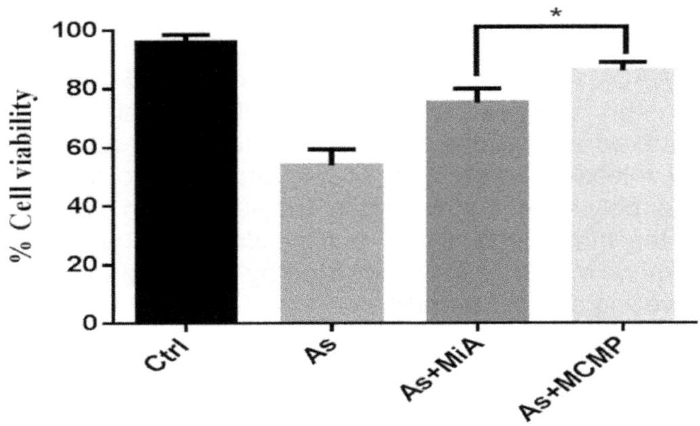

Figure 4 *MTT assay results representing % viability of Neuro-2A cell line (* represents P ≤ 0.05).*

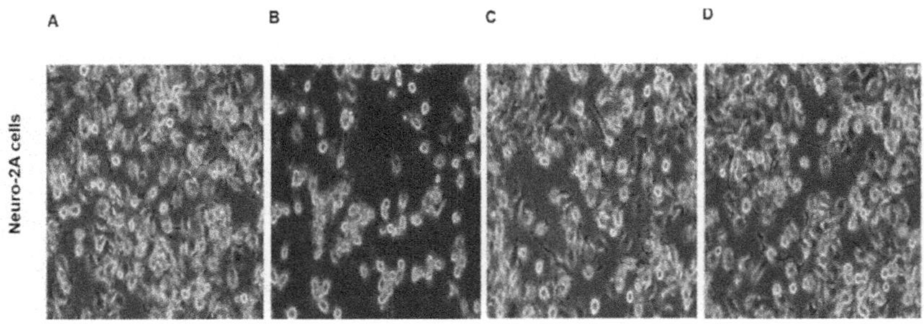

Figure 5 *Morphological changes in Neuro-2A cell line depicted as A) control cells without any exposure, B) Arsenic exposed cells (10 μM), C) plain MiADMSA treated cells, D) MCMP treated cells (Inverted microscope, Magnification 20X).*

3. CONCLUSION

Stable MiADMSA loaded microparticles were successfully prepared with average particle size, d (0.9) of 3.0 μ and characterized using FTIR and DSC. The loading capacity of prepared MCMP was found to be 23.59 % and release data showedcontrolled release of drug as compared toMiADMSA plain drug. The *in vitro* cell line study reveals improvement of viability. This finding suggests that chitosan microparticles loaded with MiADMSA could be more effective treatment for arsenic poisoning as compared to MiADMSA plain drug. Further investigation with quantitative determination oxidative stress parameters is the future perspective of study.

References

1. M. Rinaudo, Chitin and chitosan: Properties and applications, *Progress in Polymer Science,* Oxford, 2006.
2. S. P. Chawla, S. R. Kanatt and A. K. Sharma, Chitosan, in *Polysaccharides: Bioactivity and Biotechnology*, 2015.
3. A. Yadav, R. Mathur, M. Samim, V. Lomash, P. Kushwaha and U. Pathak, *et al.*, Nanoencapsulation of DMSA monoester for better therapeutic efficacy of the chelating agent against arsenic toxicity, *Nanomedicine*, 2014, **9**(4), DOI: 10.2217/nnm.13.17.
4. M. T. Yen, J. H. Yang and J. L Mau, Antioxidant properties of chitosan from crab shells, *Carbohydr. Polym.*, 2008, **74**(4), 840–844.
5. D. H. Ngo and S. K. Kim, Antioxidant effects of chitin, chitosan, and their derivatives, in *Advances in Food and Nutrition Research*, 2014.
6. P. Da Silva Mira, L. Souza-Flamini, D. Da Costa Guedes and A. Da Cruz-Filho, Evaluation of the chelating effect of chitosan solubilized in different acids, *J. Conserv. Dent.*, 2017, **20**(5), 297–301.
7. S. J. S. Flora, Arsenicals, in *Handbook of Toxicology of Chemical Warfare Agents*, 2015.
8. M. Ram Kumar, S. J. S. Flora and G. R. Reddy, Monoisoamyl 2,3-dimercaptosuccinic acid attenuates arsenic induced toxicity: Behavioral and neurochemical approach, *Environ. Toxicol. Pharmacol.*, 2013, **36**(1), 231–242.
9. Y. Fu and W. J. Kao, Drug release kinetics and transport mechanisms of non-degradable and degradable polymeric delivery systems, *Expert Opin. Drug Delivery*, 2010, 429–444.

Part 6: Proteomics Analysis

EXPLORATIVE PROTEOMIC CHARACTERIZATION OF PLATELET RICH FIBRIN AND PLATELET RICH FIBRIN MATRIX AND THEIR POTENTIAL CONTRIBUTION IN REPAIR AND REGENERATION – A PILOT STUDY

S.Nagaraja[a], S.Mathew [a], H. R.Sowmya [b] and S.Chaki [c]

[a] Faculty of Dental Sciences, MS Ramaiah University of Applied Sciences, Bangalore, India; [b] Research Scholar, Department of Biochemistry, Kuvempu University, Mysore, India; [c] Faculty of Life and Allied Health Sciences, MS Ramaiah University of Applied Sciences, Bangalore, India

Corresponding Author Email: nagaraja.shruthi@gmail.com

1 INTRODUCTION

Dental caries and pulpal diseases are serious global oral health care burden. Regenerative procedures are receiving increasing research attention in an attempt to restore lost tissue structure and function contributing to improved patient care. Many proteins, cytokines and growth factors are involved in tissue regeneration and act in orchestrated steps to restore tissue function[1]. Platelet concentrates such as Platelet Rich Fibrin (PRF) and Platelet Rich Fibrin Matrix (PRFM) are blood derivatives rich in growth factors, cytokines, and signaling molecules[2] and have evolved as a pioneering regenerative bioactive materials. PRF is rich in platelets and leucocytes with loose fibrin matrices, whereas PRFM a variant of PRF family is devoid of leucocytes. Though the main constituent of these concentrates is the fibrin scaffold, little is known about the proteins interacting in these and their role in the healing process. Identification of proteins in PRF and PRFM could deepen understanding of underlying molecular mechanisms in tissue regeneration, optimize formulation and dosage in regenerative therapies. It could enable us to further explore and unravel better therapeutic strategies to facilitate repair and regeneration, and improvise scope of translational research. Hence this study aimed to characterize the proteins enmeshed in PRF and PRFM scaffolds. (Figure 1).

2 METHODS AND RESULTS

2.1 Donors

The study was performed following principles of the declaration of Helsinki. Healthy consenting single volunteer donated blood for isolation of both PRF & PRFM.

2.2 Scaffold fabrication and sample preparation

PRF and PRFM samples were processed as per the protocol prescribed by Choukroun et al [2]and Lucarelli et al [3]. Fibrin clots were transferred into a tube without separating the liquid part. The sample was centrifuged at 10,000 rpm for 30 min at 4oC to separate fibrin proteins and to extract Fibrin clots release. The resulting supernatant with Fibrin clots release was transferred into a new tube and used for further analysis. The separated protein was quantified using the Bradford Assay. The proteins were further loaded on 12% SDS-PAGE and then taken for LCMS study as shown in the workflow (Figure 2). One sample of each group was processed independently and triplicated.

2.3 1D SDS-PAGE analysis and Tryptic digestion

A known quantity of protein extract (100µg) was used for digestion with Trypsin. The sample was treated with 100mM DTT at 95oC for 1 hr followed by 250mM IDA at room temperature in dark for 45 min. The sample was then digested with Trypsin by incubating overnight at 37°C. The peptides were extracted in 0.1% formic acid and incubated at 37°C for 45 minutes. The solution was centrifuged at 10,000 rpm and the supernatant was collected into a separate tube. The resulting sample was vacuum dried and dissolved in 20µl of 0.1% formic acid in water. The 10µL volume of sample was injected on to a C18 UPLC column for separation of peptides. The peptides separated on the column were directed to Waters Synapt G2 Q-TOF instrument for MS and MSMS analysis.

2.4 LC-MS/MS analysis:

The raw data acquired from the instrument was processed using PLGS software 3.0.2 within which data processing and database search were performed. The source of the sample being human protein PRF and PRFM, the sequence in FASTA format was downloaded by checking the homology against the proteins available in the Swiss-prot database and were used for identifying the peptides present in the samples.

2.5 Gene Ontology (GO) study

The GO of the identified proteins was deduced by comparing against the IPI Human database to determine the molecular function of the identified proteins[1]. GO analysis was performed using Panther software version 14.1.

2.6 Results

2.6.1 Identification of differential expressed proteins

A known quantity of proteins from both PRF and PRFM were loaded on to an SDS-PAGE gel (Figure 3) to compare the protein content in each of the samples. Further, the samples were digested using the enzyme 'trypsin' for identification of proteins by LC-MS/MS analysis (Figure 4). This led to the identification of 3,506 total proteins (TP) in both groups (Figure. 5a). Among these, 435 proteins were exclusively fibrin proteins (FP) (Figure 5b). Of all the identified proteins belonging to the fibrin group, FP-PRF had 269 proteins while FP-PRFM had 166 proteins. Among them, 83 proteins were present exclusively in FP-PRF, while only 36 proteins were present in FP-PRFM (Figure 5c).

2.6.2 Gene ontology study

Identified proteins were clustered according to GO hierarchy. The proteins present exclusively in the PRF and PRFM group along with the overlapping proteins which belonged to both the groups were analyzed for their biological function. Among all the PRF proteins identified, 52% were those involved in cellular processes such as cell metabolism (38.5%), cellular response to a stimulus (35.9%), signal transduction (33.3%), cellular component organization (25.6%) and cell communication (10 %). 48.4% of the identified PRFM fibrin proteins were associated with the cellular process such cell metabolism (46.7%), cellular response to a stimulus (46.7%), signal transduction (46.7%) and cellular component organization (33.3%). However, proteins involved in cell communication were not observed in PRFM group.

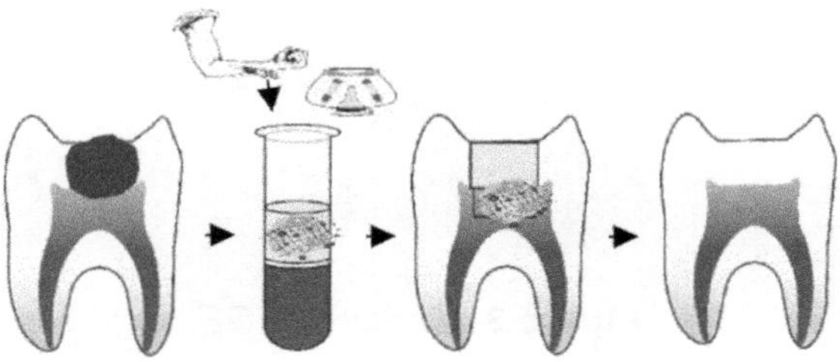

Figure 1 *Dental pulp Regenerative Therapy.*

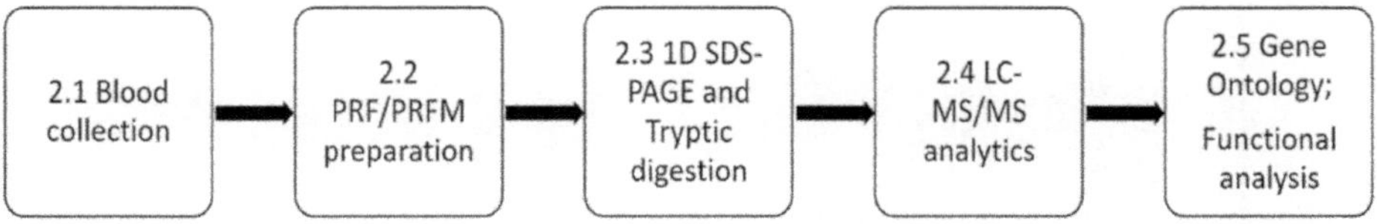

Figure 2 *Proteomics workflow.*

Table 1 *List of a few identified proteins.*

Actin
A1 Antitrypsin
A1 anti chymotrypsin
C-C motif chemokine 7 & 23
Rho guanine nucleotide exchange factor 2
TGF- β
BMP7
VEGF
Fibronectin type III
Vascular cell adhesion protein

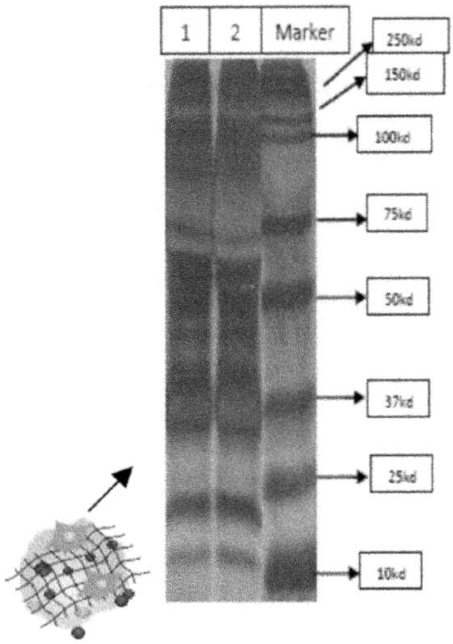

Figure 3 *1D SDS-PAGE.*

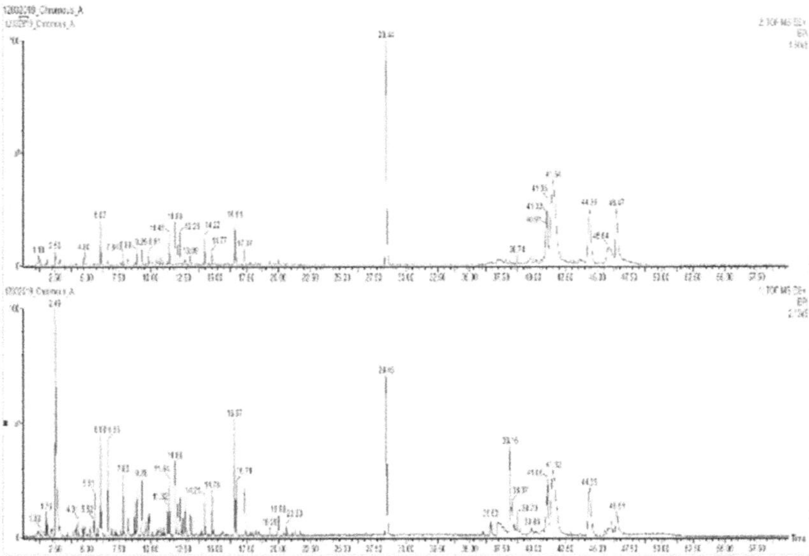

Figure 4 *LC-MS/MS Chromatogram electrophoresis.*

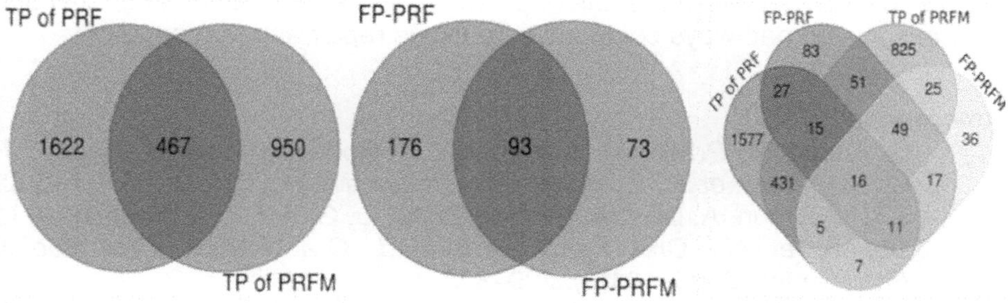

Figure 5 *Distinct and overlap proteins; Left: Total proteins, Centre: Fibrin proteins, Right: cumulative proteins. (TP-PRF: Total Human proteins in PRF, TP-PRFM: Total Human proteins in PRFM, FP-PRF: Fibrin proteins in PRF, FP-PRFM: Fibrin proteins in PRFM.)*

3. DISCUSSION

This pilot study led to the identification of several proteins associated with tissue regeneration. Major proteins identified are listed in Table 1. The therapeutic action of these proteins has been extensively demonstrated in an experimental wound model[4]. Several chemokines such as C-C motif chemokine 7, chemokine 23, a macrophage inflammatory proteins and growth factor-like TGF β were also identified. Similar results were reported in the study of PRF proteomics wherein they studied only 35 proteins [4]. Major proteins identified in the PRFM group include a) Rho guanine nucleotide exchange factor 2, which play a significant role in cell adhesion, b) BMP7, which has a role in osteo/odontogenesis, c) actin and actin-related protein, which known to anchor cells for proliferation and migration, and d) Tyrosine Protein kinase which is known for its role in cell differentiation. Some of the identified proteins such as Bone Morphogen Proteins (BMPs) are currently being used as therapeutic agents in clinical studies whereas few others are being explored for their potential clinical application[5].

This study has provided a catalog of scaffold proteins, which could have a potential role in repair and regeneration. Analysis of identified proteins are associated with cell adhesion, cellular proliferation, angiogenesis, migration, etc. Further characterization of proteins involved in these processes can shed light into the mechanism by which PRF & PRFM can facilitate regeneration.

4. CONCLUSION

The PRF & PRFM proteins harbor a rich and diverse group of proteins, which plays a significant role in repair and regeneration. The present information will help plan further studies on the molecular mechanism of proteins present in the scaffold to facilitate tissue repair and regeneration.

Future scope: To analyze the role of differentially expressed unique proteins in their associated pathways contributing to tissue repair and regeneration.

References

1. E. Anitua, R. Prado, M. Azkargorta, E. Rodriguez-Suárez, I. Iloro, J. Casado-Vela, *et al.*, *J. Tissue Eng. Regen. Med.*, 2015, **9**(11), E1–E12.
2. J. Choukroun, A. Diss, A. Simonpieri, M. O. Girard, C. Schoeffler, S. L. Dohan, *et al.*, *Oral Surg., Oral Med. Oral Pathol. Oral Radiol. Endodontol.*, 2006, **101**(3), 299–303.
3. E. Lucarelli, R. Beretta, B. Dozza, P. L. Tazzari, S. M. O'Connell, F. Ricci, *et al.*, *Eur. Cells Mater.*, 2010, **20**, 13–23.
4. E. Yaprak, M. Kasap, G. Akpinar, E. E. Islek, A. Sinanoglu, *J. Dent. Sci.*, 2018, **13**(4), 386–395.
5. J. Ilić, K. Radović, J. Roganović, B. Brković and D. Stojić, *J. Endod.*, 2012, **38**(6), 764–768.

PROTEOMIC ANALYSIS OF HUMAN PLACENTAL EXTRACT AN ALLOGENIC BIOMATERIAL FOR VITAL PULP THERAPY

A. K. Shetty,[a] S. Hegde[b] and A. J. Rai [c]

[a] Department of Pedodontics and Preventive Dentistry, Faculty of Dental Sciences, M S Ramaiah University of Applied Sciences, Bangalore-560054, India; [b] Department of Conservative and Endodontics, Faculty of Dental Sciences, M S Ramaiah University of Applied Sciences, Bangalore-560054, India; [c] Department of Oral and Maxillofacial Surgery, AB Shetty College of Dental Sciences, NITTE University, Mangalore-575018, India

1 INTRODUCTION

Dental pulp infections and trauma to pulpal tissue could apparently cause inflammation and eventually cause regeneration of pulp in cellular level as process of physiological events. Pulp response to infection is similar to any other tissue response to infection in the body.[1,2] Due to poor lymphatic drainage and limited tissue space; individuals experience immense pain and discomfort.

The release on molecular mediators like cytokines and chemokine are observed as defense mechanism. Eventually the invading bacteria and tissue debris is removed in the process of cell repair. In early stages of tooth development odontoblast were involved in the release of molecular mediators and in the later stage other cell in the pulp like fibroblast, mesenchyme cell and endothelial cell respond to sequence of events.[3] The generation of functional dentine by effective stimulation of odontoblast has not yet been achieved though there are numerous strategies been employed in regenerative dentistry.[3,4,5] Need for creating bioactive material that enables the pulpal tissue to heal itself instead of permanently removing the injured tissue and placing prostheses.

Biologically attained medicine plays a major role in curing and preventing disease process. One such naturally obtained remedy derived from traditional medicine with numerous amount of therapeutic potential is the extract obtained from human placenta. The human placenta ensures mutual exchange of nutrients of mother and fetus, determining growth and development of the fetus thus making placenta an exceptionally a temporary organ.[6] Placental is believed to be allogenic in nature and its geno type could be entirely foreign.[7] The human placental organ possesses an extensive collection of bioactive proteins which are necessary for growth and development of fetus. Placental extract acts as a stimulant for wound healing activity.[8] Additionally small peptides or small molecules present in preparation were biologically active. Biomaterial based therapies have not been entirely new in dentistry, but always there is a quest for innovative formulations which is nontoxic and effective. The major aim of this study was to extract low molecular weight proteins effectively and characterize for best clinical use in regeneration of dental pulp cells.

2 METHODS AND RESULTS

2.1 Preparation of Samples

Placenta was collected immediately following C-section deliveries from mothers with no medical complication (n=4) after obtaining informed consent. A homogenised paste was obtained by using a food processor. A broad-spectrum proteolytic enzyme inhibitor cocktail (Sigma Aldrich, catalogue number P9599) was added before processing. To the homogenate 2:1 (v/v) acetonitrile is added and vortex briefly, this is allowed to stand for 30 minutes to precipitate high molecular weight proteins and high abundant proteins. This mixture was centrifuged at 14,000 rpm (13107xg) for 10 min at 4°C to get supernatant (~550 µL).[9] This is separated and transferred to a new eppendorf tube and stored at -20^{0}C (Figure 1).

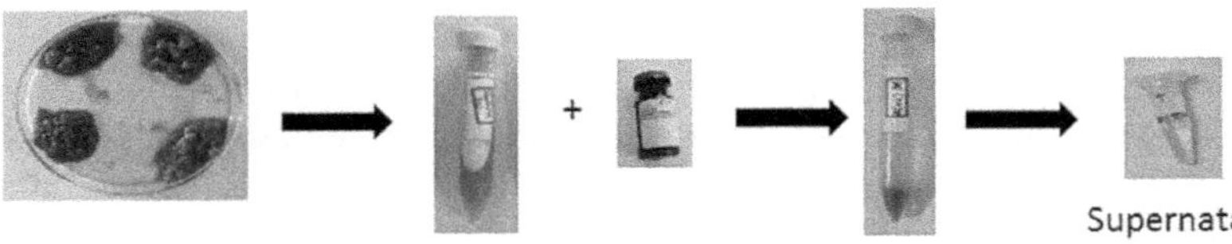

Figure 1 *Preparation of Human placental extract.*

2.2 Characterization of proteins

Mass Spectrometric analysis (200µl) of final extract was subjected to UltrafleXtreme MALDI TOF/TOF (Bruker Daltonics) and Bruker Daltonics ESI Q TOF-(Maxis Impact) Mass Spectroscopy at Molecular Biophysics Unit, Indian Institute of Sciences, Bangalore. In-solution digestion was carried out by addition of 1µL Trypsin to the extract and incubated overnight at 30°C prior to analysis. MASCOT and SwissProt database was used for protein sequencing and protein identification. About six proteins were characterized by MALDI TOF/TOF (Table 1) and six proteins by ESI Q TOF (Table 2) Mass Spectroscopy analysis.

Table 1 *Proteins identified by MALDI-TOF/TOF.*

Protein	Mw(kDa)	Score	Matched peaks
EAW50873.1	22.542	38	7
Lysine-specific histone demethylase 1A KDM1A_HUMAN	93.358	30	5
CCL24_HUMAN, C-C motif chemokine 24	13.467	39	5
AAB51135.1, MPIF-2	13.453	39	5
SURP and G-patch domain-containing protein 1 isoform X1	53.370	31	7
SURP and G-patch domain-containing protein	49.543	30	7

Table 2 *Proteins identified by ESI-Q-TOF.*

Protein	Mw(kDa)	Score	Matched peaks
SURP and G-patch domain-containing protein 1 isoform X1 [Homo sapiens]	53.370	31	7
SURP and G-patch domain-containing protein 1 isoform X2	49.543	30	7
RNA-binding protein splice variant a [Homo sapiens]	72.080	24	7
Splicing factor 4 protein	72.321	24	7
SURP and G-patch domain-containing protein 1	72.540	23	7
Splicing factor 4	72.599	23	7

2.3 Assessment of Biological activity

The information on the biological function was obtained from UniProt database.[10] Proteins identified showing associated with cell regulation were Lysine-specific histone demethylase 1A, C-C motif chemokine 24, SURP and G-patch domain-containing protein 1 and Splicing factor 4 (Table 3).

Table 3: *Biological functions of interested protein according to UniProt database.*

Mass	Gene	Protein	Biological pathways (Uniprot)
93.358	KDM1A_HUMAN	Lysine-specific histone demethylase 1A	Helps in neuron maturation, neuroblast proliferation and positive regulation of cell size.
13.467	CCL24_HUMAN	C-C motif chemokine 24	Chemotactic for resting T-lymphocytes, and eosinophils. Cell-cell signaling. Cellular response to TNF and interleukin 1, positive response to angiogenesis and cell migration.
72.540	SUGP1_HUMAN	SURP and G-patch domain-containing protein 1	Plays a role in pre-mRNA splicing.
72.599	SRSF4	Splicing factor 4	RNA binding and RNA processing

3 CONCLUSION

Human placental extract containing low molecular weight proteins has possible roles of in neuron maturation and proliferation along with cell signaling properties. The bio functional properties are supportive of cellular differentiation in microenvironment of dental pulp tissue. Thus mass spectrometric characterization of the HPE hypothesized the presence inductive biological factors for differentiation of dental pulp cells. In view of developing new biomaterial in pulpal repair human placental extract would be suggestive of promoting immunomodulation with a foundation of molecular and epigenetic strategies in future endodontic treatments.

References

1. L. Bjørndal and T. Darvann, *Caries Res.*, 1999, **33**, 50–60.
2. L. Bjørndal, T. Darvann and A. Thylstrup, *Caries Res.*, 1998, **32**, 59–69.
3. M. T. Albuquerque, S. J. Ryan, E. A. Münchow, M. M. Kamocka, R. L. Gregory, M. C. Valera and M. C. Bottino, *J. Endod.*, 2015, **41**, 1337–1343.
4. K. M .Galler, R. N. D'souza, J. D. Hartgerink and G. Schmalz, *Adv. Dent. Res.*, 2011, **23**, 333–339.
5. X. Yang, F. Yang, X. F. Walboomers, Z. Bian, M. Fan and J. A Jansen, *J. Biomed. Mater. Res., Part A*, 2010, **93**, 247–257.
6. B. Huppertz. *Hypertension*, 2008, **51**, 970–975.
7. A. L. Veenstra van Nieuwenhoven, M. J. Heineman and M. M. Faas, *Hum. Reprod. Update,* 2003, **9**, 347–357.
8. S. I. Togashi, N. Takahashi, S. Watanabe, A. Ishiguro and T. Fukui, *J. Health Sci.*, 2000, **46**,126–131.
9. K. Kedia, C. A. Nichols, C. D. Thulin and S. W. Graves, *Anal. Bioanal. Chem.*, 2015, **407**, 8543–8556.
10. UniProt Consortium, *Nucleic Acids Res.*, 2018, **47**, D506–515.

HUMAN AMNIOTIC MEMBRANE EXTRACT AS AN ANTIMICROBIAL BIOMATERIAL AGAINST STREPTOCOCCUS MUTANS AND LACTOBACILLUS ACIDOPHILUS IN MANAGEMENT OF DEEP DENTINAL CARIES

A.K. Shetty[a], A.J. Rai[b], S. Shivani[a] and S. Bharadwaj[a]

[a] Dept of Pedodontics and Preventive Dentistry, Faculty of Dental Sciences, M.S. Ramaiah University of Applied Sciences, Bangalore-560054, India

[b] Dept of Oral and Maxillofacial Surgery, AB Shetty College of Dental Sciences, NITTE University, Mangalore-575018, India

1 INTRODUCTION

Dental caries comprises of highly complex, variable and uncultivable microbial population with dental caries, even though prime organisms are commonly known to be allied with disease succession. Association of mutans group Streptococci and Lactobacilli with the initiation of caries and early caries development have been well documented.[1]Studies have shown transition of early caries causing gram-positive primary facultative bacteria to anaerobic gram-positive rods and cocci and gram-negative rods as the lesion progresses to deep carious lesion. The permeation of microorganisms into the soft dentine maintains the acidity and consequently jeopardizes the pulp vitality. Excavation of soft dentin will further lead to pulp exposure, and therefore it is imperious to preserve the pulp vitality by a thorough and conservative approach with the use of efficacious unguents instead of resorting to endodontic therapy.

Commonly used material in view of pulp preservation is Mineral Trioxide Aggregate (MTA), a calcium silicate based material that has been widely used in treatment of deep dentinal caries, including root-end filling, direct pulp capping, and regenerative endodontic procedures because of its resourceful properties like antibacterial activity, anti-inflammation, biocompatibility and osteo/odontogenicity. Previous studies have shown pronounced anti-bacterial efficiency of MTA against Lactobacillus acidophilus (LB).[2]

Complete removal of bacterial contaminants becomes critical in long-term success of vital pulp therapy. Hence, substitutes to traditional antibiotics, antibacterial peptides have been greatly studied for more than three decades for various therapeutic uses. Peptides, such as nisin have demonstrated biological and/or physical properties that may be apt for a more target centred approach for discerning regulation or elimination of recognized caries causing pathogens without causing harmful effects like irritation or staining. There is a need for more potential antibacterial peptides for their role in the management and deterrence of dental caries.[3]

The amniotic membrane is the deepest stratum of the placenta. It is 0.02 to 0.05 thick and comprises of a basement membrane, an epithelial stratum and an avascular stroma. Human amniotic membrane (hAM) has been used effectively for various applications for over a period of 70 years. The human amniotic membrane (hAM) holds diverse characteristics such as angiogenic modulation, low immunogenicity, antimicrobial activity and anti-inflammatory properties.[4]It also stimulates re-epithelization and averts scarring.[5]Hence, it has been utilized as a potential biomaterial in the fields of dermatology, as a biological dressing for skin burns, wounds and chronic leg ulcers and also as a graft material in plastic surgery and skin transplants.[6]Several studies have also demonstrated the antimicrobial effects of hAM against a few bacterial strains such as Escherichia coli, Group A streptococci, Pseudomonas aeruginosa and Staphylococcus aureus that are predominant in periodontal diseases.[7] Hence, aim of this study is to assess the antimicrobial property of hAM extract against bacterial strains such as Streptococcus mutans (SM) and Lactobacillus acidophilus.

2 METHODS AND RESULTS

2.1 Isolation and preparation of amniotic membrane

Ethical clearance was obtained from the Institute Ethical Board (UECHT/2016-18/PhD_01). Informed consent was obtained from the mothers enduring a C-section and they were screened for HIV 1&2, hepatitis B, hepatitis C, syphilis and malaria. Amniotic membrane was separated and washed with phosphate buffer solution. Tissue homogenate was prepared using food processor. 1mg of hAM extract and proteolytic inhibitor were subjected to centrifugation at 8000rpm at -4^{0}C.The supernatant was subjected to aceto-nitrile protein depletion and was again centrifuged at 14,000rpm at -4^{0}C [8]. The low molecular weight proteins were obtained after depletion. Protein estimation was done using Himedia Bradford assay kit. The protein concentration of 143µg/ml was estimated by means of Bradford Assay.

2.2 Evaluation of antimicrobial activity by Well Diffusion Method

The Streptococcus mutans (ATCC 25175) and Lactobacillus acidophilus (ATCC 314) were the test organisms used in this study. Ciprofloxacin was used as a Standard (S) and MTA was used as a control (C), hAM extract was the test compound (S2) (Figure 1). Cell suspensions of the test organisms were prepared and grown on Peptone broth, the suspensions were adjusted to 1-2 x10^5 cells/ml and the cultures were incubated at 37°C for 24-48 hours. 90 µl of Trypsin broth was mixed with 10 µl of inoculum as a control. The trays with Streptococcus mutans and Lactobacillus acidophilus (in an anaerobic chamber) were incubated at 37^{0}C for 24-48 hours. After incubation, 20 µl of Resazurin dye was added to each well and incubated for 4 hours to confirm viability of the culture. A Tecan plate reader at 590nm was used to measure the Optical density (OD). The minimal inhibitory concentration (MIC) of the drug giving 50% inhibition of the OD was determined. MIC can also be based on the color reduction of Resazurin that is, pink is viable and blue is non-viable. The protein concentration of 143µg/ml was estimated by means of Bradford Assay. The MIC of Ciprofloxacin against

SM and LB was 0.25 µg/ml and 0.125 µg/ml respectively. MIC of MTA(C) against SM and LB to be 10 µg/ml and 2.5 µg/ml respectively (Table 1). The MIC of hAM against LB was 35.75 µg/ml concentration and against SM was not determined (Table 2).

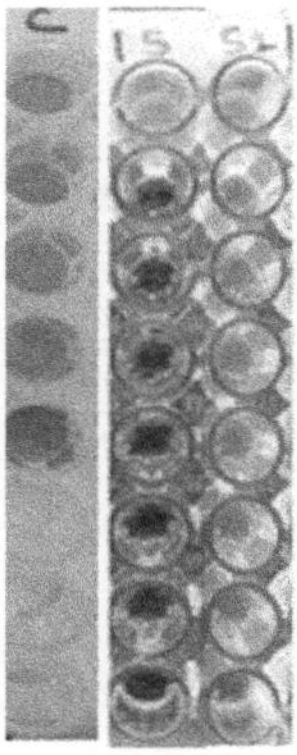

Figure 1 *Determination of antimicrobial activity by micro broth dilution of Streptococcus mutans by C –MTA sample, S-Standard ciprofloxacin and S2 – hAM extract Sample.*

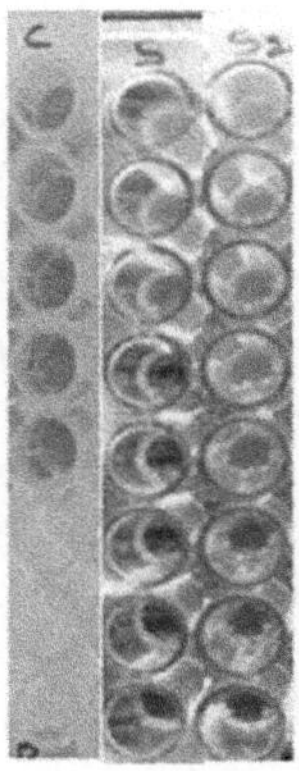

Figure 2 *Determination of antimicrobial activity by micro broth dilution of Lactobacillus acidophilus by C –MTA sample, S-Standard ciprofloxacin and S2 – hAM extract Sample.*

Table 1 *Minimal inhibitory concentration of Ciprofloxacin against SM and LB.*

Standard	Streptococcus mutans		Lactobacillus acidophilus	
Conc (µg/ml)	ABS	%Inhibition	ABS	%Inhibition
0	1.1272	0.00	1.2603	0
0.063	0.8974	20.39	0.8162	35.24
0.125	0.7596	32.61	0.6285	50.13
0.25	0.5493	51.27	0.3551	71.82
0.5	0.0539	95.22	0.1216	90.35
MIC (ug/ml)	**0.25**		**0.125**	

3 DISCUSSION

The inhibition of the growth of bacteria was the result of the composition of the antimicrobial test materials, on contact with bacteria. Though there was an increased antimicrobial activity of all the three materials with increasing concentration of the powder, the results revealed in this study were obtained when the powder concentration was 12.5 mg/mL with a clear comparison in between the materials. MTA, showed a decrease in absorbance (ABS), with increase concentration, i.e, 55.38% (~50%) inhibition was achieved at a minimal concentration of 0.6092 µg/ml against SM and 54.11% (~50%) of inhibition at a minimal concentration of 0.8672 µg/ml against LB (Table 2).
Anitmicrobial peptides have the property of an effective pathogen clearing by both immunological modulation and direct antimicrobial action, thus instigating angiogenesis and complementing wound healing[9]. The results in this study reveal a unswerving decrease in the Lactobacillus acidophilus count with increase in concentration of hAM. It can be proposed that antimicrobial peptide acts by binding to the Techoic acid in Gram positive bacteria[10]. And then, it is presumed that it reaches the cytoplasmic membrane, where it exercises its effect by putrefying the cytoplasmic membrane[11]. Human amniotic membrane extract inhibited the growth of Lactobacillus principally associated with dental diseases. Although MTA shows minimal inhibition at lower concentration, hAM being a natural extract also shows potent antimicrobial sensitivity against this anaerobic bacteria, which would allow accomplishing significant clinical effects with a reduced toxicity. Hence, human amniotic membrane could counterpart synthetic drugs in the combat against bacterial diseases.

Table 2 *Minimal inhibitory concentration of MTA against SM and LB.*

Standard	Streptococcus mutans		Lactobacillus acidophilus	
Conc (µg/ml)	**ABS**	**%Inhibition**	**ABS**	**%Inhibition**
0	1.3653	0.00	0.7353	0.00
1.25	1.1046	19.09	0.573	22.07
2.5	0.8672	36.48	0.3374	54.11
5	0.7049	48.37	0.1297	82.36
10	0.6092	55.38	0.1193	83.78
MIC (µg/mL)	**10**		**2.5**	

Table 3 *Minimal inhibitory concentration of hAM extract against SM and LB.*

Standard	Streptococcus mutans		Lactobacillus acidophilus	
Conc (µg/ml)	**ABS**	**%Inhibition**	**ABS**	**%Inhibition**
0	1.16	0.00	0.756	0
2.23	1.13	1.84	0.642	15.10
4.46	1.09	5.78	0.594	21.45
8.93	1.01	12.50	0.511	32.43
17.87	0.94	18.51	0.427	43.51
35.75	0.83	27.76	0.331	56.18
71.5	0.75	35.48	0.240	68.32
143	0.71	38.70	0.198	73.87
MIC (µg/mL)	**–**		**35.75**	

4. CONCLUSION

Human amniotic membrane extract showed no anti-bacterial effect on Streptococcus mutans but substantially inhibition of Lactobacillus acidophilus growth which is one of the principal organisms in caries progression. This novel strategy of using hAM extract has proven to be potentially antimicrobial sensitive against this gram-positive facultative anaerobic bacteria. Since hAM an allogenic material, might permit in accomplishing major clinical results with no possible toxicity. However, the antimicrobial protein or peptide that is mainly responsible can be determined by a proteomic analysis of hAM.

References

1. N. Takahashi and B. Nyvad , *J. Dent. Res.*, 2011, **90**, 294.
2. P. M. Melo Júnior, A. P. Sobral, G. C. Sampaio, I. M. Pinto and N. K. Shinohara, *RGO, Rev. Gaucha Odontol.*, 2015, **63**, 181.
3. Pepperney and M. L. Chikindas, *Probiotics Antimicrob. Proteins*, 2011, **3**, 68.
4. F. A. Tehrani, K. Modaresifar, S. Azizian and H. Niknejad, *Sci. Rep.,* 2017, **7**, 17022.
5. H. P. Lorenz and N. S. Adzick, *West. J. Med.*, 1993, **159**, 350.
6. R. Sripriya and R. Kumar. *Prog. Biomater.*, 2016, **5**, 161.
7. S. Heidarzadeh, A. Ghasemian, Z. Kalafi, F. Nikkhahi and M. M. Dallal, *Immunopathol. Persa*, 2018, **4**.
8. K. Kedia, C. A. Nichols, C. D. Thulin and S. W. Graves, *Anal. Bioanal. Chem.,* 2015, **407**, 8543.
9. J. Battersby, J. Khara, V. J. Wright, O. Levy and B. Kampmann, *Front. Imunol.*, 2016, **7**, 309.
10. L. H. Vorland, *Apmis*, 1999, **107**,971.
11. N. Orsi, *Biometals,* 2004, **17**, 189.

HOMOLOGY MODELING, MOLECULAR DYNAMICS AND DOCKING OF A NOVEL ANTI-CANCER PROTEIN OBTAINED FROM EARTHWORM COELOMIC FLUID

Dominic Augustine,[a] Roopa S. Rao, [a] Jayaraman Anbu,[b] Saraswathy G. R.,[c,e] K. N. Chidambara Murthy[d] and Hema Sree G. N. S.[e]

[a] Department of Oral Pathology, Faculty of Dental Sciences, Ramaiah University of Applied Sciences, Bangalore – 54, India; [b] Department of Pharmacology, Faculty of Pharmacy, Ramaiah University of Applied Sciences, Bangalore – 54, India; [c] Department of Pharmacy Practice, Faculty of Pharmacy, Ramaiah University of Applied Sciences, Bangalore – 54, India; [d] Principal Scientist, Central Research Laboratory, Ramaiah Medical College, Bangalore – 54, India; [e] Pharmacological Modelling and Simulation Centre, Faculty of Pharmacy, Ramaiah University of Applied Sciences, Bangalore – 54, India

1. INTRODUCTION

The specific problem encountered in combating cancer is the uncontrolled proliferation of cancer cells. Recently, concepts of using naturally available extracts to inhibit division of cancer cells have emerged, naturally occurring substances such as extracts from helminths have assumed importance in preventing the replication and division of cancer cells *in vitro* and *in vivo*.[1] This research work is designed to identify biomolecules from Earthworm Coelomic Fluid (ECF) of *Eudrilus eugeniae* that can inhibit cancer cells from proliferating. This would be highly relevant in the current scenario since cancer cell division is one of the prime reasons for failure of anticancer therapy. Homology modeling refers to constructing an atomic-resolution model of the "target" protein from its amino acid sequence. Using simulation tools, the constructed model can be docked with known receptors to analyze their interactions. Identification of the 3D structure of these biomolecules and information about interaction with receptors could be an initial step towards drug discovery and development.

2. METHODS & RESULTS

2.1 Isolation by SDS-PAGE & Initial Characterization by MALDITOF & MS-MS Sequencing

Initial isolation and purification of the crude protein from ECF was performed by multiple SDS-PAGE runs which resulted in a specific 18kDa band. The band was isolated and subjected to MALDI-TOF & MS-MS Sequencing. The masses obtained by fragmentation of the peptides in MS & MS/MS were submitted for Mascot search against the given sequence. The details of the Mascot search results are shown below, the number 96 represents the highest matching score.

Database : Lumbricina 20180918 (4677 sequences; 992615 residues)
Timestamp : 18 Sep 2018 at 10:40:31 GMT
Top Score : 96 for A0A0P4VL51_9ANNE, Extracellular globin OS=Glossoscolex Paulistus OX=1046353 GN=HgBp PE=3 SV=1

This data was insufficient as several amino acid details were missing; hence a complete amino acid sequencing was done to derive information regarding all the amino acids present in the protein

2.2 Nano-LCMS based amino acid sequencing

The Gel bands were used to extract the peptide and digest them using Trypsin. 10 µl of the 18kDa protein of ECF was injected on BEH C18 Ultra-Performance Liquid Chromatography column for separation of peptides followed by analysis on the Q-TOF instrument for MS and MS-MS. The raw data was processed by Mass Lynx 4.1 WATERS, peptide editor software to get the complete integrated sequence of the sample. The individual peptides MS-MS spectra were matched to the database sequence for amino acid sequence and protein identification on PLGS software, WATERS. **[Figure 1].** This enabled the identification of a total of 158 amino acids in the protein with the correct order and sequence.

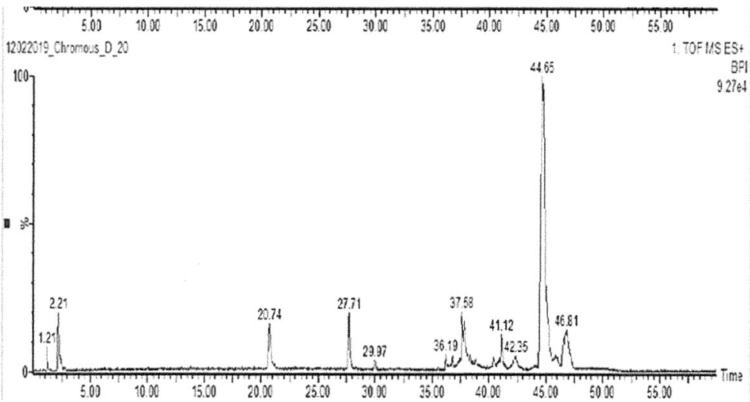

Figure 1 *Separation of peptides by UPLC.*

2.3 Homology Modeling

Due to unavailability of 3D structure in Protein Data Bank (PDB), the protein had to be modelled via energy-based methods using Prime module (Schrödinger Release 2019-3: Prime, Schrödinger, LLC, New York, NY, 2019.). This was done in order to dock the protein of interest to the Caspase-3 and Caspase-8 receptors.

Protein sequence was subjected to Basic Local Alignment Search Tool for protein (*BLASTp*) analysis in Uniprot, which revealed highest similarity with p-chain belonging to protein ''giant hemoglobin'' from species *"Glossoscolex Paulistus"* (PDB ID: 4U8U) at threshold level 0.01 using the matrix Blosum-62. The protein was modelled using reference PDB obtained through BLASTp analysis by retaining the co-crystallized structures [**Figure 2A**]. Finally, the Modelled Protein (MP) was validated by Ramachandran Plot [**Figure 2B**].

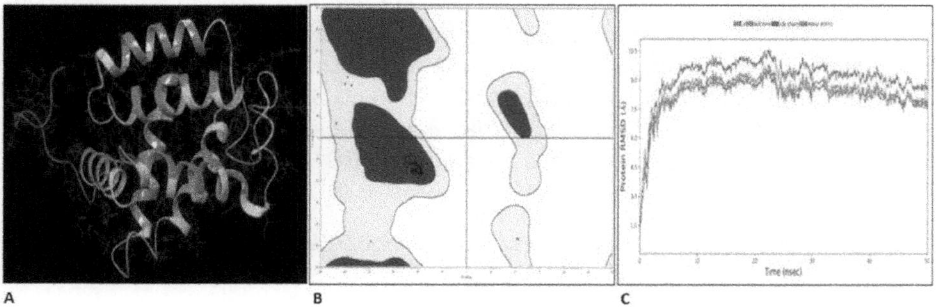

Figure 2 ***A***. *The modelled protein.* ***B***. *Ramachandran plot and* ***C.*** *The plot shows the RMSD evolution of the protein (left Y-axis).*

2.4. Molecular Dynamics Simulations

Dynamic behaviour of the modelled protein was studied using MD. The system was prepared using TIP3P solvent model under Normal Pressure (1 bar) and Temperature (300K) ensemble for 50 ns using Desmond tool of Schrödinger (2019-2) (Desmond Molecular Dynamics System, D. E. Shaw Research, New York, NY, 2019) with OPLS3e force field. Qualitative analysis predicting pressure, volume, temperature, total energy and potential energy were calculated using Simulation Quality Analysis tool. Qualitative and quantitative assessments of modelled protein was studied using simulation event analysis tools. Stability of modelled protein was described in terms of Root Mean Square Deviation (RMSD) and Protein-Root Mean Square Fluctuations (P-RMSF) [**Figure 2C**]. MP exhibited stability by 10 ns and no major fluctuations observed in RMSD values throughout the simulation period (Figure 2). P-RMSF graph exhibited high fluctuations in initial 20 residues as expected due to terminal residues. The residues involved in formation of secondary structure exhibited less fluctuations. More than half of residues (53.45%) of MP were found to be involved in α-helix formation.

2.5 Protein-Protein Docking (PPD) with pro-apoptotic receptors and Binding Energy Calculations

PPD and energy calculations were performed after MD analysis. PPD was done using Bioluminate tool (**Schrödinger Release 2019-3**: BioLuminate, Schrödinger, LLC, New York, NY, 2019) with Caspase-3 (PDB ID: 6BDV) and Caspase-8 (PDB ID: 5JQE) to analyse the affinity of the modelled protein towards selected anti-cancer targets [**Figure 3A & 3B**]. Sitemap analysis was performed for the selected proteins to identify the binding sites for docking and site with highest site score (~1) was used for generation of receptor grid. Docking analysis with 5JQE and 6BDV revealed PIPER pose score (resembles the docking score for every pose between the 2 proteins) ranging from -278.548 to 364.946 and -606.115 to -82.549. Top 5 poses exhibiting high PIPER score were subjected to Molecular Mechanics/Generalized Born Surface Area (MM-GBSA) energy calculations using Prime tool [**Table 1**].

Table 1 *The Prime Energy and bind score calculation for the top 5 poses obtained in PPD.*

PPD POSE	**PIPER Score**	**ΔBind energy**	**Δ coloumb**	**ΔG Covalent**	**ΔG H bond**	**ΔG-vdW**
MM-GBSA6bdv – Caspase 3 Receptor						
pose_18	-606.115	-59.38	167.432	30.97	3.7	-116.11
pose_3	-407.64	-69.91	5.81	14.80	-6.04	-72.48
pose_6	-351.607	-72.07	85.38	24.50	-7.48	-88.20
pose_14	-323.082	-41.94	47.58	38.66	-2.95	-87.55
pose_28	-305.217	-93.73	-11.60	25	-4.47	-97.81
MM-GBSA-5jqe – Caspase 8 Receptor						
pose_8	-278.548	-60.62	653.81	9.83	-6.15	-154.04
pose_28	-272.288	-103.21	80.28	26.01	-6.19	-100.93
pose_4	-230.801	1773.77	424.85	974.81	-8.56	792.83
pose_12	-221.794	-69.89	47.87	26.51	-3.38	-91.59
pose_6	-219.31	-23.8	304.4482	-2.48767	-0.77	-49.38

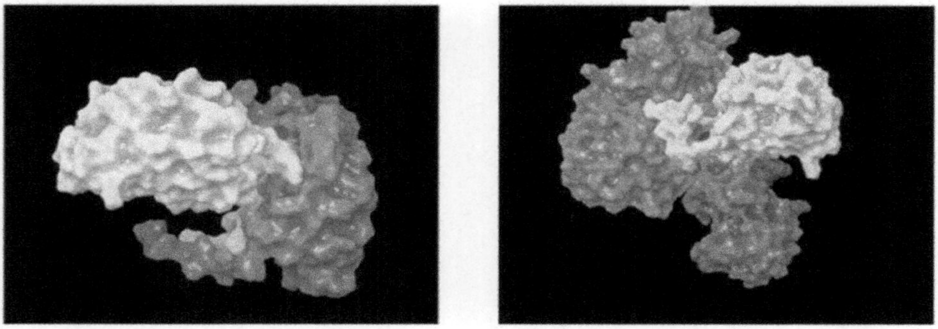

A B

Figure 3 *Top docking poses of the MP (light grey) with (A) Caspase – 3 (PDB ID: 6BDV) (dark grey) and (B) Caspase – 8-5jqe receptor (dark grey).*

3. CONCLUSION

The sequence of novel anticancer protein from ECF was identified through Nano-LCMS AA-Sequencing, 3D structure was modelled by homology modelling and stability was studied through molecular dynamic simulations. The modelled protein has been docked with pro-apoptotic cancer receptors to study the binding affinity. Results of the current study exhibited good binding affinity with pro-apoptotic genes Caspase - 3 and Caspase – 8 responsible for cancer cell apoptosis. The future scope of this study is to validate the upregulation of Caspase - 3 and Caspase – 8 genes by 18kDa protein of ECF in cancer cell lines through experimental techniques such as RT-PCR and qPCR. These findings may lead to the development of naturally available drugs to combat cancer.

References

1. M. Abraham, D. Augustine and R. S. Rao, *et al.*, *J. Evid. Based Complementary Altern. Med.*, 2017, **22**(4), 870–878.
2. E. Lionta, G. Spyrou, D. K. Vassilatis and Z. Cournia, *Curr. Top. Med. Chem.*, 2014, **14**(16), 1923–1938.

Part 7: Updates in Analytical Techniques

ESTIMATION OF SALIVARY GLUCOSE BY VARIOUS PAPER-BASED COLOURIMETRIC METHODS

Sumedha S. K.[a] and Sandhya K. V. [b]

[a]Department of Pharmaceutics, FPH-Ramaiah University of Applied Sciences, Bangalore 54, India; [b]Assistant Professor, Department of Pharmaceutics, FPH-Ramaiah University Applied Sciences, Bangalore 54, India

1. INTRODUCTION

Diabetes Mellitus is a chronic metabolic disorder affecting the blood glucose levels in both the young and old. The impaired glucose levels if not carefully monitored in the blood affects multiple systems in the body leading to further complications like loss of vision and limbs, failure of important organs like the kidney, pancreas etc., Regular monitoring of blood glucose levels of patients who pose higher risk of damage to the eyes, kidneys can facilitate early treatment. Most of the present methods used to monitor blood glucose levels are invasive techniques, which require a single drop of blood preferably from the fingertip that causes a lot of inconvenience and discomfort to diabetic patients who require blood glucose to be monitored 4-5 times a day. Since continuous glucose, monitoring is critical and to overcome the disadvantages posed by the current invasive techniques, the present study aims to develop non-invasive colorimetric detection of glucose using paper based medium in saliva.

2. METHODS

Non-invasive glucose monitoring techniques using various biological fluids like saliva, tears, sweat are gaining significance. The present study uses saliva since it shows significant correlation with respect to blood glucose. Different colorimetric methods were developed to estimate the amount of glucose in saliva and further paper was used as a medium for the development of colour using different chromogenic substrates along with different concentration of glucose (10-50µg/ml). Three colorimetric methods were developed using chromogenic substrates **potassium iodide** (Method 1-yellow colour),a combination **4 Amino anti-pyrine & phenol** (Method 2-pink colour), **methyl red** (Method 3-a pH sensitive dye) for estimation of glucose colorimetrically in saliva using UV spectrophotometer. A stable colour (linearity observed) formed among the three methods was used for further development on a paper-based medium. Three paper based medium: Whatman filter paper, Rayon fibre paper, Nitrocellulose paper was used for the development of colour using different concentration of glucose (10-50µg/ml) using simulated saliva.

BASIC PRINCIPLE: The present approach is used identify the minute quantity of glucose present in biological fluids saliva

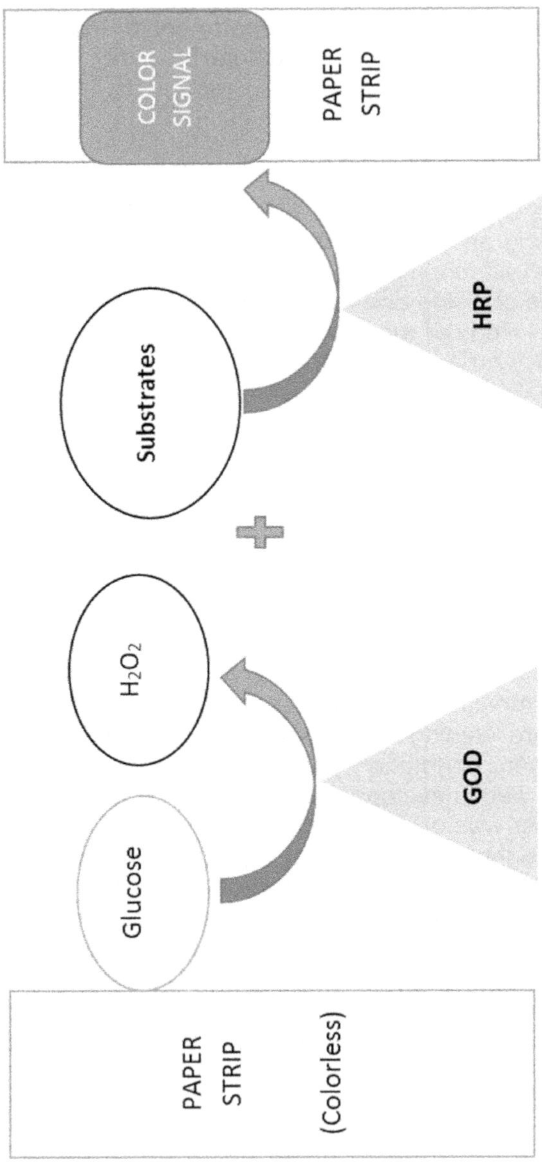

Figure 1 *Basic principle for detection of glucose non-invasively. GOD: glucose oxidase, HRP: horseradish peroxidase.*

3. RESULTS

Table 1 *Estimation of glucose using KI*

Glucose conc. (µg/ml)	GOX (µL)	HRP (µL)	KI (ml)	Colour	Absorbance* (λmax430nm)
10	600	60	1	Light yellow	0.1247 ± 0.0483
20	600	60	1	Light yellow	0.2228 ± 0.0060
30	600	60	1	Yellow	0.3124 ± 0.0040
40	600	60	1	Yellow	0.4439 ± 0.0226
50	600	60	1	Yellow	0.5472 ± 0.0269

*Average of six readings

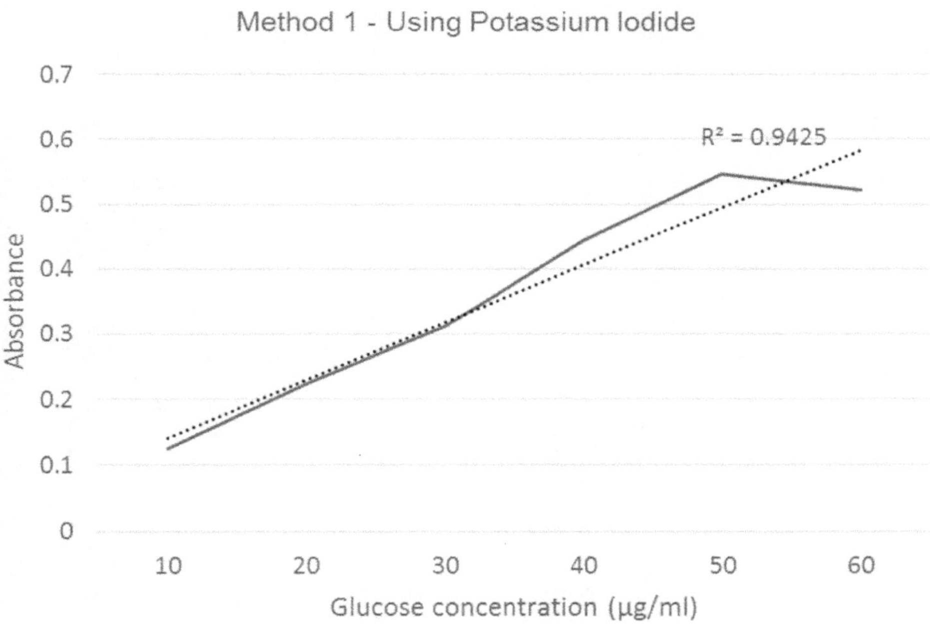

Figure 2 *R^2 value (Method 1).*

Table 2 *Estimation of glucose using 4AAP and phenol.*

Glucose conc. (µg/ml)	GOX (µL)	HRP (µL)	4AAP (ml) (4mM)	Phenol (µL)	Colour	Absorbance* (λmax510nm)
10	1000	100	2	250	Pale pink	0.1361 ± 0.0211
20	1000	100	2	250		0.2620 ± 0.0255
30	1000	100	2	250		0.3621 ± 0.0313
40	1000	100	2	250		0.4492 ± 0.0346
50	1000	100	2	250	Dark pink	0.5730 ± 0.0380

*Average of six readings

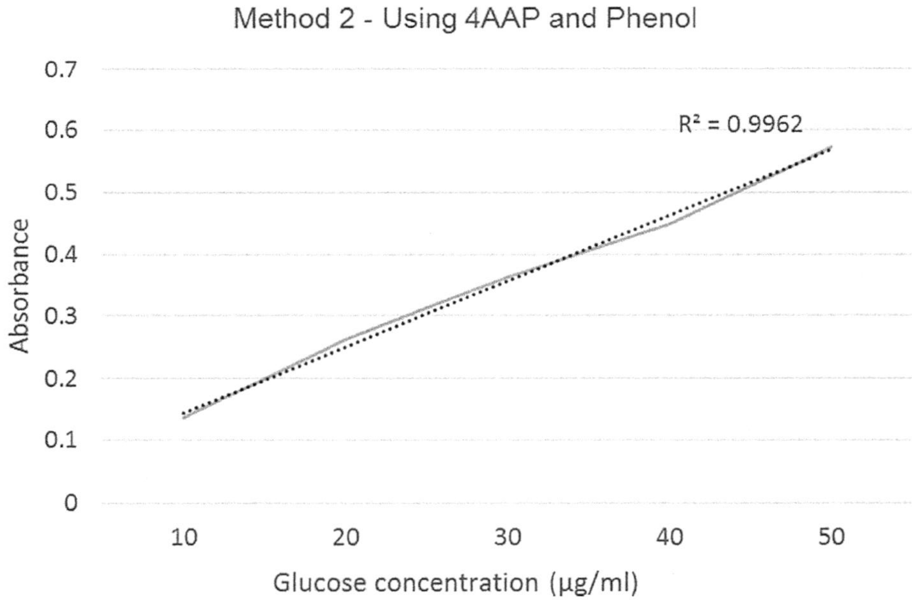

Figure 3 *R^2 value (Method 2).*

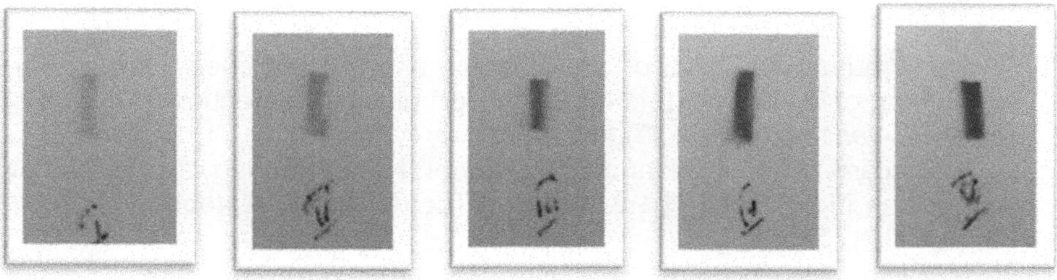

Figure 4 *Color development on Whatman filter paper using KI (10–50 μg/ml glucose).*

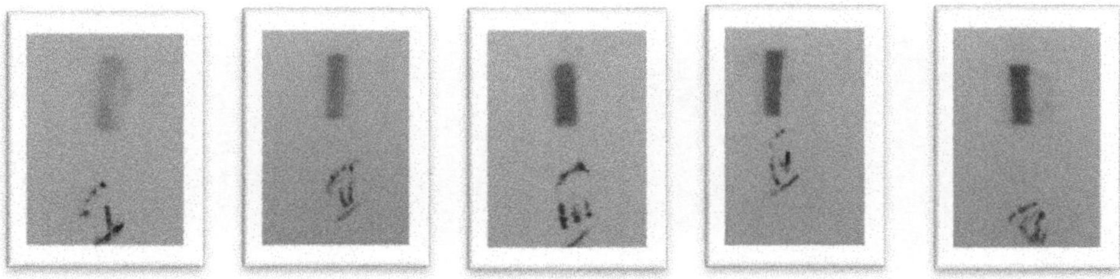

Figure 5 *Color development on Whatman filter paper using 4 AAP and phenol (10–50 μg/ml glucose).*

4. CONCLUSIONS

The colorimetric method of estimation was based on the reaction of the primary enzymes **Glucose oxidase (GOD) and Horse radish peroxidase (HRP)** with glucose in presence of chromogenic substrates leading to the formation of a coloured dye whose intensity increased as the concentration of glucose increased from 10 -50 μg/ml

Among the three different methods, Method 1 and Method 2 resulted in the formation of a stable colour which was further developed on a paper based medium

Among the three different paper mediums used, Whatman filter paper showed a linear increase in the intensity of colour and uniform dispersion of colour along the entire length of the paper whereas Nitrocellulose paper and rayon fibre paper showed saturation of the enzyme at a single position and non-uniform dispersion of colour.

References

1. R. Dominguez, M. Orozco, G. Chávez and A. Márquez-Lucero, The evaluation of a low-cost colorimeter for glucose detection in salivary samples, *Sensors.*, 2017, **17**(11), 2495.
2. C. M. Riccardi, R. M. Kasi and C. V. Kumar, Nanoarmoring of enzymes by interlocking in cellulose fibers with poly (acrylic acid), *Methods Enzymol.*, 2017, **590**, 475–500.
3. A. Soni and S. K. Jha, A paper strip based non-invasive glucose biosensor for salivary analysis, *Biosens. Bioelectron.*, 2015, **67**, 763–768.

ANALYSIS OF INTERACTION BETWEEN CHLORINE DIOXIDE AND BISPYRIDINE-BASED IRRIGATING SOLUTION – AN OBSERVATIONAL STUDY

S. Rajatha,[a] K. Dinesh,[a] M. S. Lakshmi,[b] N. Shruthi[a] and M. Sylvia[a]

[a]Department of Conservative Dentistry and Endodontics , Faculty of Dental Sciences, M S Ramaiah University of Applied Sciences, Bengaluru, India; [b]Department of Pharmaceutical Chemistry, Faculty of Pharmacy, M S Ramaiah University of Applied Sciences, Bengaluru, India

1 INTRODUCTION

Microbial infections are the most common cause of an endodontic disease. To critically reduce bacterial loads, endodontic treatment regimens combine mechanical instrumentation with chemicals, namely irrigation and medication. An optimum agent for endodontic irrigation should have antimicrobial activity, dissolve necrotic tissue, aid in the debridement of the canal system and nontoxic to the periradicular tissues. None of the irrigating solutions used till date possess all the ideal properties. They are used in succession to enhance their individual properties but it is a concern that their interaction could result in reactionary products. This is especially evident when chlorhexidine (CHX) is used in succession with sodium hypochlorite (NaOCl) producing a brown precipitate of parachloroaniline (PCA), which is said to be toxic and carcinogenic and can occlude dentinal tubules.[1] Hence, alternate combinations need to be explored for enhanced properties with no deleterious by-products.

Octinisept (OCT; Sch€ulke & Mayr, Nordersdedt, Germany) is an antiseptic containing 0.1% octenidine hydrochloride and 2% phenoxyethanol (PE). Octenidine is a positively charged bis-pyridinamine with a broad spectrum of antibacterial, antifungal and some antiviral properties.[2] However, its potential as a stand-alone irrigant is limited because of its poor tissue-dissolving properties.[2] Chlorine dioxide (ClO_2) has been researched recently as a substitute for hypochlorite. It has pulp tissue dissolution property, less toxic, has antibacterial activity and is biocompatible.[3] A combination of OCT with ClO_2 as an endodontic irrigation regimen could have added beneficial effects but has not been researched yet. Therefore, study was aimed to evaluate the interaction between 0.1% OCT and 5% ClO_2 and identification of the final compound.

2 METHODS AND RESULTS

2.1 Chemicals and reagents: Octenisept (OCT; Schulke & Mayr, Nordersdedt, Germany), Chlorine dioxide (Kresko Projects Pvt Ltd, chemical manufacturer in Ahmedabad, Gujarat) was freshly prepared as manufacturer's instructions, HPLC grade water & acetonitrile (Sigma Aldrich Chemicals Pvt Ltd from Bengaluru, Karnataka, India)

2.2 Study design: Study was divided into 2 parts: Firstly, evaluation of colour change and precipitate formation. Secondly, High Performance Liquid Chromatography (HPLC) analysis of the final product.

All samples were procured and samples were divided into three different groups: Group 1- 0.1% OCT Group 2- 5% ClO_2 Group 3- 0.1% OCT and 5%ClO_2 (test solution).

2.3 Evaluation of colour change and precipitate formation

A sterile polystyrene round-bottom tube was filled with 2 mL OCT alone-Group1. Group 2- 2 mL ClO_2 and Group 3 received 1 mL OCT and 1 mL ClO_2. All tubes were stored at 36.5°C and 95% humidity for one week . The solutions were observed for any color change or precipitate formation by an independent observer immediately, 12 hrs, 24 hrs, 1 week.

2.4 High performance liquid chromatography (HPLC) analysis

This was performed to investigate any interaction between 0.1% OCT and 5% ClO_2. A preliminary study design was used in which OCT and ClO_2 were injected separately followed by the test solution under following condition: Mobile phase: Acetonitrile 80%: Water 20%, flow rate of 1ml/min, λmax- 270 nm and injection volume of 20 µL. 5 µg/ml of OCT and ClO_2 were injected separately followed by the test solution (OCT & ClO_2) to get HPLC chromatogram of sample.

On evaluation of colour change- Group 1 and 2 became clearer on standing whereas Group 3 (OCT and ClO_2) immediately turned cloudy and turbid on mixing, which remained unchanged for 12 hours then by 24 hrs it turned transparent with no precipitate. According to the results obtained by HPLC analysis, there is no interaction seen with the combination of OCT (retention time t_R = 3.0 minutes) and ClO_2 in test solution (t_R = 2.5 minutes) as the HPLC chromatogram of the test solutions were similar to those obtained when irrigants were injected alone (Figure 1).

Previous studies on the precipitate formed with CHX-NaOCl have resulted in varying interpretation of the compound present in the precipitate.[4] Although the precipitate being insoluble raises questions about leaching of PCA, shown to be cytotoxic, genotoxic. However, these peach-brown precipitates do penetrate deeper into the dentinal tubules and may further act as a reservoir within the tubules even if it is removed from the main canal by passive ultrasonic irrigation.[5] It may prevent diffusion of intracanal medicaments, interfere with the sealing ability of obturation materials and stain the dental structures. A previous study evaluated interaction between OCT with NaOCl. It was observed that it resulted in a whitish precipitate, phenoxyethanol (PE), a compound already present in OCT. In the present study, combining OCT with ClO_2 did not result in any precipitate and HPLC analysis also confirmed the same. HPLC is non-destructive analytical technique used to separate, identify and quantify each component in a mixture. Considering the advantages of OCT and ClO_2, a combination of these solutions can be considered as an alternate irrigation regimen. Further studies on its antimicrobial efficacy, effect on root dentin, interaction with endodontic sealers need to be evaluated before it can be recommended for clinical use as a routine irrigation protocol.

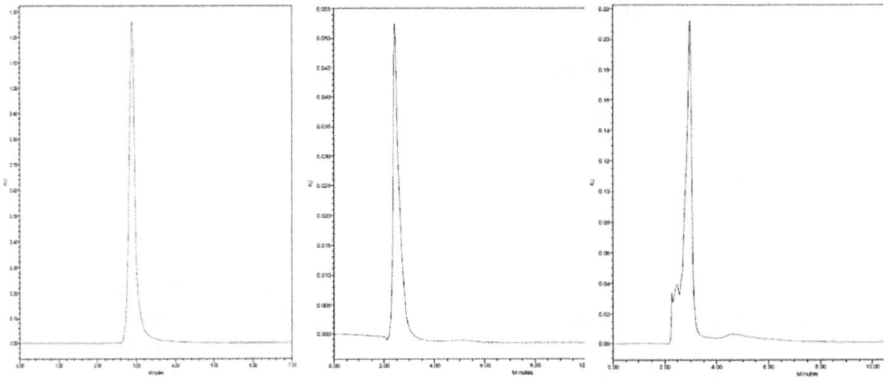

Figure 1 *HPLC chromatogram of (left) OCT only, (centre) ClO_2 and (right) Test solution.*

3. CONCLUSION

In this in vitro study, there is no interaction seen on mixing OCT and ClO_2. They do not lead to formation of any precipitate which might block the dentinal tubules, which was also confirmed by HPLC analysis. Hence, they can be used as potential endodontic irrigants in combination as an alternative to NaOCl and CHX. Further studies evaluating other aspects of this combination needs to be assessed.

References

1. T. B. Bui, J. C. Baumgartner and J. C. Mitchell, *J. Endodont.,* 2008, **34**(2), 181–185.
2. K. A. Thaha, R. L. Varma, M. G. Nair, V. G. S. Joseph and U. Krishnan, *J. Endodont.,* 2017, **43**(1), 135–140.
3. F. K. Cobankara and H. B. Ozkan, *J. Endodont.,* 2010, **36**(2), 272–274.
4. B. R. Basrani, S. Manek and R. N. S. Sodhi, *J. Endodont.,* 2007, **33**(8), 966–969.
5. K. P. Kolosowski, R. N. S. Sodhi, A. Kishen and B. R. Basrani, *J. Endodont.,* 2014, **40**(12), 2036–2040.

EVALUATION OF PROCESS RELATED IMPURITIES OF APREMILAST

P. Ravi Kumar[a] and K. Sesha Maheswaramma[b]

[a] Research Scholar, JNTUA, Ananthapuramu, India; [b] Department of Chemistry, JNTUA College of Engineering (Autonomous), Pulivendula-516390, Andhra Pradesh, India
Email: kallurumahi.chemistry@jntua.ac.in

1 INTRODUCTION

Apremilast is an inhibitor of Phosphodiesterase type 4 an enzyme which breaks down cyclic adenosine monophosphate while using the treatment of Psoriasis and arthritis. This is one of advanced oral drug with low side effects as comparatively other psoriatic drugs like adalimumab, etanercept, infliximab, ustekinumab, secukinumab, and cyclosporine.

2 METHODS AND RESULTS

The present work related to identification and synthesis of probable impurities during synthesis of Apremilast drug substance based on the route of synthesis. The identified impurities were characterised by corresponding spectral data (IR, LC-MS and NMR) and consequently evaluated by novel HPLC method and LCMS method (genotoxic impurity) as per current ICH guidelines.[1,2,3] The importance of impurities synthesis is to verify the method specificity, relative response factors to each impurity, the exact percentage of composition in drug substance.[4] Total eleven impurities were identified as carry over from starting materials and process related; these were characterized and evaluated to meet the regulatory requirements.

The impurities are N-(2-(1-(3,4-dihydroxyphenyl)-2-(methylsulfonyl)ethyl)-1,3-dioxoisoindolin-4-yl)acetamide (A), N-(2-(1-(3,4-dimethoxyphenyl)-2-(methylsulfonyl)ethyl)-1,3-dioxoisoindolin-4-yl)acetamide (B), N-(2-(1-(4-ethoxy-3-methoxyphenyl)-2-(methylsulfonyl)ethyl)-1,3-dioxoisoindolin-4-yl)-acetamide (C), N-(2-(1-(3,4-diethoxyphenyl)-2-(methylsulfonyl)ethyl)-1,3-dioxoisoindolin-4-yl)acetamide (D), N-(2-(1-(3-ethoxy-4-methoxyphenyl)-2-(methylsulfonyl)ethyl)-1,3-dioxoisoindolin-4-yl)-butyramide (E), N-(2-(1-(3-ethoxy-4-methoxyphenyl)-2-(methylsulfonyl)ethyl)-1,3-dioxoisoindolin-4-yl)-propionamide (F), 4-amino-2-(1-(3-ethoxy-4-methoxyphenyl)-2-(methylsulfonyl)ethyl)isoindoline-1,3-dione (G), 4-amino-2'-(1-(3-ethoxy-4-methoxyphenyl)-2-(methyl sulfonyl)ethyl)-2,4-biisoindoline-1,1',3,3'-tetrone (H), N-(2'-(1-(3-ethoxy-4-methoxyphenyl)-2-(methylsulfonyl)ethyl)-1,,1',3,3'-tetraoxo[2,4'-biisoindolin]-4-yl)acetamide (II), 2-(1-(3-ethoxy-4-methoxyphenyl)-2-(methyl sulfonyl)ethyl)-4-nitroisoindoline-1,3-dione (J), N-

(1-(3-ethoxy-4-methoxyphenyl)-2-(methylsulfonyl)ethyl)acetamide (K). Among those A, B, C, D, E, F, G and K are carryover impurities, H and I are process impurities, J is carryover and genotoxic impurity.

2.1 Synthesis of Apremilast

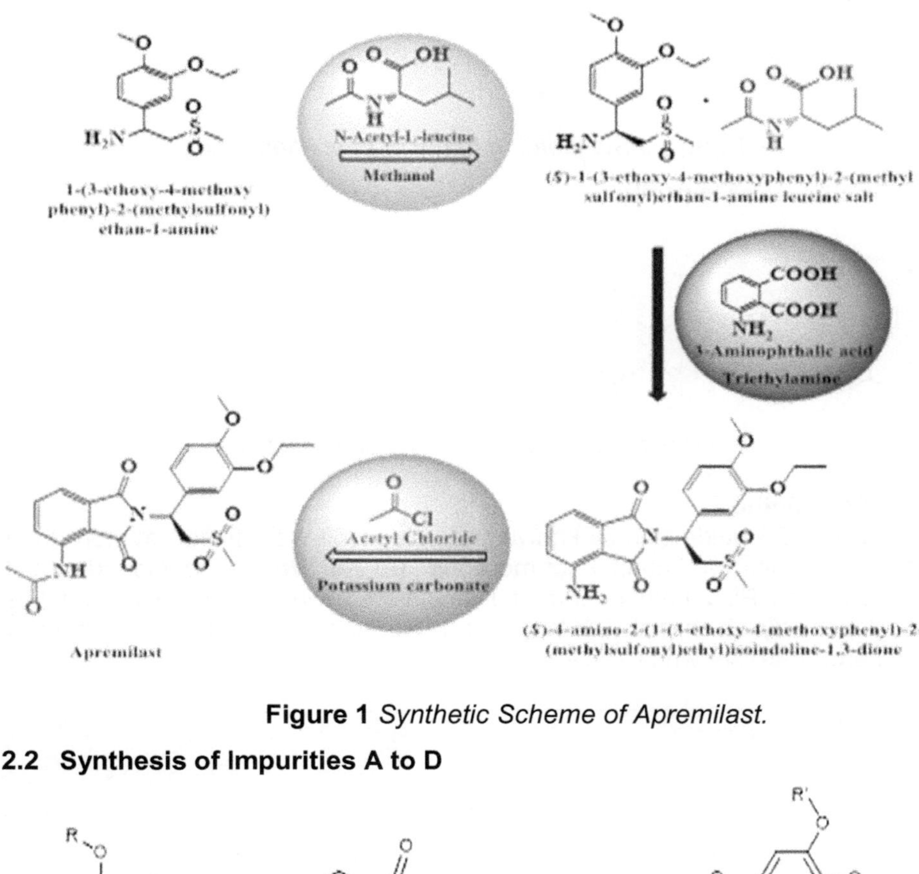

Figure 1 *Synthetic Scheme of Apremilast.*

2.2 Synthesis of Impurities A to D

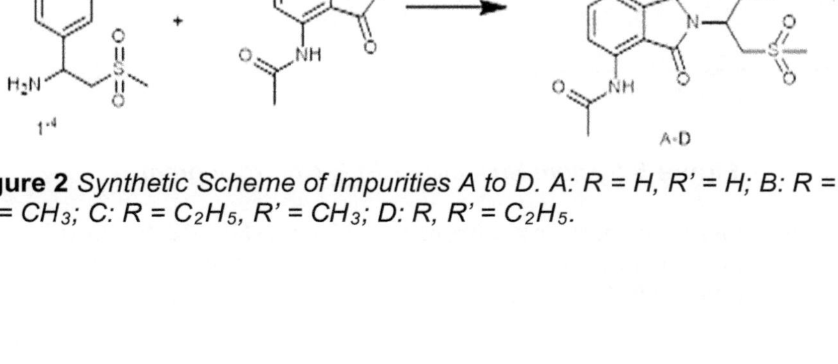

Figure 2 *Synthetic Scheme of Impurities A to D. A: R = H, R' = H; B: R = CH_3, R' = CH_3; C: R = C_2H_5, R' = CH_3; D: R, R' = C_2H_5.*

2.3 Synthesis of Impurities E and F

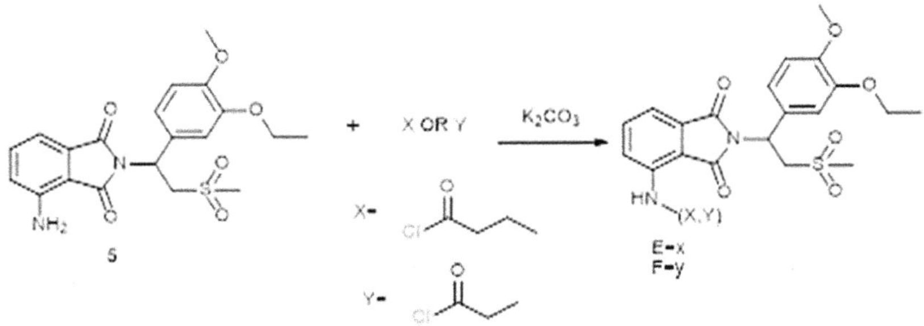

Figure 3 *Synthetic Scheme of Impurities E and F.*

2.4 Synthesis of Impurity J

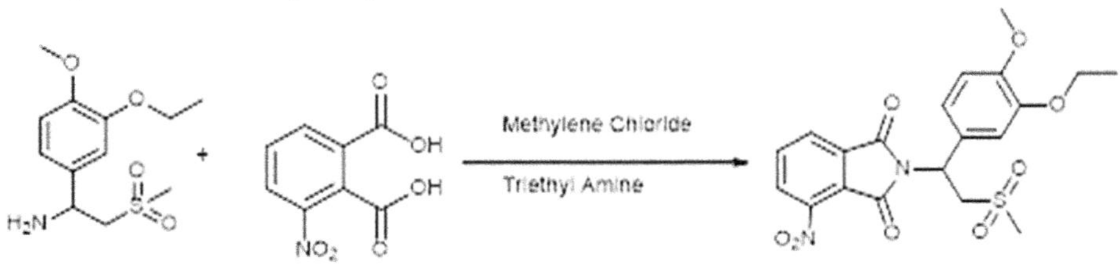

Figure 4 *Synthetic Scheme of Impurity J.*

2.5 HPLC Method

As per ICH Q3 guidelines all known impurities limit were fixed not more than 0.15% and unknown impurity not more than 0.10%. We had developed robust and accurate HPLC method for quantification of known and unknown impurities simultaneously which is shown in Figure 5.[1,2,3] The validation results of impurity A to I and K are shown in Table 1.

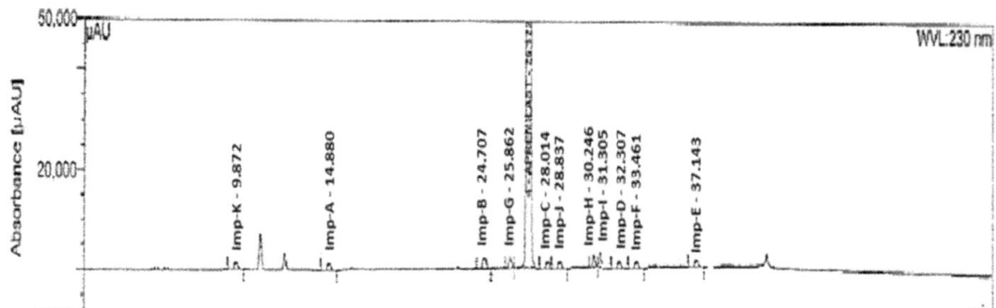

Figure 5 *System Suitability HPLC Chromatogram.*

2.6 Trace analysis of genotoxic Impurity -J by LC-MS/MS Method

Impurity-J was structural alert and it was confirmed from DEREK nexus software and positive in Ames test. Based on ICH M7 guidelines limit was fixed as 25 ppm and monitored by LC-MS/MS method. The validation results of impurity J is shown in Table 2. The specificity chromatogram for impurity-J and Apremilast are shown in Figures 6 and 7.

Table 1 *HPLC method validation results of impurity A to I and K.*

Impurities	LOD (%)	LOQ (%)	Repeatability[a] (%RSD)	Accuracy [b]	Linearity and range (%)
A	0.004	0.011	0.51	96.7– 102.1	0.01 – 0.25
B	0.004	0.011	0.27	97.5– 102.8	0.01 – 0.25
C	0.004	0.011	0.29	98.0 - 103.2	0.01 – 0.25
D	0.004	0.008	0.48	98.4 - 102.7	0.01 – 0.25
E	0.004	0.011	0.77	97.0 – 101.6	0.01 – 0.25
F	0.004	0.011	0.62	95.8 – 104.8	0.01 – 0.25
G	0.012	0.030	0.39	98.6 –14.3	0.03 – 0.25
H	0.012	0.030	0.83	96.3 – 103.6	0.03 – 0.25
I	0.012	0.031	0.44	97.8 – 103.4	0.03 – 0.25
K	0.010	0.030	0.86	98.6 – 104.6	0.03 – 0.25

LOD: Limit of Detection; LOQ: Limit of Quantification; RSD: Rate of Standard Deviation; [a] n = 6; [b]Demonstrated at LOQ 50%, limit level (100%) and 200%, n = 6; C R2 >0.9991

Table 2 *LC-MS/MS method validation results of impurity – J.*

Parameter	Impurity-J
Limit Of Detection (ppm)	1
Limit Of Quantification (ppm)	4
Slope	472.3
Intercept	823.0
Correlation coefficient	0.999
Linearity range (ppm)	4-50
System precision (%RSD)	0.6
Method precision (%RSD)	1.1
Intermediate precision (%RSD)	0.3
% Accuracy at 4.0 ppm (LOQ)	94.1-97.8
% Accuracy at 12.5 ppm (50%)	98.6-102.7
% Accuracy at 25 ppm (100%)	100.1-103.1
% Accuracy at 50 ppm (200%)	97.4-99.6

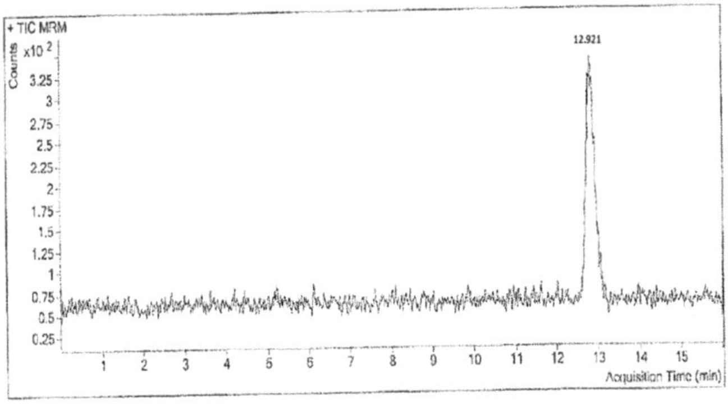

Figure 6 *Specificity chromatogram for impurity-J.*

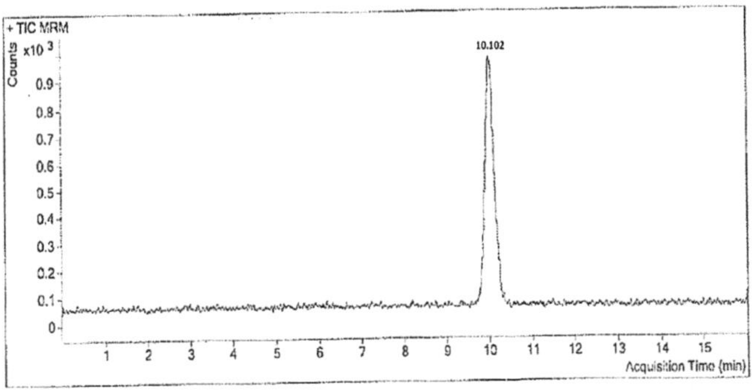

Figure 7 *Specificity chromatogram for Apremilast.*

3 CONCLUSION

Chemical synthesis was carried out for the identified probable carry over and process related impurities of apremilast and characterized using analytical tools such as HPLC, HNMR, Mass and IR. The acceptable level of known related compound impurity should be less than 0.15% and unknown impurity should be less than 0.10% respectively in a drug substance as per ICH Q3 guidelines. If the impurity is a potential genotoxic, threshold limit of toxicological value in the drug as per ICH and European guidelines is 1.5 µg/day. The impurity A to I and K were less than 0.15% and impurity-J,a potential genotoxic was less than 25 ppm as per regulatory requirements.

References

1. ICH (Q 11) Guidelines, 2012.
2. ICH Hormonised Tripartite ($Q_3A(R2)$) Guidelines, 2016.
3. ICH Hormonised (M7(R1)) Guidelines, 2017.
4. S. Lakshmana Prabu and T. N. K. Suriyaprakash, *Int. J. Pharm. Sci. Rev. Res.*, 2010, **3**, 2.

Part 8: Full-length Articles

IN SILICO DRUG DESIGN, MOLECULAR MODELLING, SYNTHESIS AND CHARACTERIZATION OF NOVEL BENZIMIDAZOLES

R. Sreerama[a] and C. Kaushik*[a]

[a] Department of Chemistry, School of Advanced Sciences, Vellore Institute of Technology, Vellore – 632 014, Tamil Nadu, India
Email: chandakaushik1@gmail.com

1 INTRODUCTION

Benzimidazole and its scaffolds are considered as an important heterocyclic motif containing two annular nitrogen atoms of bicyclic compound fused with benzene to imidazole ring established and obtained in many synthetic and natural compounds. In recent years, a few marketed drugs containing 1,2,5-trisubstituted benzimidazoles have shown extensive interest with anticancer (Bendamustine), antifungal (Benomyl), analgesic (Etonitazine), antihypertensive (Telmisartan) and antihistaminic (Norastemizole) activities etc, as shown in Figure 1. It also exhibits varied pharmaceutical applications resembling antibacterial,[1] antifungal,[2] anti-inflammatory,[3] antitumor,[4] anthelminthic,[5] antihypertensive,[6] anti-HIVs,[7] antiulcer,[8] anti-convulsant,[9] and anti-diabetic[10] activities. In addition, most of the drugs have substitution at 2nd and 5th position at the benzimidazole core. Moreover; at these two positions, substitutions of the benzimidazole moiety were recognized as fundamental for potency of receptor binding and receptor activation.[11]

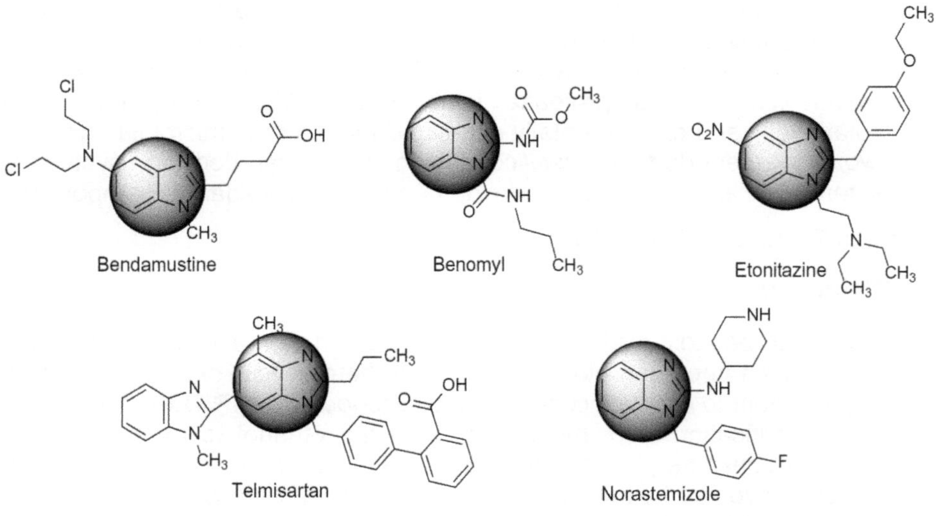

Figure 1 *Chemical structure of biologically active benzimidazole derivatives containing marketed drugs.*

In recent trends, some newly synthetic strategies have been reported for the synthesis of 1,2,5-trisubstituted benzimidazole derivatives via N-C-N bond formation reaction. In the initial method, for the benzimidazole derivatives via the N-C-N bond formation which engage the reaction between *o*-phenylenediamaine and carboxylic acids under different conditions in presence of various catalysts (Scheme **1**) BF_3-Et_2O in CH_2Cl_2,[12] $Bu_4N^+Cl^-$ in H_2O and toluene[13] and vanadyl acetylacetonate[14], useful for the synthesis of targeted compounds. Recently, Sitwala and his co-workers accomplished the liquid-phase combinatorial synthesis of 1,2,5-trisubstituted benzimidazoles, where the soluble polymer supported PEG 5000 reacted with the 4-floro-3-nitro-benzoic acid and amines to form polymer supported amines followed by S_NAr reaction, nitro group reduction and cyclization using aldehydes.[15] Thus, the improvement of the benzimidazole scaffold bearing a diverse range of substituent's at the N-1, C-2 and the C-5 positions pose a synthetic challenge. Benzimidazole derivatives are widely investigated for antibacterial activity but failure in reporting the candidates with *in-vivo* and certainly *in-vitro* efficacy could be either unsuccessful may be due to the incapability of the compounds to go through the bacteria or the proper binding of ligand to the target site.[16]

Based on the various synthetic approaches and biological profiles of the benzimidazole and also in extension of our research work on synthesis of novel benzimidazoles, the objective of study is to carry out the molecular docking studies to elucidate the interactions between the benzimidazole derivatives as a protease inhibitor for severe acute respiratory syndrome (SARS) coronavirus and to verify the synthesized compounds with anti-microbial activity by using Auto dock program. Molecular docking is a computational method that can afford information about intermolecular interactions of proteins, nucleic acids, lipids, and ligands.[17] Function of molecular docking is to attain optimized confirmation of proteins and ligands and also to obtain a relative orientation between proteins and ligands through the minimized energy free system. Binding affinity of 1,2,5-trisubstituted benzimidazoles towards protease inhibitor of SARS coronavirus were computationally and experimentally examined for safer antimicrobial agent. In this work, we reported the *in-silico* design, molecular docking, synthesis, characterization and microbial activity of the novel benzimidazole compounds.

2 RESULTS AND DISCUSSION

2.1 Chemistry

4-chloro-3-nitrobenzoic acid is a wide marketable multi reactive building block that can play an important starting material in heterocyclic oriented synthesis (HOS) prominent to different condensed nitrogenous cycle.[18] Our-work started with the esterification step, by reacting with the ethanol to formation of 4-chloro-3-nitrobenzoate (**1**). The new base-free route for the relevant C-N bond forming step involved by the nucleophilic displacement reaction of 4-chloro-3-nitrobenzoate with aliphatic/aromatic amines in continuous stirring with dichloromethane as solvent for overnight to obtain the N-alkyl-3-nitrobenzoate (**2**). Conversely, to accelerate the process, reduction of the nitro group was the next step in the synthesis sequence, this was achieved using Zn dust with

HCOONH4 in H_2O: IPA as solvent medium at 80 °C temperature to form an intermediate ethyl-3-amino-4-substituted amino benzoate (*o*-phenylenediamaine precursors) (**3**). The core nucleus 2-aryl-1-alkyl/aryl-1*H*-benzo[*d*]imidazole-5-carboxylate (**4**) was prepared by reacting with various substituted aromatic aldehydes in the presence of H_2O: IPA as solvent medium at 80 °C without any catalyst for heterocyclization reaction. The scheme of synthesis in this research is presented in Scheme **2**.

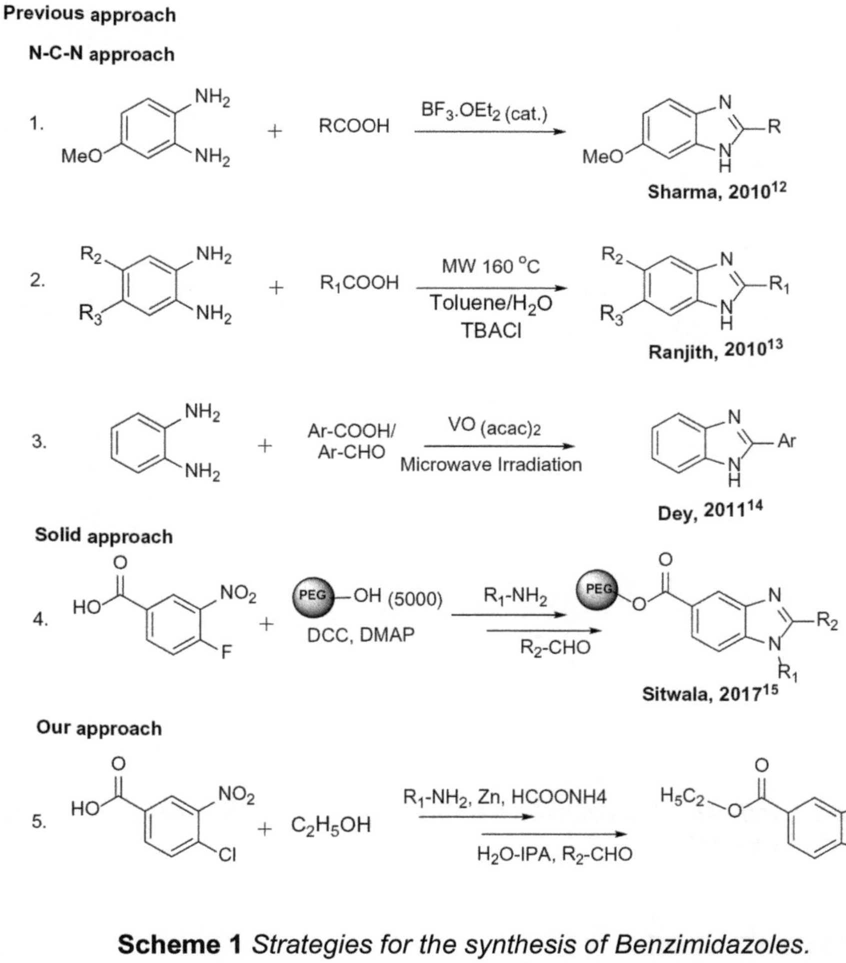

Scheme 1 *Strategies for the synthesis of Benzimidazoles.*

The formation of synthesized 1,2,5-trisubstituted benzimidazoles was characterized by ^{1}H NMR, ^{13}C NMR and high resolution mass spectral analysis. The ^{1}H NMR spectra of benzimidazoles showed two doublets (d) in the region δ 7.66 - 7.97 and δ 7.22 – 7.24 ppm with coupling constant 10.4 Hz that corresponds to the ketone group (-C=O-C-Ar) and aromatic form. The other aromatic protons emerged as multiplet (m) in the aromatic region δ 7.30

- 7.86 ppm. The formation of benzimidazole derivatives were further confirmed by HRMS analysis.

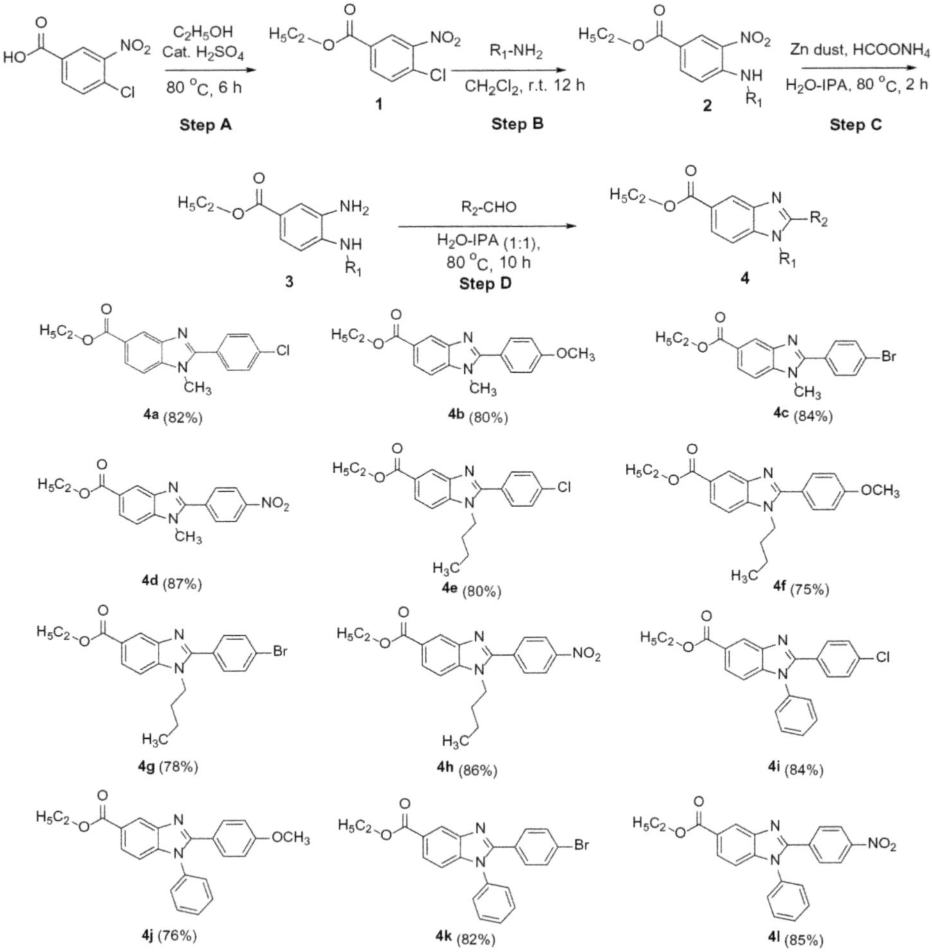

Scheme 2 *Synthesis of benzimidazoles from 4-chloro-3-nitrobenzoic acid and its additional examples.*

2.2. Biological Evaluation

The antimicrobial activity of the synthesized derivatives **4a** - **4l** was evaluated against the selected pathogenic organism two gram-negative (*Escherichia coli, Pseudomonas aeruginosa*) and two gram-positive (*Staphylococcus aureus, Bacillus subtilis*) bacterial strains by agar-well diffusion method[19]. The results were compared with standard drug Ciprofloxacin. The zone of inhibition of the compounds was depicted in Table **1**. The results indicate that all the synthesized compounds are good to moderately active against tested pathogenic organisms. Among these, the derivatives, **4d** and **4h** have emerged as significant anti-microbial activity.

2.3 Molecular Docking Simulation

The receptor is the X-ray crystallography derived SARS-Targeting Coronavirus Main Proteases (CoV Mpro) in complex with an inhibitor N1 from Protein Data Bank with PDB ID: 1WOF. The ligand was separated from receptor structure using BIOVIA Discovery Studio 2017 R2 Client software. The 3D structures of benzimidazole and its derivatives as ligand were optimized by Chem Draw Ultra 8.0 (PerkinElmer Inc.). The SARS- CoV Mpro receptor and ligands were prepared for molecular docking simulation using Auto Dock Tools 1.5.6. Further, receptors and ligands are protonated, receptor as macromolecule added the Kollman charges while the ligands added the Gasteiger charges. The grid parameter file was according to the grid box that comprised of 40 x 40 x 40 points with 0.375 Å space and was centred on the active site of SARS- CoV Mpro (x = -40.957, y = -51.293, z = -22.318). The ligand conformation with the lowest free binding energy (ΔG) was chosen from the most favoured cluster. The best ligand conformation was used for the further step of analysis. The receptor-ligand complexes from docking simulation were visualized using EduPyMOL 1.7 and BIOVIA Discovery Studio Visualizer 2017. The predicted best binding domain of SARS-CoV Mpro receptor with a ligands were in a similar way as shown in Figure **2**.

2.4 Prediction of Computational docking method in interaction sites

In an elucidation of binding interaction of proteins and interacting ligands of the nucleic acid/protein with 1,2,5-trisubstituted benzimidazoles were executed by docking studies. The compounds **4d** and **4h** were established to have strong affinity with SARS-Targeting Coronavirus Main Proteases.[20] The intermolecular energy involvement of the compounds **4d** and **4h** to the total energy was found -8.78 and -8.87 kcal/mol (Table **2**) respectively. The interaction complex connecting the protein with compounds **4d** and **4h** had been stabilized by non-covalent interactions, such as hydrophobic and electrostatic, further, hydrogen bonds to achieve proper binding. The benzimidazole moiety in compound **4d** was positioned to form π-π interaction with the residue ARG188 and in compound **4h** it formed a π-interaction with the residue HIS41 (Figure **2**).

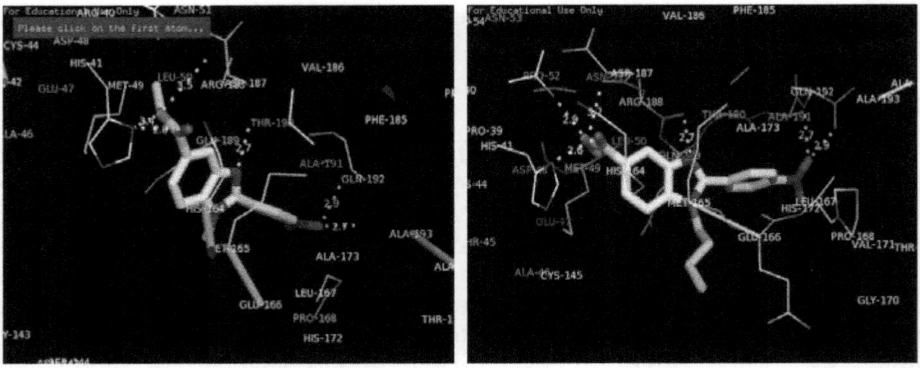

Figure 2 *3D structure of the best docked pose of SARS- CoV Mpro receptor that shows the hydrophobic interactions with binding pocket of compounds **4d** and **4h**.*

Table 1 *The antimicrobial activity of the synthesized derivatives depicted as zone of inhibition in millimetres against tested pathogenic organism*

Compound	Zone of inhibition (mm)							
	Escherichia coli MTCC 1302 (µg/mL)		*Pseudomonas aeruginosa* MTCC 741 (µg/mL)		*Staphylococcus aureus* MTCC 740 (µg/mL)		*Bacillus subtilis* MTCC 121 (µg/mL)	
	100	50	100	50	100	50	100	50
4a	49	20	49	19	35	20	39	15
4b	47	19	46	22	36	19	38	17
4c	41	20	45	17	34	18	44	19
4d	52	30	52	24	46	24	56	22
4e	40	20	31	20	32	16	46	18
4f	39	21	35	15	38	15	43	17
4g	46	23	34	15	29	14	37	18
4h	56	26	44	21	45	19	52	23
4i	44	20	35	17	30	16	42	20
4j	50	29	46	26	49	23	54	26
4k	37	20	34	15	37	16	39	16
4l	40	23	39	18	34	17	45	17
Standard#	60	36	58	32	56	30	62	30
Control	-	-	-	-	-	-	-	-

#Standard used: Ciprofloxacin; control: DMSO.

2.4 Prediction of Computational docking method in interaction sites

In an elucidation of binding interaction of proteins and interacting ligands of the nucleic acid/protein with 1,2,5-trisubstituted benzimidazoles were executed by docking studies. The compounds **4d** and **4h** were established to have strong affinity with SARS-Targeting Coronavirus Main Proteases.[20] The intermolecular energy involvement of the compounds **4d** and **4h** to the total energy was found -8.78 and -8.87 kcal/mol (Table **2**) respectively. The interaction complex connecting the protein with compounds **4d** and **4h** had been stabilized by non-covalent interactions, such as hydrophobic and electrostatic, further, hydrogen bonds to achieve proper binding. The benzimidazole moiety in compound **4d** was positioned to form π-π interaction with the residue ARG188 and in compound **4h** it formed a π-interaction with the residue HIS41 (Figure **2**).

2.5 *In silico* physicochemical properties

The estimation of drug-likeliness nature of active compounds in pharmaceutical research plays an important role in lead generation and lead optimization. Lipinski's "rule of five" helps to develop bioactive drug candidates which are decided by the ADME (absorption, distribution, metabolism and elimination) profile.[21] In our present study, the physicochemical properties of the potent compounds were assessed to determine the ADME properties by using the Swiss ADME property calculation toolkit and the results are summarized in Table **2**. All the target compounds satisfied all the parameters.

3 MATERIALS AND METHODS

3.1. General Methods

All the chemicals and solvents used were of laboratory commercial reagent grade, procured from Avra, Merck, Sigma-Aldrich Co. and were used without further purification. Purity of the compounds was accomplished on thin layer chromatography (TLC) plates in the solvent system hexane-ethylacetate (8:2; 9:1). The spots were located under UV light of short (254 nm)/long (365 nm). The final synthesized compounds were chemically described by thin layer chromatography (TLC), Fourier-transform infrared (FT-IR), nuclear magnetic resonance (^{1}H and ^{13}C NMR), and high-resolution mass (GC-MS) spectroscopy studies. Melting points were determined by open capillary tubes in Gallenkamp (Electronic) melting point apparatus (Hicon, New Delhi, India) and were uncorrected. Fourier-transform infrared spectroscopy (FTIR) spectra were reported on Bruker model Tensor 27 FT-IR spectrometer using KBr pellets; υ_{max} values are given in cm^{-1}. The ^{1}H and ^{13}C NMR were recorded on Bruker model DRX-400 MHz NMR spectrometer (^{1}H at 400 MHz, ^{13}C at 100 MHz) in $CDCl_3$ with tetramethylsilane as internal reference. In NMR spectroscopy, chemical shifts (δ) were accomplished in ppm and coupling constants (*J* values) were mentioned in Hz. Mass spectra were carried out on Perkin Elmer model no. Clarus 680 (GC), presented as *m*/*z*. The purity of reactants, intermediates and products were monitored by TLC on silica-gel aluminium plates from Merck Company (Merck 60 F_{254}). Thin-layer chromatography (TLC) was run throughout the reaction to optimize the reaction for purity and completion. Column chromatography purification was done in solvent system hexane-ethylacetate (9:1) using Silica Gel 60 (0.063-0.200 mm).

Table 2 *In silico physicochemical properties of selected target compounds.* [a]

Code	MW	MR	LogP	HBA	HBD	RotB	TPSA	BS	DS
4a	314.7	87.53	3.26	3	0	4	44.12	0.55	-9.28
4b	281.2	77.94	2.18	4	0	3	67.18	0.56	-9.29
4c	359.2	90.22	3.29	3	0	4	44.12	0.55	-9.39
4d	325.3	91.34	2.55	5	0	5	89.94	0.55	-8.78
4e	356.8	101.95	3.75	3	0	7	44.12	0.55	-9.26
4f	352.4	103.43	3.76	4	0	8	53.35	0.55	-9.02
4g	401.3	104.64	3.99	3	0	7	44.12	0.55	-9.25
4h	367.4	105.76	3.19	5	0	8	89.94	0.55	-8.87
4i	376.8	107.60	3.86	3	0	5	44.12	0.55	-10.58
4j	372.4	109.08	3.91	4	0	6	53.35	0.55	-9.97
4k	421.2	110.29	3.99	3	0	5	44.12	0.55	-10.64
4l	387.3	111.41	3.36	5	0	6	89.94	0.55	-9.77

[a] MW: molecular weight; MR: Molar refractivity; LogP: logarithm of partition coefficient between n-octanol and water; HBA: number of hydrogen bond acceptors; HBD: number of hydrogen bond donors; RotB: number of rotatable bonds; TPSA ($Å^2$): topological polar surface area; BS: Bioavailability score; DS: Docking score.

3.2. Molecular docking

The 3D structures of CoV Mpro (PDB code: 1WOF) was selected as a optimal target protein for docking with 1,2,5-trisubstituted benzimidazole derivatives. Naturally occurring tripeptide, pyrrolidine-2-ones and isoxazoles were occupied as standard ligands.

Protein preparation

The particular PDB files of CoV Mpro (PDB code: 1WOF) was applied for docking studies were collected from the RSCB site (www.rscb.org) and pre-evaluated for subsequent to the docking calculations with the help of BIOVIA Discovery Studio Visualizer 2017 program. The protein preparation was followed according to the manual of Auto Dock Tools 1.5.6. Protein structures were analyzed and handled with Pymol 1.6.0 (The PyMOL Molecular Graphics System, Version 1.6.0.0 Schrödinger, LLC). Chem BioDraw Ultra 11.0 software graphical interface in Chem BioOffice 2008 was used for preparing the chemical structures. Open Babel 2.3.2 software, with MMFF94s force field was used to generate the lowest energy conformations. Docking study has been performed with AutoGrid 4.2 software, employing Auto Dock Tools (ADT) graphical interface in MGL Tools 1.5.6. For the protein and the ligand, Kollman and Gasteiger charges were computed respectively and the non-polar hydrogen atoms were merged with ADT. The Lamarckian genetic algorithm (LGA) was used as a search engine. The active site was defined using AutoGrid 4.2.

3.2.1 Studies of ADME properties

Swiss ADME software was used to compute, and affords the information about the absorption, distribution, metabolism and excretion (ADME) properties of the ligand molecules.

3.3. General Procedure

3.3.1 Synthesis of Ethyl-4-chloro-3-nitrobenzoate (1).

A 100 mL round-bottomed flask was alleged with 4-chloro-3-nitrobenzoic acid (5 g, 2 mmol), ethanol (30 mL) and few drops of catalyst concentrated sulphuric acid refluxed for 6 h. After completion of the esterification process (as evident by TLC), the solvent was evaporated under reduced pressure. The reaction mixture was neutralized with aqueous $NaHCO_3$; further aqueous layer was extracted with ethyl acetate (10 mL x 3). The resulting organic layer was pre-washed with brine (5 mL), dried over anhydrous Na_2SO_4 and concentrated by rotary evaporation, dehydrated, under reduced pressure to yield compound **1** as cream-yellow powder, 98% yield; m.p.: 102-104 °C. ^1H NMR ($CDCl_3$) δ 8.73-8.71 (d, J = 7.1 Hz, 1H), 8.35 (m, 1H), 7.40 (t, J = 8.6 Hz, 1H), 4.53-4.43 (t, 2H, CH_2), 1.43 (s, 3H, CH_3) ppm; ^{13}C NMR (100 MHz, $CDCl_3$) δ: 163.58, 159.31, 156.62, 137.23, 136.46, 127.62, 118.81, 62.08, 14.20 ppm. FT-IR υ (cm^{-1}, KBR): 3742, 3071, 2948, 1695, 764. HRMS (ESI, *m/z*) calcd for $C_9H_8ClNO_4$: 229.0142; found: 228.0854 $[M+H]^+$.

Synthesis of Ethyl-N-alkyl/aryl-3-nitrobenzoate (2).

In a 100 mL vessel, to a mixture of Ethyl-4-chloro-3-nitrobenzoate, **1** (1 g, 4 mmol) and amine (methyl; butyl; benzyl) (3.2 mmol) were mixed in dichloromethane, CH_2Cl_2 (5 mL) stirred to 12 h at room temperature. After completion of the reaction (evident from TLC), the reaction mixture was washed with 1N HCl; further aqueous layer was extracted with ethyl acetate (10 mL x 2). The organic layer was dried over anhydrous Na_2SO_4 under reduced pressure to yield compound **2**.

(**2a**): A pale yellow powder, 82% yield; ^{1}H NMR ($CDCl_3$) δ 8.87 (d, J = 2.3 Hz, 1H), 8.35 (q, J = 5.1 Hz, 1H), 8.09 (d, J = 8.2 Hz, 1H), 6.87 (d, J = 8.2 Hz, 1H), 4.37 (q, J = 6.4 Hz, 2H), 3.09 (d, J = 5.1 Hz, 3H), 1.39 (t, J = 6.4 Hz, 3H) ppm; ^{13}C NMR (100 MHz, $CDCl_3$) δ: 165.20, 148.48, 136.45, 131.31, 129.33, 117.53, 113.09, 61.03, 29.93, 14.39 ppm. HRMS (ESI, *m/z*) calcd for $C_{10}H_{12}N_2O_4$: 224.2133; found: 224.1226 $[M]^+$.

(**2b**): Light yellow powder, 88% yield; ^{1}H NMR ($CDCl_3$) δ 8.86 (d, J = 2.3 Hz, 1H), 8.34 (q, J = 8.1 Hz, 1H), 8.06 (d, J = 4.9 Hz, 1H), 6.87 (d, J = 8.1 Hz, 1H), 4.36 (q, J = 6.4 Hz, 2H), 3.35 (t, J = 5.7 Hz, 2H), 1.74 (p, J = 5.9 Hz, 2H), 1.50 (d, J = 6.2 Hz, 2H), 1.39 (t, J = 6.4 Hz, 3H), 1.00 (t, J = 6.7 Hz, 3H) ppm; ^{13}C NMR (100 MHz, $CDCl_3$) δ: 165.20, 147.75, 136.33, 131.12, 129.45, 117.33, 113.47, 60.98, 42.96, 30.85, 20.19, 14.39, 13.73 ppm. HRMS (ESI, *m/z*) calcd for $C_{13}H_{18}N_2O_4$: 266.2930; found: 267.3610 $[M+H]^+$.

(**2c**): Light yellow powder, 90% yield; ^{1}H NMR ($CDCl_3$) δ 9.71 (s, 1H), 8.90 (d, 1H, J = 2.2 Hz), 8.32 (d, J = 8.1 Hz, 1H), 7.42 (d, J = 7.9 Hz, 1H), 7.27 – 7.12 (m, 2H), 6.80 (t, J = 7.0Hz, 1H), 6.68 (m, 2H), 4.42 (q, J = 6.4 Hz, 2H), 1.41 (t, J = 6.4 Hz, 3H) ppm; ^{13}C NMR (100 MHz, $CDCl_3$) δ: 164.75, 163.61, 159.35, 156.66, 145.53, 137.25, 136.2, 132.4, 130.1, 129.14, 127.6, 118.62, 115.38, 62.11, 21.04, 14.35 ppm. HRMS (ESI, *m/z*) calcd for $C_{15}H_{14}N_2O_4$: 286.2827; found: 287.0361 $[M+H]^+$.

3.3.2 Synthesis of Ethyl-N-alkyl/aryl-3-aminobenzoate (or) o-Phenylenediamaine precursors (3).

To a solution of ethyl-N-alkyl/aryl-3-nitrobenzoate, **2** (0.9 g, 2 mmol) was added with Zn dust (0.6 g, 5 mmol) were mixed in $HCOONH_4$ and H_2O:IPA (5 mL, 1:1) as solvent. The content of flask was refluxed for 2 h at 80 °C until the completion (solution turned colourless) to obtain the diamine derivatives. Further, the reaction mixture was filtered to remove zinc and the filtrate was dried under the reduced pressure. The filtrate was extracted in ethylacetate and dried over the anhydrous Na_2SO_4 under reduced pressure to yield compound **3**.

(**3a**): Colourless powder, 84% yield; ^{1}H NMR ($CDCl_3$) δ 7.41 (d, J = 8.6 Hz, 1H), 7.26 (d, J = 2.2 Hz, 1H), 6.57 (d, J = 8.5 Hz, 1H), 4.32 (q, J = 6.4 Hz,

2H), 4.00 (q, J = 5.1 Hz, 1H), 3.26 (s, 2H), 2.91 (d, J = 5.1 Hz, 3H), 1.36 (t, J = 6.4 Hz, 3H) ppm; ^{13}C NMR (100 MHz, $CDCl_3$) δ: 167.1, 148.4, 143.9, 132.1, 129.3, 124.3, 119.2, 117.8, 108.8, 60.2, 30.4, 14.4 ppm. HRMS (ESI, *m/z*) calcd for $C_{10}H_{14}N_2O_2$: 194.2304; found: 194.0830 $[M]^+$.

(**3b**): Colourless powder, 88% yield; ^{1}H NMR ($CDCl_3$) δ 7.60 (d, J = 8.4 Hz, 1H), 7.41 (d, J = 2.2 Hz, 1H), 6.59 (d, J = 8.4 Hz, 1H), 4.31 – 4.29 (m, 2H), 3.91 (s, 1H), 3.24 (q, J = 5.6 Hz, 2H), 1.66 (d, J = 5.9 Hz, 2H), 1.46 (d, J = 23.1 Hz, 2H), 1.35 (d, J = 8.1 Hz, 3H), 0.97 (t, J = 6.7 Hz, 3H) ppm; ^{13}C NMR (100 MHz, $CDCl_3$) δ: 167.1, 143.2, 131.9, 124.3, 118.9, 109.2, 60.2, 43.4, 31.5, 20.3, 14.4, 13.9. HRMS (ESI, *m/z*) calcd for $C_{13}H_{20}N_2O_2$: 236.3101; found: 236.1498 $[M]^+$.

(**3c**): Colourless powder, 78% yield; ^{1}H NMR ($CDCl_3$) δ 8.91 (d, J = 8.3, 2.3 Hz, 1H), 7.44 (m, 1H), 7.35 (s, 1H), 7.26 – 7.13 (m, 2H), 7.03 (d, J = 8.3 Hz, 1H), 6.82 (t, J = 7.0 Hz, 1H), 4.44-4.42 (m, 4H), 4.36 (q, 2H), 1.42 (t, J = 6.4 Hz, 3H) ppm; ^{13}C NMR (100 MHz, $CDCl_3$) δ: 164.7, 163.6, 159.3, 156.6, 145.5, 136.5, 132.1, 130.1, 127.7, 126.3, 118.8, 115.3, 62.1, 14.2 ppm. HRMS (ESI, *m/z*) calcd for $C_{15}H_{16}N_2O_2$: 256.2997; found: 256.3498 $[M]^+$.

3.3.3 Synthesis of 2-aryl-1-alkyl/aryl-1H-benzo[d]imidazole-5-carboxylate (4).

Ethyl-N-alkyl/aryl-3-aminobenzoate (**3**) (0.5 g, 2.5 mmol) in H_2O:IPA (5 mL, 1:1) was mixed with different substitute aromatic aldehydes (2.5 mmol) and the reaction mixture was heated for 10 h at 80 °C. Terminus of the reaction was progressed by TLC. Further, the reaction mixture was cooled to room temperature and extracted with ethylacetate (10 mL x 2). The combined organic layer was subjected to evaporation, dried over the anhydrous Na_2SO_4 under reduced pressure and concentrated to give the crude product, which was purified by column chromatography using 10% ethylacetate: hexane as the solvent to obtain the products **4a-l**.

(**4a**): Brown liquid, 82% yield; ^{1}H NMR ($CDCl_3$) δ 7.97 (d, J = 2.3 Hz, 1H), 7.92 – 7.86 (m, 2H), 7.82 (d, J = 8.2 Hz, 1H), 7.49 – 7.43 (m, 2H), 7.22 (d, J = 8.2 Hz, 1H), 4.42 (q, J = 6.4 Hz, 2H), 4.05 (s, 3H), 1.40 (t, J = 6.4 Hz, 3H) ppm; ^{13}C NMR (100 MHz, $CDCl_3$) δ: 166.3, 153.3, 138.8, 138.6, 134.3, 129.4, 128.9, 124.9, 118.6, 115.2, 60.9, 31.9, 14.1 ppm. HRMS (ESI, *m/z*) calcd for $C_{17}H_{15}ClN_2O_2$: 314.7662; found: 315.2144 $[M+H]^+$.

(**4b**): Yellow crystals, 80% yield; ^{1}H NMR ($CDCl_3$) δ 7.97 (d, J = 2.3 Hz, 1H), 7.85 – 7.77 (m, 3H), 7.22 (d, J = 8.2 Hz, 1H), 7.01 – 6.95 (m, 2H), 4.42 (q, J = 6.4 Hz, 2H), 4.05 (s, 3H), 3.82 (s, 3H), 1.40 (t, J = 6.4 Hz, 3H) ppm; ^{13}C NMR (100 MHz, $CDCl_3$) δ: 167.3, 160.7, 153.3, 138.8, 138.6, 128.5, 125.2, 124.9, 123.0, 118.6, 115.2, 114.8, 60.7, 55.9, 31.9, 14.1 ppm. HRMS (ESI, *m/z*) calcd for $C_{18}H_{18}N_2O_3$: 310.3471; found: 310.4826 $[M]^+$.

(**4c**): Light orange crystals, 84% yield; ^1H NMR ($CDCl_3$) δ 8.13 – 8.07 (m, 2H), 7.97 (d, J = 2.3 Hz, 1H), 7.82 (d, J = 8.2 Hz, 1H), 7.78 – 7.72 (m, 2H), 7.22 (d, J = 8.2 Hz, 1H), 4.42 (q, J = 6.4 Hz, 2H), 4.05 (s, 3H), 1.40 (t, J = 6.4 Hz, 3H) ppm; ^{13}C NMR (100 MHz, $CDCl_3$) δ: 166.0, 153.3, 132.2, 129.7, 125.2, 124.9, 123.1, 118.6, 115.2, 60.8, 31.9, 14.1 ppm. HRMS (ESI, *m*/*z*) calcd for $C_{17}H_{15}BrN_2O_2$: 359.2172; found: 360.0461 $[M+H]^+$.

(**4d**): Orange crystals, 87% yield; ^1H NMR ($CDCl_3$) δ 8.31 – 8.25 (m, 2H), 8.03 – 7.95 (m, 3H), 7.82 (d, J = 8.2 Hz, 1H), 7.22 (d, J = 8.2 Hz, 1H), 4.42 (q, J = 6.4 Hz, 2H), 4.05 (s, 3H), 1.40 (t, J = 6.4 Hz, 3H) ppm; ^{13}C NMR (100 MHz, $CDCl_3$) δ: 166.1, 153.3, 148.4, 138.8, 138.6, 128.4, 124.9, 121.6, 118.6, 115.2, 60.9, 31.9, 14.1 ppm. HRMS (ESI, *m*/*z*) calcd for $C_{17}H_{15}N_3O_4$: 325.3187; found: 325.1063 $[M]^+$.

(**4e**): Light brown liquid, 80% yield; ^1H NMR ($CDCl_3$) δ 7.97 (d, J = 2.3 Hz, 1H), 7.85 – 7.77 (m, 3H), 7.22 (d, J = 8.6 Hz, 1H), 7.01 – 6.95 (m, 2H), 4.42 (q, J = 6.4 Hz, 2H), 4.02 (t, J = 7.0 Hz, 2H), 3.82 (s, 3H), 2.00 – 1.93 (m, 2H), 1.40 (t, J = 6.4 Hz, 3H), 1.37 – 1.27 (m, 2H), 0.91 (t, J = 6.8 Hz, 3H) ppm; ^{13}C NMR (100 MHz, $CDCl_3$) δ: 166.0, 153.4, 138.8, 138.6, 129.4, 128.9, 125.2, 124.9, 118.6, 115.2, 60.9, 48.0, 33.1, 20.2, 14.1,13.8 ppm. HRMS (ESI, *m*/*z*) calcd for $C_{20}H_{21}ClN_2O_2$: 356.8459; found: 356.8802 $[M]^+$.

(**4f**): Yellow powder, 75% yield; ^1H NMR ($CDCl_3$) δ 7.97 (d, J = 2.3 Hz, 1H), 7.85 – 7.77 (m, 3H), 7.22 (d, J = 8.6 Hz, 1H), 7.01 – 6.95 (m, 2H), 4.42 (q, J = 6.4 Hz, 2H), 4.02 (t, J = 7.0 Hz, 2H), 3.82 (s, 3H), 2.00 – 1.93 (m, 2H), 1.40 (t, J = 6.4 Hz, 3H), 1.37 – 1.27 (m, 2H), 0.91 (t, J = 6.8 Hz, 3H) ppm; ^{13}C NMR (100 MHz, $CDCl_3$) δ: 166.0, 153.3, 138.8, 138.6, 128.5, 125.2, 124.9, 123.0, 118.6, 115.2, 114.8, 60.9, 55.9, 48.0, 33.1, 20.2, 14.1, 13.8 ppm. HRMS (ESI, *m*/*z*) calcd for $C_{21}H_{24}N_2O_3$: 352.4269; found: 353.5802 $[M+H]^+$.

(**4g**): Orange crystal, 78% yield; ^1H NMR ($CDCl_3$) δ 8.13 – 8.07 (m, 2H), 7.97 (d, J = 2.3 Hz, 1H), 7.82 (d, J = 8.6 Hz, 1H), 7.78 – 7.72 (m, 2H), 7.22 (d, J = 8.6 Hz, 1H), 4.42 (q, J = 6.4 Hz, 2H), 4.02 (t, J = 7.0 Hz, 2H), 1.97 (t, J = 6.9, 6.2 Hz, 2H), 1.40 (t, J = 6.4 Hz, 3H), 1.37 – 1.27 (m, 2H), 0.91 (t, J = 6.8 Hz, 3H) ppm; ^{13}C NMR (100 MHz, $CDCl_3$) δ: 166.1, 153.3, 138.8, 138.6, 132.2, 129.7, 125.2, 124.9, 123.1, 118.6, 115.2, 60.9, 48.0, 33.1, 20.2, 14.1, 13.8 ppm. HRMS (ESI, *m*/*z*) calcd for $C_{20}H_{21}BrN_2O_2$: 401.2969; found: 402.1842 $[M+H]^+$.

(**4h**): Light orange powder, 86% yield; ^1H NMR ($CDCl_3$) δ 8.31 – 8.25 (m, 2H), 8.03 – 7.95 (m, 3H), 7.82 (d, J = 8.6 Hz, 1H), 7.24 (d, J = 8.6 Hz, 1H), 4.42 (q, J = 6.4 Hz, 2H), 4.02 (t, J = 7.0 Hz, 2H), 1.97 (t, J = 6.9, 6.2 Hz, 2H), 1.40 (t, J = 6.4 Hz, 3H), 1.37 – 1.27 (m, 2H), 0.91 (t, J = 6.8 Hz, 3H) ppm; ^{13}C NMR (100 MHz, $CDCl_3$) δ: 166.4, 153.4, 148.4, 138.8, 138.6, 136.6, 128.4, 125.2, 124.9, 121.6, 118.6, 115.2, 60.9, 48.2, 33.1, 20.2, 14.3

ppm. HRMS (ESI, *m/z*) calcd for $C_{20}H_{21}N_3O_4$: 367.3984; found: 367.4208 $[M]^+$.

(**4i**): Brown liquid, 84% yield; ^1H NMR ($CDCl_3$) δ 7.97 (d, J = 2.3 Hz, 1H), 7.92 – 7.86 (m, 2H), 7.83 (d, J = 8.2 Hz, 1H), 7.66 (d, J = 8.2 Hz, 1H), 7.49 – 7.37 (m, 5H), 7.37 – 7.31 (m, 2H), 4.42 (q, J = 6.4 Hz, 2H), 1.40 (t, J = 6.4 Hz, 3H) ppm; ^{13}C NMR (100 MHz, $CDCl_3$) δ: 166.0, 149.4, 139.5, 138.8, 137.4, 134.4, 129.7, 129.4, 128.8, 125.2, 122.4, 118.6, 115.2, 60.9, 14.1 ppm. HRMS (ESI, *m/z*) calcd for $C_{22}H_{17}ClN_2O_2$: 376.8356; found: 377.2081 $[M]^+$.

(**4j**): Light yellow powder, 76% yield; ^1H NMR ($CDCl_3$) δ 7.97 (d, J = 2.3 Hz, 1H), 7.86 – 7.80 (m, 2H), 7.79 (d, J = 1.3 Hz, 1H), 7.66 (d, J = 8.3 Hz, 1H), 7.47 – 7.37 (m, 3H), 7.37 – 7.31 (m, 2H), 7.01 – 6.95 (m, 2H), 4.42 (q, J = 6.4 Hz, 2H), 3.82 (s, 3H), 1.40 (t, J = 6.4 Hz, 3H) ppm; ^{13}C NMR (100 MHz, $CDCl_3$) δ: 166.2, 160.7, 149.4, 139.5, 138.8, 137.4, 129.7, 128.5, 125.2, 124.8, 123.0, 122.4, 118.6, 115.2, 114.8, 60.9, 55.8, 14.1 ppm. HRMS (ESI, *m/z*) calcd for $C_{23}H_{20}N_2O_3$: 372.4165; found: 374.0285 $[M+1]^+$.

(**4k**): Light orange crystal, 82% yield; ^1H NMR ($CDCl_3$) δ 8.13 – 8.07 (m, 2H), 7.97 (d, J = 2.3 Hz, 1H), 7.83 (d, J = 8.2 Hz, 1H), 7.78 – 7.72 (m, 2H), 7.66 (d, J = 8.2 Hz, 1H), 7.47 – 7.37 (m, 3H), 7.37 – 7.31 (m, 2H), 4.42 (q, J = 6.4 Hz, 2H), 1.40 (t, J = 6.4 Hz, 3H) ppm; ^{13}C NMR (100 MHz, $CDCl_3$) δ: 165.8, 149.4, 139.5, 138.8, 137.9, 132.2, 125.2, 124.9, 123.1, 122.4, 118.6, 60.9, 14.1 ppm. HRMS (ESI, *m/z*) calcd for $C_{22}H_{17}BrN_2O_2$: 421.2866; found: 422.0453 $[M+1]^+$.

(**4l**): Orange powder, 85% yield; ^1H NMR ($CDCl_3$) δ 8.10 – 8.04 (m, 2H), 7.97 (d, J = 2.3 Hz, 1H), 7.87 – 7.80 (m, 3H), 7.66 (d, J = 8.2 Hz, 1H), 7.47 – 7.37 (m, 3H), 7.37 – 7.31 (m, 2H), 4.42 (q, J = 6.4 Hz, 2H), 1.40 (d, J = 12.8 Hz, 3H) ppm; ^{13}C NMR (100 MHz, $CDCl_3$) δ: 166.0, 149.5, 148.4, 139.6, 138.7, 137.5, 136.8, 128.4, 125.4, 124.9, 121.6, 118.6, 115.2, 60.8, 14.1 ppm. HRMS (ESI, *m/z*) calcd for $C_{22}H_{17}N_3O_4$: 387.3881; found: 388.1219 $[M+1]^+$.

3.4. Biological evaluation

Anti-microbial assay

The synthesized compounds **4a-4l** were screened *in vitro* for antimicrobial activity against two gram-negative (*Escherichia coli (MTCC-1302) and Pseudomonas aeruginosa (MTCC-741)*) and two gram-positive (*Staphylococcus aureus (MTCC-740) and Bacillus subtilis (MTCC-121)*) bacterial strains by the agar-well diffusion method. Ciprofloxacin was used as standard drug to compare the antibacterial activites. The wells (6 mm in diameter) were dug in the media with the help of a germ-free metallic borer with centres at least 24 mm apart. In a nutrient agar media applying two to eight hours old bacterial inoculate containing approximate 10^4-10^6 colony-forming units (CFU/mL) were spread with a help of cotton swab. The tests were accomplished at a concentration of 100 μg/mL and 50 μg/mL. After completion of incubation at 37 °C for 48 h, the zone of inhibition was measured in millimetres.

4. CONCLUSION

This work is focused on finding novel potent compounds based on benzimidazole nucleus and to achieve this, a series of twelve new compounds were designed and efficiently synthesized with good yields and characterized by various spectroscopic techniques. Amongst the synthesized benzimidazoles evaluated, the ones with electron donating substituent in the 2nd position with aromatic ring showed higher binding affinity towards SARS-Targeting Coronavirus Main Proteases. Compounds, **4d** and **4h** emerged as promising an anti-microbial agent which was identified by molecular docking simulation and all the derivatives met with all the parameters of *Lipinski's Rule* of five. Novel benzimidazole derivatives are needed for the further pharmacological investigation and for future development of anti-cancer agents.

Acknowledgements

The authors thank the Chancellor and Vice Chancellor of Vellore Institute of Technology (Deemed to be University) for providing opportunity to carry out this study. Further, the author Sreerama Rajasekhar wishes to thank the Management and Principal of Sri Venkateswara College of Pharmacy, Chittoor for providing the facilities and financial assistance to carry out this study and thanks to UGC, New Delhi, Government of India for funding through F.No.4-4/2014-15 (MRP/SEM/UGC/SERO) P.No.1412. They are also thankful to SIF, VIT, Vellore for the spectral analysis.

References

1. N. N. A. Mohammed, Y. Alias, Z. Abdullah, R. M. Shakir, E. M. Taha and A. A. Hamid, *Molecules*, 2013, **18**, 11978.
2. Y. B. Bai, A. L. Zhang, J. J. Tang and J. M. Gao, *J. Agric. Food Chem.*, 2013, **61**, 2789.
3. R.K. Arora, N. Kaur, Y. Bansal and G. Bansalm, *Acta Pharm. Sin. B.*, 2014, **4**, 368.
4. Y. Zheng, M. Zheng, X. Ling, Y. Liu, Y. Xue, L. An, N. Gu and M. Jin, *Bioorg. Med. Chem. Lett.*, 2013, **23**, 3523.
5. B. Munguia, P. Mendina, R. Espinosa, A. Lanz, J. Saldana, M. J. Andina, X. Ures, A. Lopez, E. Manta and L. Domínguez, *Lett. Drug Desig. Discov.*, 2013, **10**, 1007.
6. M. C. Sharma, D. V. Kohli, S. Sharma and A. D. Sharma, *Pharmacia Sinica*, 2010, **1**, 104.
7. J. Cheng, J. Xie and X. Luo, *Bioorg. Med. Chem. Lett.*, 2005, **15**, 267.

8. Y. Bansal and O. Silakari, *Med. Chem. Res.*, 2015, **24**, 1170.
9. R. V. Shingalapur, K. M. Hosamani, R. S. Keria and M. H. Hugar, *Eur. J. Med. Chem.*, 2010, **45**, 1753.
10. Y. E. Bakri, E. H. Anouar, I. Marmouzi, K. Sayah, Y. Ramli, M. E. A. Faouzi, E. M. Essassi and J. T. Mague, *J. Mol. Model.*, 2018, **24**, 179.
11. H. Yang, W. Xie, X. Xue, K. Yang and J. Ma, *PLOS Biol.*, 2005, **3**, e324.
12. M. C. Sharma, D. V. Kohli, S. Sharma and A. D. Sharma, *Pharm. Sin.*, 2010, **1**, 104.
13. R. P. Karuvalam, M. Siji, N. Divia and K. R. Haridas, *J. Korean Chem. Soc.*, 2010, **54**, 589.
14. M. Dey, K. Deb and S. S. Dhar, *Chin. Chem. Lett.*, 2011, **22**, 296.
15. N. D. Sitwala, V. K. Vyas, B. C. Variya, S. S. Patel, C. C. Mehta, D. N. Rana and M. D. Ghate, *Bioorg. Chem.*, 2017, **75**, 118.
16. P. Picconi, C. Hind, S. Jamshidi, K. Nahar, M. Clifford, M. E. Wand, J. M. Sutton and K. M. Rahman, *J. Med. Chem.,* 2017, **60**, 6045.
17. M. A. A. El-Sayed, N. I. Abdel-Aziz, A. A. M. Abdel-Aziz, A. S. El-Azab and K. E. H. El-Tahir, *Bioorg. Med. Chem.*, 2012, **20**, 3306.
18. S. Křupková, P. Funk, M. Soural and Jan Hlaváč, *ACS Comb. Sci.,* 2013, **15**, 20.
19. A. U. Rahman, M. I. Choudhary and W. J. Thomsen, *Bioassay Techniques for Drug Development*, Harwood Academic Publishers, Amsterdam, The Netherlands, 2001, p. 16.
20. B. Xia and X. Kang, *Protein Cell,* 2011, **2**, 282.
21. T. B. Fernandes, M. C. F. Segretti, M. C. Polli and R. Parise-Filho, *Lett. Drug Des. Discovery*, 2016, **13**, 999.

IN SILICO STUDIES OF BENZOXAZOLE DERIVATIVES USING FERRITE-*L*-CYSTEINE MAGNETIC NANOPARTICLES: GREEN SYNTHESIS

G. Neelima,[a, b*] K. Lakshmi[b] and K. Sesha Maheswaramma[c]

[a] JNTUA, Ananthapuramu, India; [b] Department of Pharmaceutical Analysis, Sri Venkateswara College of Pharmacy, RVS Nagar, Chittoor-517127, Andhra Pradesh, India ;[c] Department of Chemistry, JNTUA College of Engineering (Autonomous), Pulivendula-516390, Andhra Pradesh, India

Email: grandhe.neelima37@gmail.com

1 INTRODUCTION

Heterocyclic compounds of the benzene ring fused with heteroatom (N, O) five-member ring of benzoxazole is playing vital role as pharmaceutical agents. Benzoxazole derivatives are recognized as an important scaffold in anti-inflammatory[1], anti-oxidant[2], antitumor agents[3], cathepsin S inhibitor[4], melatonin receptor agonists[5], Rho-kinase inhibitors[6], amyloidogenesis inhibitors[7] etc. In particular, there are several drugs contain the benzoxazole moiety such as pseudopteroxazole (**A**, used for the treatment of tuberculosis), flunoxaprofen, benoxaprofen (**B**, **C** as an anti-inflammatory drugs), caboxamycin (**D**, as an antibiotic), tafamidis (**E**, for deadly neurodegenerative disease), available in the market as shown in Figure 1.

Recently, endurable modification is the challenge for the researchers to develop and design simple synthetic methods under mild and ecofriendly conditions. In 1998, Anastas and his co-workers propounded by the twelve principles of green synthesis describes usage of green solvent and catalytic reagents that will alter the phase of synthetic organic reactions and a corner stone for innovative changes in chemical process. Catalyst has prominent role in the transformation of raw material into appropriate products and gives immense scope of versatile applications in the field of organic synthesis .In recent trends, nanocatalysts are regarded as a revolution in the field of catalysis and green synthesis. But, previous researchers accomplished that metal nanocatalyst ($CuFe_2O_3$, $NiFe_2O_3$ and $CoFe_2O_3$) at high temperature become aggregate or decompose easily and are restricted to limited heterocyclic reactions only. To overcome these problems eventually metal nanoparticle has further undergone to integrate new class of catalyst i.e. organo-nanocatalyst[8-12].

Currently, greener methods have been explored ionic liquids, monolayer and multilayer coated organo- nanocatalyst having promising super catalytic properties can play a fundamental role in both industrial and academic investigators alternative to conventional catalyst. Organocatalyst are conventionally occur to be homogenous form, but when converted as nano-organocatalyst they behave like heterogeneous catalyst. Many chemists focussed on heterogeneous catalysis to overcome the environmental problems, as it possess advantages like recyclability, efficient activity, good recovery, inexpensive in preparation, more selectivity and sufficient chemical stability[13].

Heterogeneous organo-nanocatalyst consists immobilization of the amino acid anchored to the magnetic solid support and improves recyclability. It consists of metal nanoparticle coated with active organic moiety. However it has several advantages over conventional catalyst, easily isolated and separated from a reaction vessel using an advanced technique called magnetic decantation process instead of filtration and centrifugation methods. Some of the researchers developed organo-nanocatalyst like Fe_3O_4-Chitosan[14], Fe_3O_4-HAP-sulfonic acid[15], Fe_3O_4-silica[16], Fe_3O_4-proline[17], Fe_3O_4-*L*-arginine[18], Fe_3O_4-glutathione[19] and Fe_3O_4-Cysteine[20] have been tremendously studied.

In the present study, we have designed monolayer coated organo-nanocatalyst affords immobilization of thiol moiety with catalytic property of *L*-cysteine anchored to free hydroxyl functional group onto the surface of ferromagnetic activity of ferrite as a magnetic nano-core. Therefore, synthesis of benzoxazole derivatives using nanocatalyst and eco-benign green solvent under simple and moderate reaction conditions is considerably inexpensive, eco-friendly, less toxic compared to other organic solvents. The synthesis of benzoxazole with nano magnetite recyclable catalyst and lemon juice is the highlight of this work. In recent drug discovery substituted benzoxazole derivatives are the important building blocks for various pharmacological activities and so these are evaluated for *in-vitro* anti-inflammatory and anti-oxidant activity[21-24].

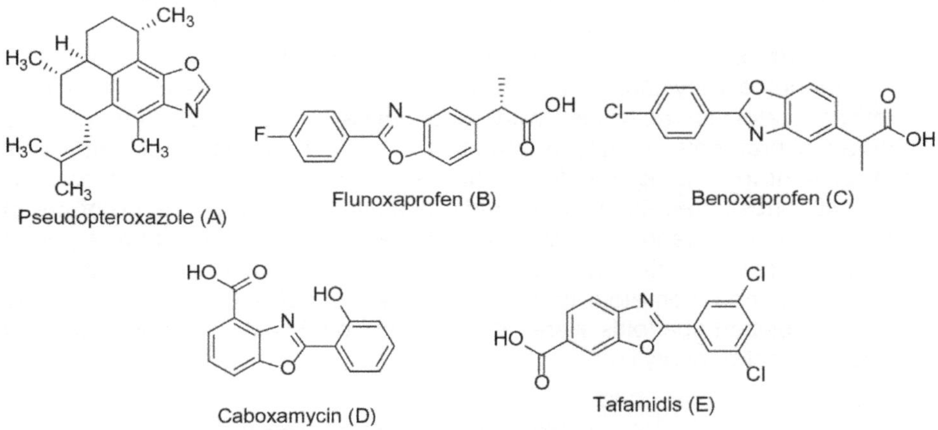

Figure 1 *Benzoxazole scaffold containing marketed drugs.*

2 EXPERIMENTAL

2.1. Materials and Instrumentation

In different circumstances indicated all the chemicals and catalytic reagents were used as obtained from different commercial suppliers as received without further purification. The progress of reactions were observed by thin layer chromatography (TLC) using silica gel-G UV 254 plates using hexane and ethylacetate (8:2 and 9:1) as mobile phase. Infrared spectra were depicted on a Bruker, FT-IR spectrometer Tensor 27 using potassium bromide pellets. ^1H NMR (400 MHz) and ^{13}C NMR (100 MHz) spectra were recorded on a Bruker DRX400 spectrometer in $CDCl_3$ and DMSO-d_6 solvents. Chemical shifts (δ) were revealed in ppm corresponding to tetramethylsilane (TMS) as internal standard, *J* values were mentioned in Hz. Powder X-ray diffraction (XRD) analysis was compiled on Bruker SMART APEX II CCD provided with a graphite monochromator. Transmission electron microscope (TEM) analysis was reported on JEOL 2100 to reveal the morphological nature of nano particles. A small quantity of diluted sample in alcohol was dripped on a TEM grid. Mass spectra were recorded on a Perkin Elmer, Calrus 680 GC-MS spectrometer.

2.2. Optimization studies

Initially the effect of catalyst loading, optimum reaction temperature, the nature of solvent and time were examined. To enhance yield of the reaction were optimized, by using of different solvents and catalysts. Maximum yield obtained when the reaction was carried out with lemon juice and ferrite cysteine nano organocatalyst. Monolayer coated organo-nanocatalyst affords immobilization of the amino acid anchored to the magnetic solid support which decrease the temperature of a chemical reaction, enhances the chemical transformation, reduces the atom economy and improve recyclability.

2.3. Synthesis of benzoxazole derivatives (3a-g)

Nano magnetite recyclable supported catalysis recently emerged as highly promising catalytic activities with economically viable precursors and preparation protocols. Acidity is maintained by lemon juice, considered as green solvent in organic transformation. 3-amino-4-hydroxy benzoic acid (**1**) reacts with substituted aliphatic/aromatic aldehydes (**2**), in presence of green solvent and nanoorgano catalyst under different conditions. Purity analysis of synthesized compounds was done by recrystallization, melting point, thin layer and column chromatography. The synthetic scheme is shown in Scheme **1**, the synthesized scaffolds were characterized by FT-IR, ^1H NMR, ^{13}C NMR and Mass spectral analysis.

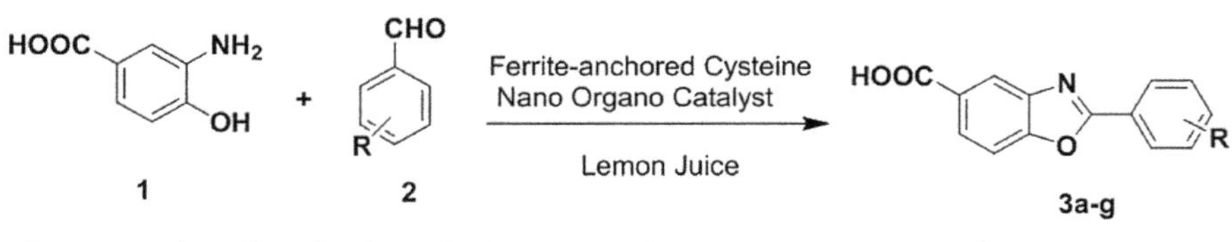

Scheme 1 *Synthetic Scheme of benzoxazoles from 3-amino-4-hydroxybenzoic acid using ferrite cysteine nano organocatalyst.*

2.4. Preparation of Ferrite magnetic nanoparticles (MNPs):

To a solution of ferric chloride hexahydrate (10.8 g) and urea (7.3 g) in 200 mL of deionized water at 85 to 90 ˚C for 2 h. Then, the reaction mixture turned to brown, added ferrous sulphate (5.6 g) and sodium hydroxide (0.1 M) was injected until to get the pH 10. Formed hydroxides were homogenized by ultrasound sonication in the sealed flask by maintaining the temperature 30 to 35 ˚C for 30 min. After vigorous stirring for 5 h, resultant black precipitate of ferrite was washed, purified and dried in a vacuum.

2.5. Preparation of Ferrite - Cysteine Nano organocatalyst:

Typically, ferrite MNPs (1 g) were added to the deionized water (20mL) and L-cysteine (1 g) dispersed in 40-50 mL methanol-water (1:1) was dissolved to above suspended ferrite MNPs. The resultant content was vigorously stirred at r.t. for 24 h (1000 RPM) in a magnetic stirrer.[25] The ferrite cysteine MNPs were separated by external magnetic field then washed with distilled water, methanol (2 spells 10 mL) and dried under a vacuum at 60 ˚C for 24 h as shown in Scheme **2**.

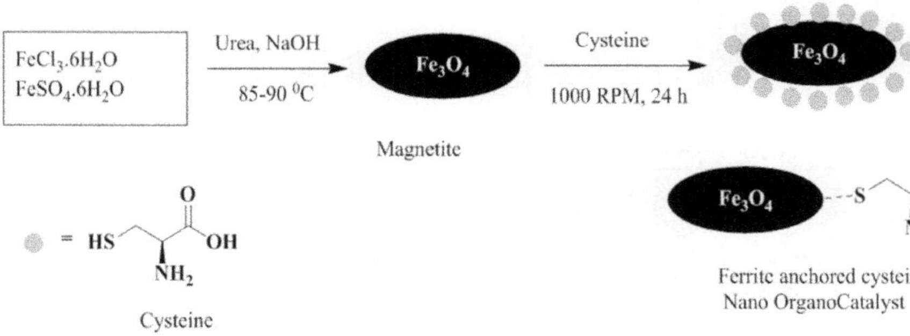

Scheme 2 *Synthesis of Ferrite Cysteine MNPs.*

2.6. General procedure for preparation of Nano organocatalyst mediated Benzoxazole derivatives:

To a mixture of 2-aminophenol (1.5 mmol), aromatic aldehydes (1.5 mmol), nano organo catalyst ferrite – cysteine MNPs (0.005 g) were dissolved in 10 ml of green solvent was refluxed at 80 ˚C for an appropriate time until the reaction content became solidified. The terminus of the reaction was progressed by TLC (n-hexane: ethyl acetate 8:2). After end of the reaction content turned clear and then the catalyst was efficiently alienated by using an external magnet and recycled for next synthesis. After elimination of catalyst, the mixture was cooled to room temperature. Resulting crude reaction was purified by recrystallization from warm ethanol to afford the final derivatives. The resultant synthesized benzoxazoles (**3**) were obtained in good to excellent yields.

Table 1 *Analytical data of synthesized benzoxazole derivatives (3a-3g).*

Code	Structure of the compound	Molecular formula	MW[a]	Yield (%)
3a	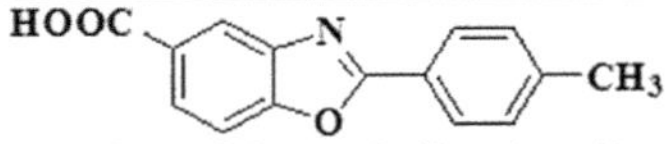 2-*p*-tolylbenzo[*d*]oxazole-5-carboxylic acid	$C_{15}H_{11}NO_3$	253.25	95.15
3b	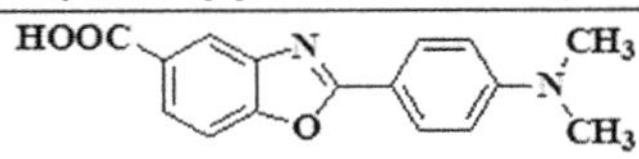 2-(4-(dimethylamino)phenyl)benzo[*d*]oxazole-5-carboxylic acid	$C_{16}H_{14}N_2O_3$	282.29	95.47
3c	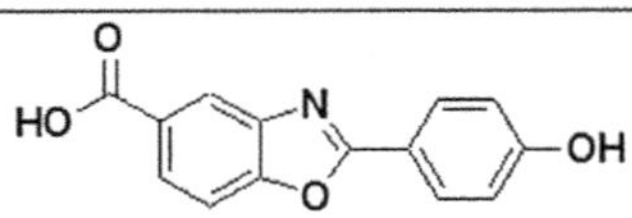 2-(4-hydroxyphenyl)benzo[d]oxazole-5-carboxylic acid	$C_{14}H_9NO_4$	255.23	94.38
3d	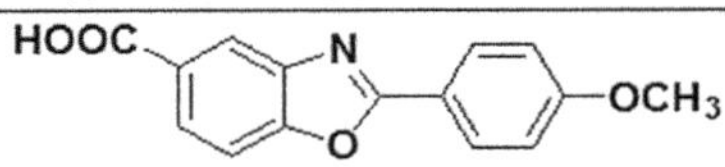 2-(4-methoxyphenyl)benzo[*d*]oxazole-5-carboxylic acid	$C_{15}H_{11}NO_4$	269.25	90.19
3e	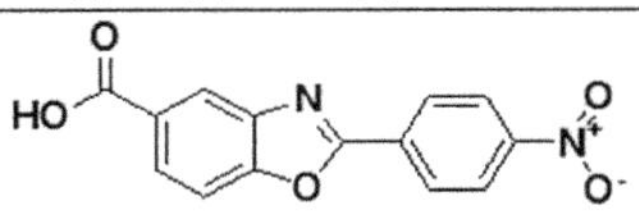 2-(4-nitro phenyl)benzo[*d*]oxazole-5-carboxylic acid	$C_{14}H_8N_2O_5$	284.22	93.48
3f	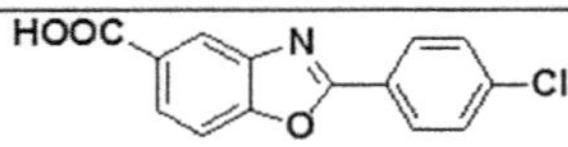 2-(4-chloro phenyl)benzo[*d*]oxazole-5-carboxylic acid	$C_{14}H_8ClNO_3$	273.67	85.26
3g	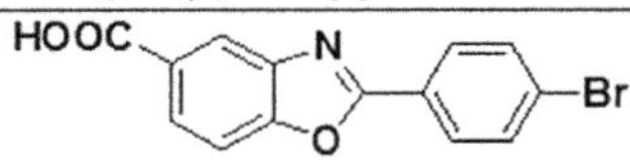 2-(4-bromo phenyl)benzo[*d*]oxazole-5-carboxylic acid	$C_{14}H_8BrNO_3$	318.12	90.28

[a]MW: molecular weight

3 RESULTS AND DISCUSSION

3.1 Characterization of Ferrite - Cysteine Nano organocatalyst

The synthesized nano catalyst characterized by FT-IR spectroscopy, transmission electron microscopy (TEM) and X-ray diffraction (XRD) techniques. The morphology of the prepared nanocatalyst was determined from the TEM image (Figure 2a.) shows that there is a weak interaction of cluster formation may be due to hydrogen bonding and vander-waals forces between the ferrite and cysteine molecules are considerably spherical in shape and the diameter range of catalyst between 100-300 nm.

The powder XRD diffraction pattern image of prepared nanocatalyst is shown in Figure 2b. The various diffraction peaks at angle 2θ = 19.3, 28.4, 35.2,

37.4, 58.6, 62.6 corresponds to the different planes. The sharp diffraction peak clearly shows the highly crystalline nature of catalyst reveals that the product image consists of *L*-cysteine and ferrite with high purity.

3.2 Optimization studies

In the current study first series of experiments, the synthesized and well characterized Ferrite Cysteine MNPs were analyzed for the O-C-N approach and catalytic cyclization by reductive elimination. The optimization of reaction by cyclization of 3-amino-4-hydroxybenzoic acid and aryl aldehydes was considered as a typical reaction in Table 2.

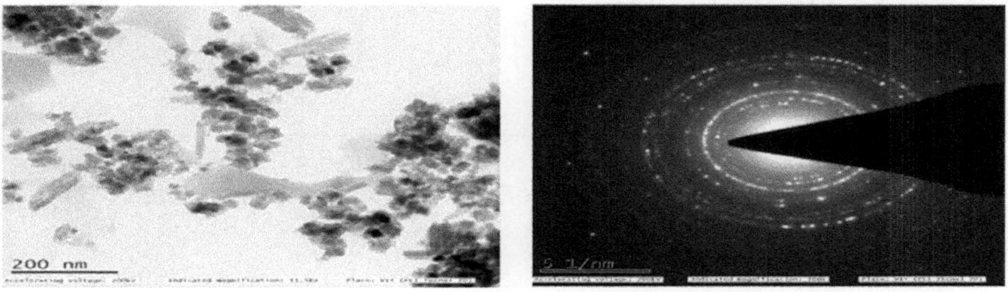

Figure 2a *TEM image of Ferrite Cysteine MNPs catalyst at 200 nm.*

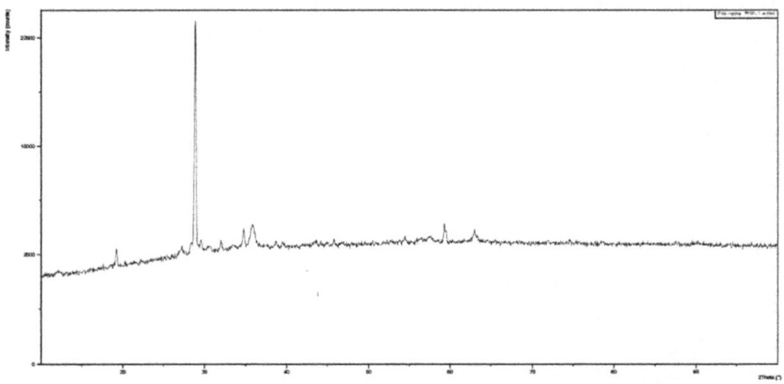

Figure 2b *Powder XRD pattern of Ferrite Cysteine MNPs catalyst.*

Table 2 *Optimization of reaction conditions.*[a]

Entry	Catalysts	Solvent	Time (min.)	Temp (°C)	Yield (%)[b]
1	-	-	240	RT	NR
2	-	-	120	80	Trace
3	Fe- Cys MNP	-	240	100	38
4	Fe- Cys MNP	-	240	60	42
5	Fe- Cys MNP	-	120	80	58
6	Fe- Cys MNP	H_2O	120	80	92
7	Fe- Cys MNP	MeOH	120	80	85
8	Fe- Cys MNP	EtOH	120	80	90

[a]Reaction conditions: 3-amino-4-hydroxybenzoic acid = 1 mmol, aldehydes = 1 mmol, Fe- Cys MNP = 50 mg, reflux. RT= room temperature, NR= no reaction.
[b]Yield of isolated product.

2.7. A tentative mechanistic pathway

According to the literature reports and results possible mechanism for the synthesis of benzoxazole analogues has been shown in Scheme 3. Initally, ferrite -*L*- Cysteine nanocatalyst react with the starting material on their surface to form intermediate 2A and then converted into intermediate 2B in the presence of aldehydes which accomplish the catalytic by the reductive elimination of the fused product.

2.8. Catalyst reusuability:

The reusability and recyclability of the magnetically separable Ferrite *L*-Cysteine catalyst was investigated in the synthesis of benzoxazole derivatives. After the completion of reaction, the nanocatalyst was collected by the influence of external magnetic field (monitored by TLC) and thoroughly washed with water, acetone and dried. Recyclability of nanocatalyst was examined that would be reused upto six successive cycles without any loss in its catalytic efficiency and also observed substantial decrease in the yield of the product (95%, 95%, 94%, 93%, 93% and 91%).[26,27]

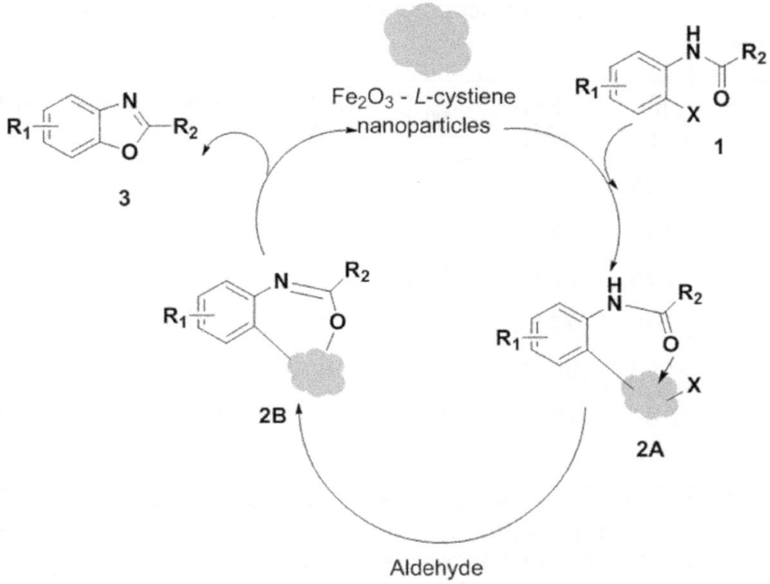

Scheme 3 *Mechanism for Ferrite Cysteine MNPs catalysed benzoxazole derivatives.*

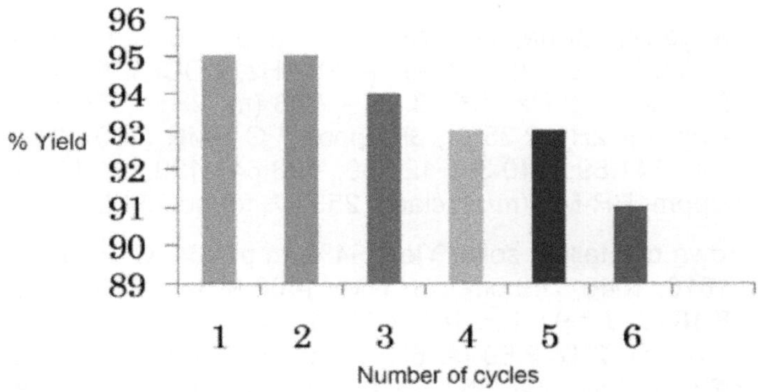

Figure 3 *Reusability plot of ferrite –L-Cysteine nanocatalyst.*

2.9. *In silico* Studies

The physicochemical properties of the synthesized compounds were screened to determine the "absorption, distribution, metabolism, and excretion" (ADME) properties by using the Swiss ADME property calculation toolkit and the results are summarized in Table 3. All the synthesized compounds satisfy the "Lipinski rule of five" with all the parameters[28-30].

Table 3 *In silico physicochemical properties of selected target compounds.*[a]

Compound	Mi Log P	TPSA (Å²)	MW	nOHBA	nOHBD	nROTB	nVioltn
3a	2.34	63.33	253.25	4	1	2	0
3b	2.44	66.57	282.29	4	1	3	0
3c	1.79	83.56	255.23	5	2	2	0
3d	2.40	72.56	269.25	5	1	3	0
3e	1.73	109.15	284.22	6	1	3	0
3f	2.44	63.33	273.67	4	1	2	0
3g	2.50	63.33	318.12	4	1	2	0

[a] miLOGP: logarithm of partition coefficient between n-octanol and water; TPSA: topological polar surface area; MW: molecular weight; nOHBA: number of hydrogen bond acceptors; nOHBD: number of hydrogen bond donors; nROTB: number of rotatable bonds; nVioltn: number of violations.

2.10. Spectral data of synthesized compounds

(3a): Light orange crystalline solid; Yield 95%; m.p. 170 ˚C; FT-IR (υ_{max}/ cm^{-1}): 1728, 1618, 1098, 766 cm^{-1}; ^{1}H NMR (400 MHz, $CDCl_3$): δ 8.62 (d, *J* = 2.3 Hz, 1H), 8.16 (d, *J* = 9.0 Hz, 1H), 8.04 – 7.98 (m, 2H), 7.77 (d, *J* = 9.0 Hz, 1H), 7.34 – 7.28 (m, 2H), 2.35 (s, 3H) ppm; ^{13}C NMR (100 MHz) δ 166.32, 163.21, 152.51, 141.89, 140.36, 129.59, 128.64, 128.28, 127.68, 122.57, 110.31, 21.42 ppm; HR-MS (m/z): clacd. 253.07, found 254.21.

(3b): Light brown crystalline solid; Yield 94%; m.p. 164 ˚C; FT-IR (υ_{max}/ cm^{-1}): 3394, 1762, 1616, 1046, 749 cm^{-1}; ^{1}H NMR (400 MHz, $CDCl_3$): δ 8.64 (d, *J* = 2.3 Hz, 1H), 8.18 (d, *J* = 9.0 Hz, 1H), 7.78 (d, *J* = 9.0 Hz, 1H), 7.70 – 7.64 (m, 2H), 6.80 – 6.74 (m, 2H), 2.89 (s, 6H) ppm; ^{13}C NMR (100 MHz) δ 166.32, 163.21, 153.22, 152.51, 140.36, 128.69, 122.57, 120.32, 111.84, 110.31, 40.43 ppm; HR-MS (m/z): clacd. 282.29, found 283.02.

(3c): Dark brown crystalline solid; Yield 90%; m.p. 183 ˚C; FT-IR (υ_{max}/ cm^{-1}): 3741, 1758, 1622, 1419, 695 cm^{-1}; ^{1}H NMR (400 MHz, $CDCl_3$): δ 8.63 (d, *J* = 2.3 Hz, 1H), 8.15 (d, *J* = 9.0 Hz, 1H), 7.76 (d, *J* = 9.0 Hz, 1H), 7.04 – 6.98 (m, 2H), 6.96 – 6.90 (m, 2H), 5.82 (s, 1H) ppm; ^{13}C NMR (100 MHz): δ 166.32, 163.21, 160.80, 152.51, 140.36, 128.65, 122.57, 120.01, 116.06, 110.31 ppm; HR-MS (m/z): clacd. 255.23, found 255.82.

(3d): Light brown crystalline solid; Yield 94%; m.p. 192 ˚C; FT-IR (υ_{max}/ cm^{-1}): 3746, 1693, 1624, 1417, 694 cm^{-1}; ^{1}H NMR (400 MHz, $CDCl_3$): δ 8.62 (d, *J* = 2.3 Hz, 1H), 8.16 (d, *J* = 9.0 Hz, 1H), 7.77 (d, *J* = 9.0 Hz, 1H), 7.13 – 7.07 (m, 2H), 7.03 – 6.97 (m, 2H), 3.82 (s, 3H) ppm; ^{13}C NMR (100 MHz): δ 166.32,

163.21, 162.98, 152.51, 140.36, 129.66, 128.64, 124.48, 122.57, 114.39, 110.31, 55.35 ppm; HR-MS (m/z): clacd. 269.25, found 269.82.

(**3e**): Dark brown crystalline solid; Yield 90%; m.p. 209 ˚C; FT-IR (υ_{max}/ cm^{-1}): 3546, 1748, 1622, 1418, 698 cm^{-1}; ^{1}H NMR (400 MHz, $CDCl_3$): δ 8.64 (d, *J* = 2.3 Hz, 1H), 8.37 – 8.31 (m, 2H), 8.25 – 8.19 (m, 2H), 8.16 (d, *J* = 9.0, 1H), 7.77 (d, *J* = 9.0 Hz, 1H) ppm; ^{13}C NMR (100 MHz): δ 166.32, 163.21, 152.51, 148.49, 140.36, 133.65, 128.64, 128.41, 124.21, 122.57, 110.31 ppm; HR-MS (m/z): clacd. 284.22, found 285.62.

(**3f**): Dark brown crystalline solid; Yield 85%; m.p. 189 ˚C; FT-IR (υ_{max}/ cm^{-1}): 3467, 1745, 1620, 1420, 695 cm^{-1}; ^{1}H NMR (400 MHz, $CDCl_3$): δ 8.62 (d, *J* = 2.3 Hz, 1H), 8.16 (d, *J* = 9.0, 2.2 Hz, 1H), 8.10 – 8.04 (m, 2H), 7.77 (d, *J* = 9.0 Hz, 1H), 7.54 – 7.48 (m, 2H) ppm; ^{13}C NMR (100 MHz): δ 166.32, 163.21, 152.51, 140.36, 136.70, 131.57, 129.41, 129.06, 128.64, 122.57, 110.31 ppm; HR-MS (m/z): clacd. 273.67, found 274.86.

(**3g**): Blackish red crystalline solid; Yield 90%; m.p. 219 ˚C; FT-IR (υ_{max}/ cm^{-1}): 3548, 1742, 1618, 1424, 696 cm^{-1}; ^{1}H NMR (400 MHz, $CDCl_3$): δ 8.62 (d, *J* = 2.4 Hz, 1H), 8.16 (d, *J* = 9.0 Hz, 1H), 7.77 (d, *J* = 9.0 Hz, 1H), 7.61 – 7.55 (m, 2H), 7.48 – 7.42 (m, 2H) ppm; ^{13}C NMR (100 MHz): δ 166.32, 163.21, 152.51, 140.36, 131.81, 129.71, 129.45, 128.64, 126.93, 122.57, 110.31 ppm; HR-MS (m/z): clacd. 318.12, found 317.86.

4 BIOLOGICAL ACTIVITY

4.1. Anti – Inflammatory activity:

The synthesized compounds were evaluated for *in-vitro* anti-inflammatory activity by using inhibition of protein denaturation method. The test sample and standard drug were diluted in minimal amount of DMF. The test and standard compounds containing of drug was diluted with 1 ml of 1% mM protein solution in saline phosphate buffer and the mixture were incubated at 37 ˚C for 15 min. Diluted with 3 ml of phosphate buffer (0.2 M, pH 7.4). Denaturation was prompted by preserving the reaction content at 70 ˚C in water bath for 15 min. After cooling, their absorbance was measured at 660 nm (SHIMADZU, UV-Spectrophotometer). Denaturation inhibition rate was depicited from the control without addition of drug (Table 4). All experiments were performed in triplicate times and averaged; diclofenac sodium was used as a standard drug.[31]

Percentage of inhibition of protein denaturation was depicited by using the formula.

$$\% \text{ of inhibition of denaturation} = 100((V_c/V_t) -1)$$

Where V_c and V_t are mean absorbance value of control and test compound.

Table 4 *Percentage inhibition of denaturation of synthesized benzoxazole derivatives (3a-3g).*

Compound code	Mean ±SD	% Inhibition of denaturation
Control	0.019	-
3a	0.012 ±0.0005	84
3b	0.003 ±0.0005	96
3c	0.011 ±0.0005	85
3d	0.004 ±0.0005	94
3e	0.0001±0.0005	87
3f	0.007±0.0005	90
3g	0.011 ±0.0005	85
Diclofenac sodium	0.001 ±0.0005	98

Results exhibited here are the mean values from triplicate times± S.D.

4.2. Anti – Oxidant activity: DPPH method

The synthesized compounds were evaluated for *in-vitro* anti-oxidant activity by using free radical scavenging activity by DPPH method. The stock solution was prepared by using DPPH (10 mg) dissolved in 10 mL dimethylsulfoxide (DMSO) and added at different concentrations in methanolic solution (10 mL). Solutions of the test compounds in varying concentrations (100, 150, 200 µg/mL) as well as of ascorbic acid as reference standard were prepared in methanol. An aliquot of each of these solutions (1.0 mL) was taken in different 10 mL volumetric flasks to which the DPPH stock solution (1.0 mL) was added and volume was made to 10 mL. The reaction vessels were incubated at 37 °C for 20 minutes and the absorbance of resulting solution was measured at 540 nm. Each experiment was performed in triplicate times and averaged (Table 5).[32]

% scavenging activity was depicited by using the formula

$$\text{Scavenging activity (\%)} = \frac{Control - Test}{Control} \times 100$$

Table 5 *Percentage DPPH scavenging activity of synthesized benzoxazole derivatives (3a-3g).*

Compound code	Concentration in µg/mL	Percentage of scavenging ± SD
3a	200	74.32 ± 0.102
	150	73.88 ± 0.364
	100	66.28 ± 0.148
3b	200	59.84 ± 0.587
	150	52.67 ± 0.452
	100	49.28 ± 0.892
3c	200	72.54 ± 0.644
	150	71.16 ± 0.528
	100	69.54 ± 0.352
3d	200	67.88 ± 0.377
	150	66.22 ± 0.852
	100	63.33 ± 0.583
3e	200	63.42 ± 0.824
	150	61.23 ± 0.432
	100	59.66 ± 0.258
3f	200	75.58 ± 0.658
	150	74.54 ± 0.912
	100	73.96 ± 0.850
3g	450	60.78 ± 0.956
	300	59.93 ± 0.932
	150	58.68 ± 0.868
Ascorbic acid	200	76.83 ± 0.542
	150	75.96 ± 0.562
	100	74.82 ± 0.256

5 CONCLUSION

Benzoxazole derivatives were successfully synthesized by simple, eco-friendly and mild reaction conditions. This technique gives significant enhancements in the reaction rates and good to excellent yield, by avoiding hazardous solvents use. Benzoxazole derivatives and organo - nanocatalyst were characterized by different spectroscopic and electron microscopy techniques respectively. The method which entangles stable, inexpensive, easily available green solvent, reusable and simple recovery of catalyst will bring much interesting in both industry and academic field. This protocol provides a new greener approach for the synthesis of heterocyclic compounds. Compound 3b containing dimethyl amino moiety exhibited significant anti-inflammatory activity and derivatives 3a, 3c and 3f showed the prominent anti- oxidant activity comparable to the standard respectively and also these compounds screened through Lipinski's rule of five. "Green Synthesis" is not only the ways to synthesize the desired products economically and environmentally congruent way but it is also helps to conserve the environment.

Acknowledgements

The authors gratefully acknowledge the Management and Principal of Sri Venkateswara College of Pharmacy, Chittoor for providing the financial assistance and analytical facilities to carry out this study. They also like to thank SIF, VIT (Deemed to be University), Vellore for the spectral analysis and other catalyst characteristic techniques.

References

1. N. Kablaoui, S. Patel, J. Shao, D. Demian, K. Hoffmaster, F. Berlioz, M. L. Vazquez, W. M. Moore and R. A. Nugent, *Bioorg. Med. Chem. Lett.*, 2013, **23**, 907.
2. R. Gudipati, R. N. Anreddy and S. Manda, *J. Enzyme Inhib. Med. Chem.*, 2011, **26**, 813.
3. S .M. Rida, F. A. Ashour, S. A. M. El-Hawash, M. M. El-Semary, M. H. Badr and M. A. Shalaby, *Eur. J. Med. Chem.*, 2005, **40**, 949.
4. M. Kim, J. Jeon, J. Song, K. H. Suh, Y. H. Kim, K. H. Min and K. O. Lee, *Bioorg. Med. Chem. Lett.,* 2013, **23**, 3140.
5. L. Q. Sun, J. Chen, M. Bruce, J. A. Deskus, J. R. Epperson, K. Takaki, G. Johnson, L. Iben, C. D. Mahle, E. Ryan and C. Xu, *Bioorg. Med. Chem. Lett.*, 2004, **14**, 3799.
6. E. H. Sessions, Y. Yin, T. D. Bannister, A. Weiser, E. Griffin, J. Pocas, M. D. Cameron, C. Ruiz, L. Lin, S. C. Schürer, T. Schröter, P. L. Grasso and Y. Feng, *Bioorg. Med. Chem. Lett.* 2008, **24**, 6390.
7. H. Razavi, S. K. Palaninathan, E. T. Powers, R. L. Wiseman, H.E. Purkey, N. N. Mohamedmohaideen, S. Deechongkit, K. Chiang, M. T. Dendle, J. C. Sacchettini and J. W. Kelly, *Angew. Chem.,* 2003, **42**, 2758.

8. P. I. Dalko and L. Moisan, *Angew. Chem.,* 2001, **113**, 3840; P. I. Dalko and L. Moisan, *Angew. Chem., Int. Ed.,* 2001, **40**, 3726; P. I. Dalko and L. Moisan, *Angew. Chem.,* 2004, **116**, 5248; P. I. Dalko and L. Moisan, *Angew. Chem., Int. Ed.,* 2004, **43**, 5138.
9. J. Jiang, H. Zou, Q. Dong, R. Wang, L .Lu, Y. Zhu and W. He, *J. Org. Chem.*, 2016, **81**, 51.
10. B. Maleki, M. Baghayeri, S. M. Mohammadzadeh and S. Akhoondhi, *RSC Adv.,* 2015, **5**, 46545.
11. M. A. Patil, P. A. Ubale, S. S. Karhale and V. B. Helavi, *Der Chemica Sinica,* 2017, **8**, 198.
12. P. T. Anastas and J. C. Warner, *Green Chemistry: Theory and Practice*, Oxford University Press, New York, 1998.
13. N. G. Singh, M. Lily, S. P. Devi, N. Rahman, A. Ahmed, A. K. Chandra and R. Nongklaw, *J. Name.,* 2013, **00**, 1.
14. A. Maleki, N. Ghamari and M. Kamalzare, *RSC Adv,* 2014, **4**, 9416.
15. M. Mohsenimehr, M. Mamaghani, F. Shirini, M. Sheykhan and F. A. Moghaddam, *Chin. Chem. Lett.,* 2014, **25**, 1387.
16. K. Sharma, Y. Monga and A. Puri. *J. Mol. Catal. A. Chem.,* 2014, **393**, 84
17. K. Azizi and A. Heydari. *RSC Adv.,* 2014, **4**, 6508.
18. K. Azizi, M. Karimi, H. R. Shaterian and A. Heydari, *RSC Adv.,* 2014, **4,** 42220.
19. A. Gupta, R. Jamatia and A. K. Pal, *New J. Chem.,* 2013, **37**, 1.
20. R. Swapnil and N. Sharad, *Current Catalysis,* 2018, **7**, 43.
21. D. Yang, X. Zhu, W. Wai, M. Jiang, N. Zhang, D. Ren, J. You and H. Wang, *Synlett.,* 2014, **28**, 729.
22. M. B. Gawande, A. Velhinho, I. D. Nogueria, C. A. A. Ghumman, O. M. N. D. Teodorod and P. S. Branco, *RSC Adv.,* 2012, **2**, 6144.
23. J. S. Ghomi and S. Zahedi, *Appl. Organometal. Chem.,* 2015, **29**, 566.
24. R. Perumal, A. Dey, R. Manavalan, K. Prakasam, E. Jayachandran and G. M. Sreenivasa, *Int. J. Chem. Sci.,* 2008, **6**, 2016.
25. D. Wang and D. Astruc, *Chem. Rev.,* 2014, **114**, 6949.
26. R. Luque, B. Baruwatib and R. S.Varma, *Green Chem.,* 2012, **12**, 1540.
27. P. Rani, K. N. Singh and A. Kaur, *J. Chem. Sci.,* 2018, 130, 160.
28. C. M. Chagas, S. Moss and L. Alisaraie, *Int. J. Pharm.*, 2019, **549**, 133.
29. J. Mahar, A. Saeed, K. D. Belfield, F. A. Larik, P. A. Channar, M. A. Kazi, Q. Abbas, M. Hassan, H. R. Raza and S. Y. Seo, *Bioorg. Chem.,* 2019, **84**, 170.
30. M. A. Abbasi, M. Hassan, A. U. Rehman, S. Z. Siddiqui, H. Raza, S. A. A. Shaah and S. Y. Seo, *Bioorgan. Med. Chem.,* 2018, **26**, 3791.
31. A. S. Gondkar, V. K. Deshmukh and S. R. Chaudhari, *Drug Invent. Today,* 2013, **5**, 175.
32. W. K. Okupnik, B. Jasiewicz, T. Pospieszny, R. Jastrza, M. Skrobanska and L. Mrowezynska. *J. Mol. Struct.,* 2018, **1169**, 130.

ANTIMYCOBACTERIAL ACTIVITY OF HEXAPEPTIDE RELATED TO APIDAECIN IA

Pandurangan Perumal[a] and Prabhu Seenivasan[b]

[a] Faculty of Pharmacy, Asian Institute of Medicine, Science and Technology (AIMST) University, 08100 Bedong, Kedah Darul Aman, Malaysia; [b] Department of Bacteriology, National Institute for Research in Tuberculosis, Chetpet, Chennai 600 031, India.
Corresponding Author E-mail: perupharma78@gmail.com

INTRODUCTION

Tuberculosis

Tuberculosis (TB) is a most treacherous disease caused by *Mycobacterium tuberculosis.* Right now the emergence of multiple drug resistance (MDR) and human immune virus (HIV) associated tuberculosis. According to World Health Organization (WHO) more than 13 million deaths related to infectious diseases, including tuberculosis have been reported worldwide. As per the global tuberculosis scenario 2013 (1-2) 3.5% of new and 20.5% of previously treated TB cases were estimated to have had MDR-TB. This translates into an estimated 480000 people having developed MDR-TB. On average, an estimated 9.0% of patients with MDRTB had extensively drug resistant TB (XDR-TB). If all notified TB patients (6.1 million, new and previously treated) had been tested for drug resistance, an estimated 300000 cases of MDR-TB would have been detected, more than half of these in three countries alone: India, China and the Russian Federation. In 2013, 136000 of the estimated 300000 MDR-TB patients who could have been detected were diagnosed and notified. This was equivalent to almost one in two (45%) and up from one in six in 2009. Progress in the detection of drug-resistant TB has been facilitated by the use of new rapid diagnostics *M. tuberculosis* affects about one third of the human population and represents the infectious condition causing more death worldwide, particularly in India nearly one third of the global population of TB about 2 million people acquire TB every year (2). Globally, there were an estimated 1.7 million deaths from HIV in 2011 and of these people an estimated 430,000 (25%) also had TB (2-3).

Drug resistance

An emerging current issue is the multiple drug resistant pathogens, nowadays increasing its occurrence. The most six scariest microbes are *Staphylococcus aureus, Escheria coli, Klebsiella, Acinetobacter baumanni, Aspergillus, Pseudomonas aeruginosa*. In this Staphylococcus aureus is the most dangerous one and it's classified under gram positive bacteria. The Infectious Diseases Society of America, having more than 8,000 infectious-disease specialists, has announcing a hit list of these six most worrisome germs facing in their clinical practice. The emergence of bacterial resistance to a number of antimicrobial agents such as β-lactam antibiotics, quinolines, vancomycin, ciprofloxacin and methicillin are becoming a major worldwide health problem (4). The urgent need for potent broad-spectrum antibiotics those are efficacious against resistant microbes including multidrug resistant strains (MDRS). In 1992, 19,000 deaths were directly caused by nosocomial infections and 28% of nosocomial infections were resistant to the preferred antibiotic treatment. More than 90% of strains of *S.aureus* in US hospitals are resistant to penicillin and β-lactam antibiotics (5**).** *Enterococci* are the most common cause of nosocomial infections and vancomycin is frequently the only effective agent. Vancomycin was previously the most effective drug for methicillin resistant *S. aureus*. However, in 1997, strains of *S. aureus* with decreased susceptibility to vancomycin were reported in Japan and United States (6).

Antimicrobial peptides (AMPs)

The harmful pathogens are exposed by living organisms through ingestion and inhalation (7). While during pathogenic invasion the first line of defence involves the innate immunity followed by acquired immunity (8). In contrast of acquired immune mechanism, endogenous peptides (which are in the GI, respiratory tract, genitourinal tract) which are induced a fast and effective defence against pathogens. This group of molecules are termed as 'antimicrobial peptides' (AMPs). These peptides are found in humans, animals, insects and plants (9). The AMPs are short peptides, generally between 12 to 50 amino acids present in the sequence of the peptides and these peptides are potent, broad spectrum antibiotics which exhibited potential as novel therapeutic agents (10&11). These peptides are also called as host defence peptides (12). AMPs demonstrated to kill gram positive and gram-negative bacteria, mycobacteria (including *M. tuberculosis*), fungus, viruses and cancer cells (13&14) and contains more than 50% of hydrophobic residues and positively charged residues (2). AMPs are believed to have a mechanism of action entirely distinct from those of current clinically used antibiotics, and there is a great interest in their development for treatment of drug-resistant infections (15).

Apidaecins

Apidaecin IA is a heat stable, 18 amino acid containing peptide originally isolated from honeybees' source *Apis mellifera* L, which is lethal for many gram-negative bacteria (16-18). It is a proline rich peptide having 39% hydrophobic residue and exists in four isoforms that are each 18 amino acids long, commonly including three arginine and six proline residues (19). Enzymatic inhibition is the basic mechanism of this peptide and the secondary structure of these peptide is a non-helical form. Killing occurs very rapidly, through a not as yet understood mechanism (16). The most unique feature about its mode of action is a total lack of pore-forming activity. AP was recognized as a potential therapeutic agent as compared with antibiotics because of their immediate effect, nontoxicity and little or no bacterial resistance (20). Although it does not display membrane- disrupting activity (bacteriolytic action), AP exhibits bacteriostatic action specifically toward gram-negative bacteria such as *Escherichia coli* (18). Mechanism for the action of AP was binds initially to lipopolysaccharide in the cell membrane and subsequently to the heat shock protein DnaK and related chaperones of *E. coli* in a specific manner. In addition, all D-enantiomers are devoid of any measurable function, indicating a significant degree of stereospecificity (18). The functional role of each amino acid in the Apidaecin structure, consist of constant regions, conferring general antibacterial and variable regions, determining specificity of the spectrum. Collectively, the data suggest that Apidaecin antibacterial function may be transporter-based type. In the present study designed the short peptide from Apidaecin-IA by splicing method, synthesized by SPPS using Fmoc chemistry protocol. Characterized by Mass spectroscopy and amino acid sequence analysis. Evaluated the antibacterial, antifungal and antimycobacterial activity of these peptides and discussed.

MATERIALS AND METHODS

MATERIALS

Chemicals and solvents

All amino acids, polymer bound resin, cleaving reagents and assay kits were purchased from Sigma Aldrich, Malaysia.

METHOD

The peptides were synthesized by SPPS using Fmoc chemistry protocols (21-22).

Procedure

Design the peptide sequence

Designed the proline rich anti-microbial hexapeptide from Apidaecin-IA by splicing method. The C-terminal portion of six amino acids was spliced from Apidaecin-IA and synthesis of the modified amino acid (replace by alanine in the place of proline) containing peptide (AIA3).

Synthesis of the hexapeptide related to Apidaecin-IA (AIA3)

Sequence

AIA2- A R A A H A- $CONH_2$

Resin activation (2)

About 500 mg rink amide MBHA resin was allowed to swell in dry DMF for about an hour and then the excess DMF was decanted (**Figure 1**). The resin was mixed with 10 mL of 20% piperidine solution in DMF swirled for 8 minutes and then decanted. This procedure was used twice to ensure complete removal of the Fmoc protective group. The resin was washed with 6 mL of dry DMF (six times with 2-minute interval between the consecutive washes) before attaching the first amino acid to the resin.

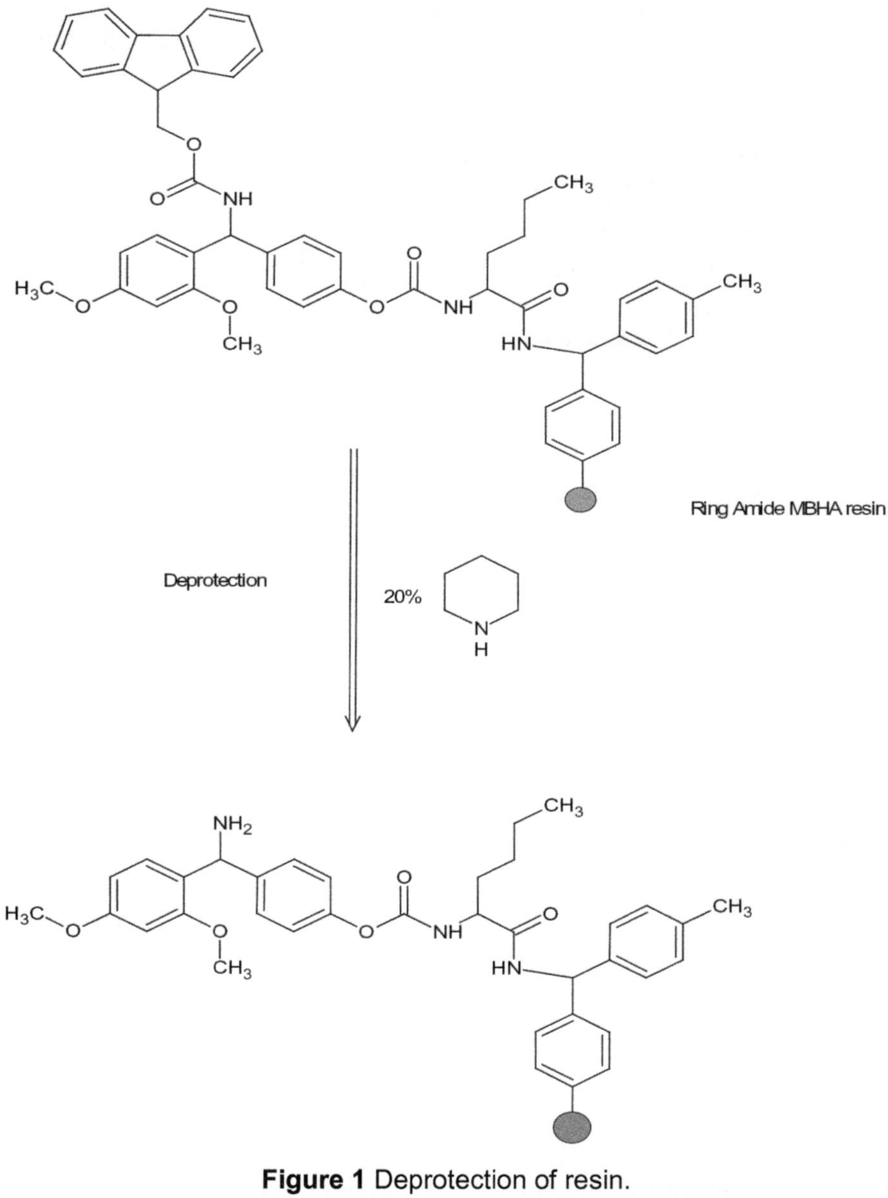

Figure 1 Deprotection of resin.

First aminoacid attachment (2)

Fmoc-L-Alanine was mixed with HBTU and HOBT in 1 mL DMF. DIPEA was added to the above milky colloidal solution and sonicated to get a clear solution. The mixture became transparent yellow colour solution after one minute. Then the contents were mixed with the activated resin and gently shaken for 90 minutes continuously (Figure 2). Then the resin was washed with 6 mL of dry DMF (six times with 2 minutes interval between the consecutive washes) to remove the excess of unreacted amino acid active ester. The negative result of ninhydrin test indicated 100% attached to the resin.

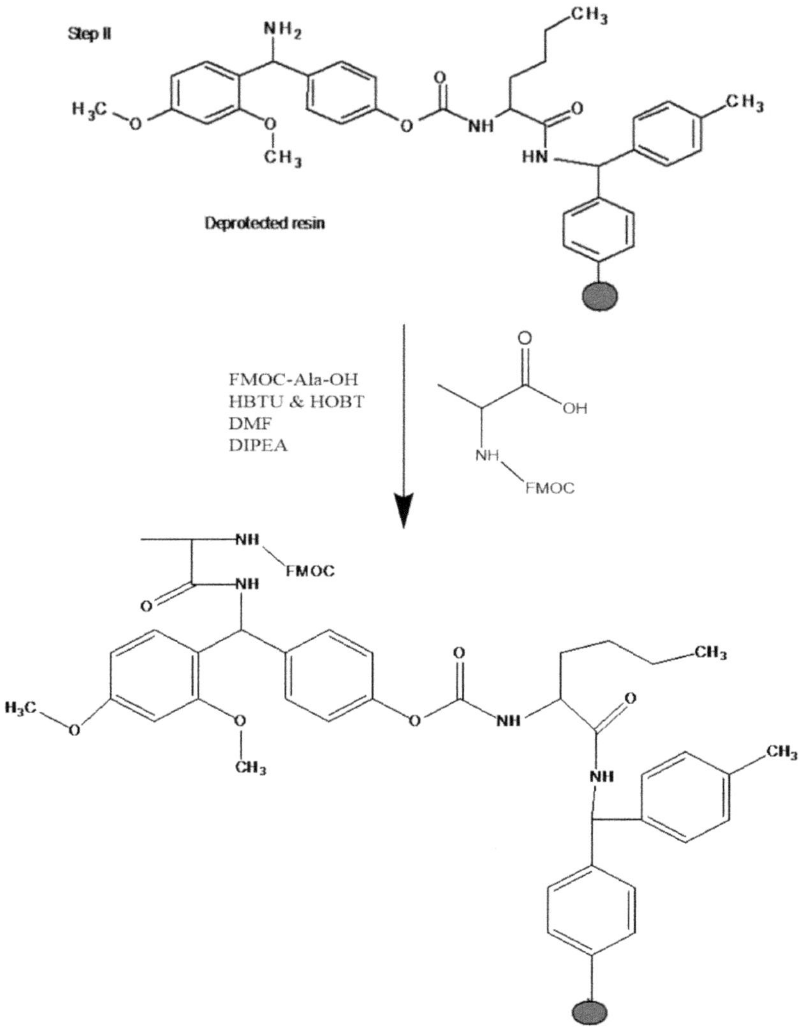

Figure 2 First amino acid attachment with resin.

Fmoc deprotection and coupling of remaining aminoacid (2)

After washing the resin with dry DMF, using 10 mL of 20% piperidine solution was added and shaken for 8 minutes and this was used twice. Finally, the resin was washed with 6 mL of dry DMF (Figure 3) (six times with 2 minutes interval between the consecutive washes). The positive result of ninhydrin test indicated 100% removal of Fmoc from the resin. AIA2 was synthesized by SPPS using a manual peptide synthesizer. The flow chart protocol for the synthesis of AIA2 is listed in Table 1.

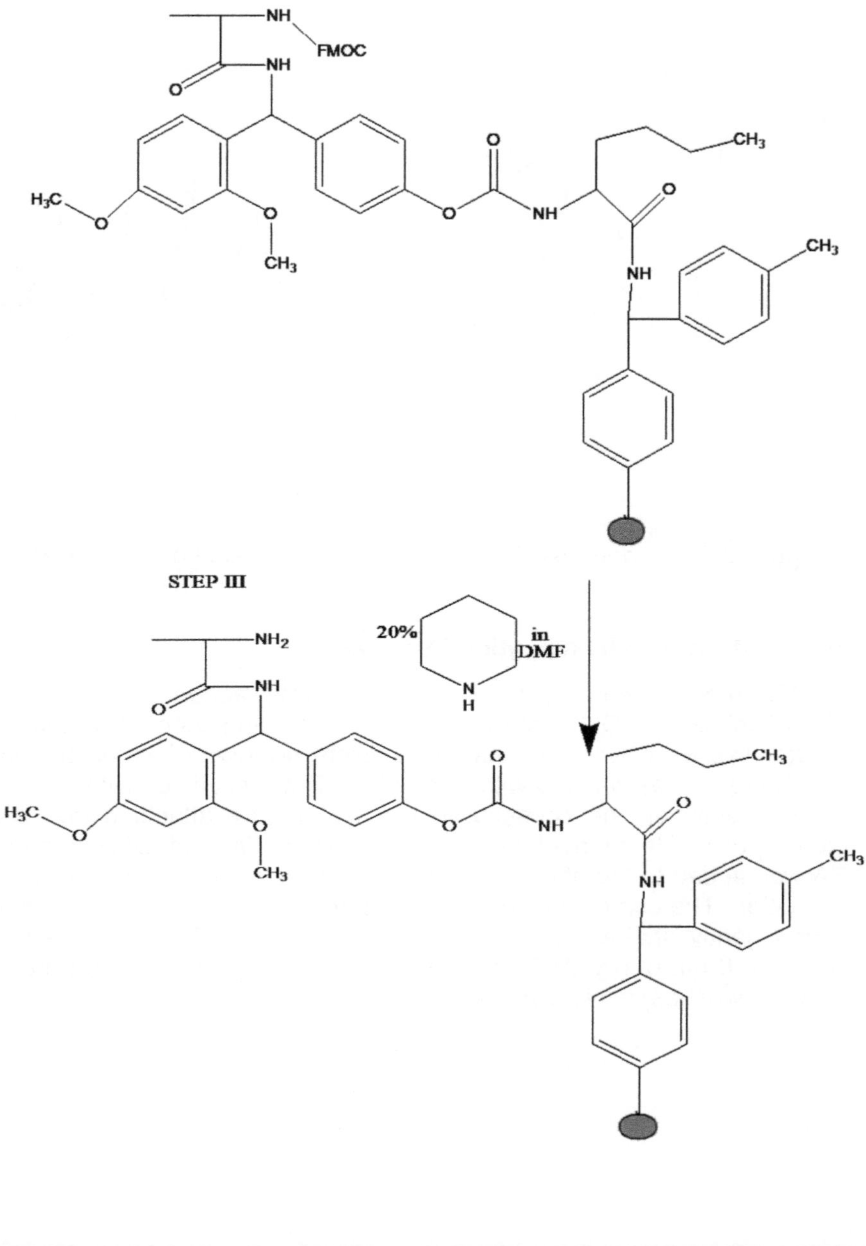

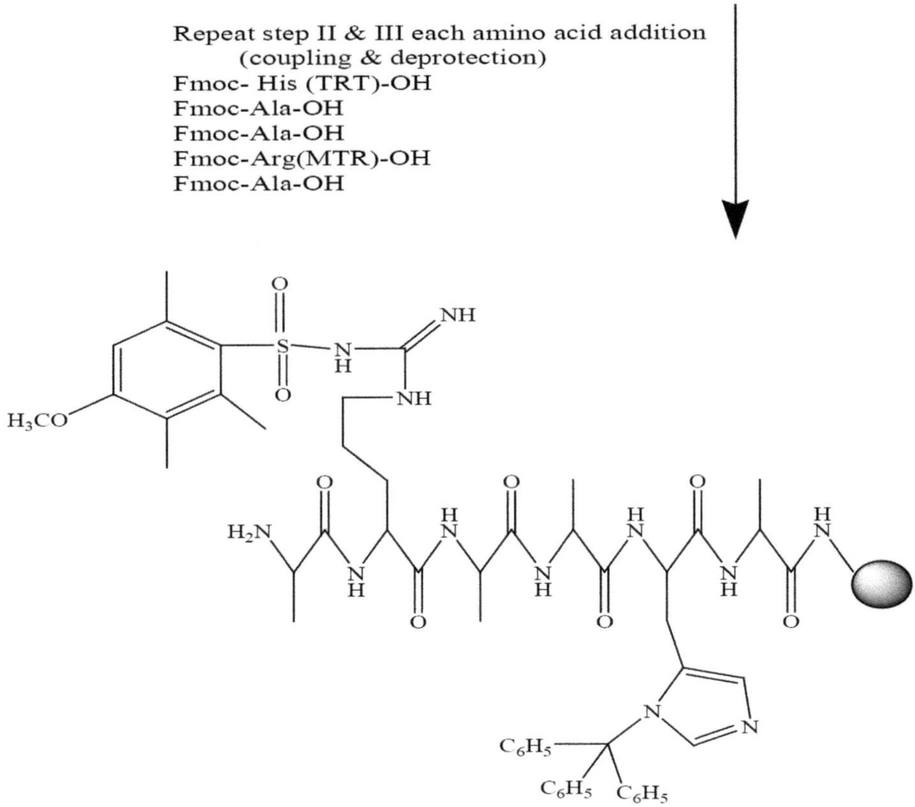

Figure 3 *Fmoc deprotection and coupling of remaining amino acid.*

Synthesis of the acyl hexapeptide (AIA3) (2)

About 480 mg of resin bound hexapeptide was added 10 mL of 20% piperidine solution in DMF swirled for 8 minutes and then decanted. This procedure was used twice to ensure complete removal of Fmoc protective group. The resin was washed with 6 mL of dry DMF (six times with 2 minutes the interval between the consecutive washes) before attaching the stearic acid to the resin. Stearic acid was mixed with HBTU in 1 mL DMF (Figure 4). DIPEA was added to the above milky colloidal solution and sonicated to get a clear solution. The contents were mixed with the modified hexapeptide bound resin and gently shaken for 60 minutes continuously. Then the resin was washed with 6 mL of dry DMF (six times with 2 minutes interval between the consecutive washes) to remove the unreacted amino acid active ester.

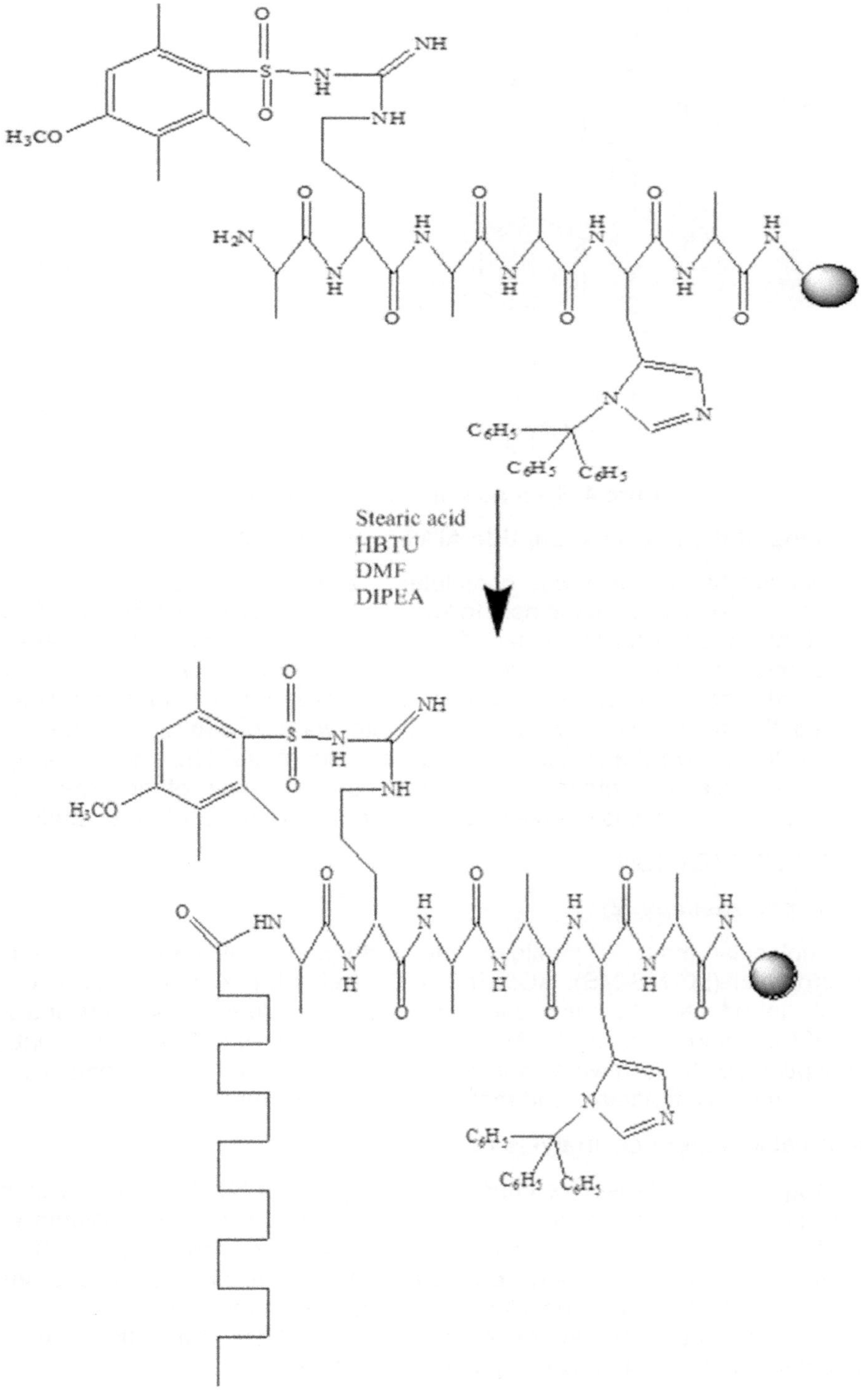
H₃CO
NH
NH
H₂N
Stearic acid
HBTU
DMF
DIPEA
HN

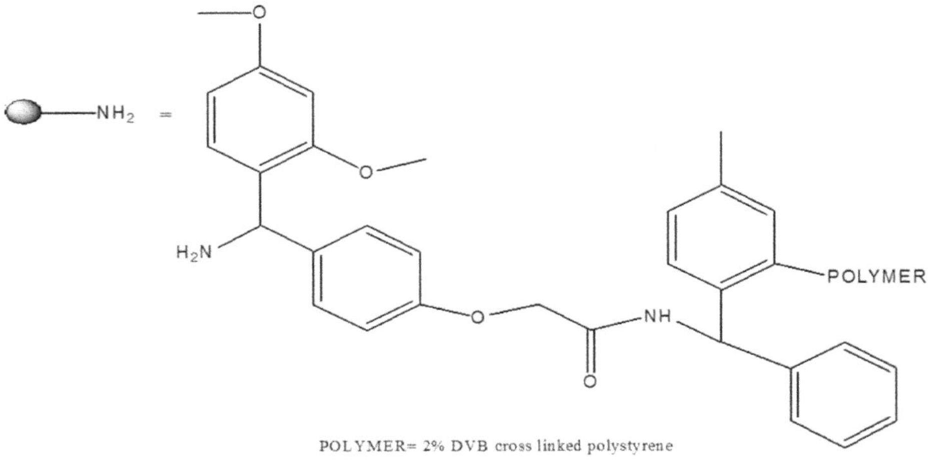

Figure 4 *Synthesis of the acyl hexapeptide.*

Cleavage of the acyl hexapeptide AIA3 from resin (2)

The peptide bound resin was completely washed two times with 10 mL of DCM, 10 mL of glacial acetic acid followed by 10 mL of DCM twice and finally with 10 mL of ether for two times. Then the peptide bound resin was treated with a mixture of 0.2 mL ethane-dithiol, 0.2 mL m-cresol, and 7 mL TFA (Figure 5). The contents were occasionally shaken for one hour and filtered, washed the resin with TFA and the filtrate was concentrated under high vacuum to remove the residual TFA to get solid mass. The solid mass was mixed with ice-cold ether and centrifuged to decant ether. The white precipitate obtained was dried and stored in a refrigerator for further study.

CHARACTERIZATION

Mass spectroscopy (2)

The molecular mass of peptides was performed in API-3000 sciex mass spectrometer (LC-MS-MS). AGILENT 1200 series Binary pump was used in sample introduction to mass spectrometer. C18 column was used and the size of the column was AGILENT XDB 150 mm, 4.6 mm, 5 µ. About 10µL of the peptide solutions in water was injected into the mobile phase composed of 10 mM ammonium acetate and methanol (50:50) mixture.

Amino acid sequence analysis (2)

The sequence analysis was characterized by AGILENT 1200 series binary pump and C18 column was used (LC-MS-MS). The size of the column was AGILENT XDB 150 mm, 4.6 mm, 5 µ. 0.3 mg of sample taken in 10 mL volume of container and adds 6N HCl solution in 5 mL and passes small volume of nitrogen gas and close the container immediately with the cap sealed and kept in hot air oven for 24 hours. After completion take the container out and kept for cooling at room temperature.

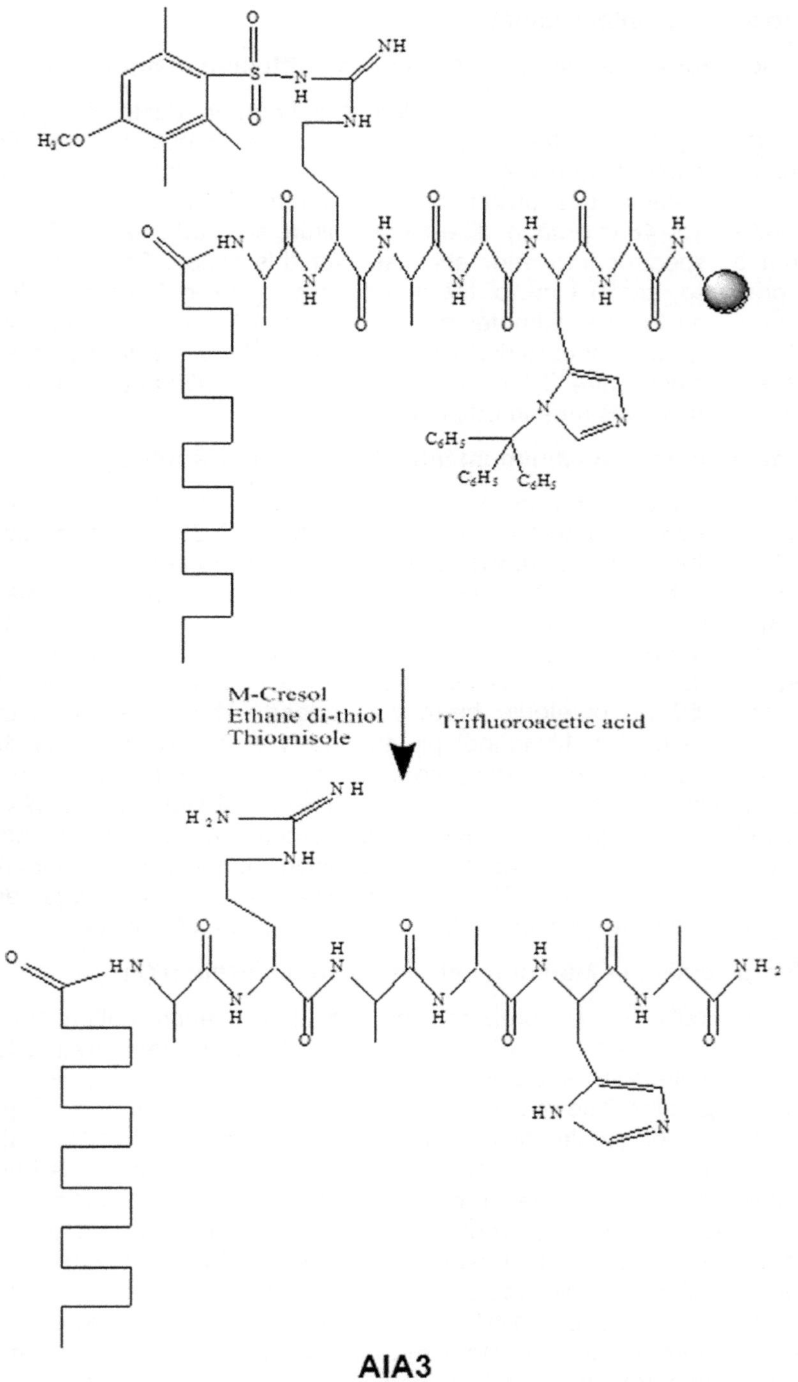

AIA3

Figure 5 *Cleavage of the acyl hexapeptide.*

In vitro antimicrobial evaluation

Antifungal activity-zone of inhibition (well diffusion method) (2).

Culture on the receipt was sub cultured sabouraud dextrose agar (SDA) plates (Fungi) and further stored in slants as stock culture. For the experiment, stock culture was prepared by inoculating culture from slants to the flask in sterile sabouraud dextrose broth (SDB) incubated at room temperature for 48hr (Fungi). The stock culture was adjusted to 0.11 OD at 650 nm by spectrophotometer and used for this assay. Sterile SDA plates were prepared, and 0.1 mL of the inoculum was blowout consistently. Wells were prepared for the diameter of 10 mm and 100 µL of the test peptides (AIA2 & AIA3) and standard (ketoconazole) was added in well separately. The plates were placed at 4^0C for 1 h and incubated at optimal temperature. The zone of inhibition was measured in mm.

Antibacterial Activity-Minimum inhibitory concentration (2)

Cultures was subcultured on nutrient agar (NA) and further stored in slants. Stock culture was prepared in sterile nutrient broth (NB) and incubated at 37^0 C for 24hr. The stock culture was adjusted to 0.5 McFarland standard turbidity and used for the assay. The resazurin solution was prepared by dissolving 270 mg in 40 mL of sterile distilled water and plates were prepared under aseptic conditions and a sterile 96 well plate was labelled. A volume of 100 µL of test material in DMSO was pipetted into the first line of the plate and all other wells 50 µL of sterile broth was added. The serial dilutions were performed using a multichannel pipette, each well had 50 µL of the test material in serially descending concentrations and each well 10 µL of resazurin indicator solution was added. The 30 µL of sterile broth was added by a pipette and 10 µL of 0.5 McFarland microbial suspension was added to each well and each plate has a set of positive, negative and standard. The plates were incubated at 37°C for 48 h. The colour change was recorded visually and the lowest concentration was taken as the MIC value.

Antifungal activity - Minimum inhibitory concentration (2)

The stock culture was prepared by inoculating each culture in sterile sabouraud dextrose broth (SDB) and incubated at room temperature for 48h. The stock culture was adjusted and used to 0.11 OD at 650 nm by spectrophotometer. The resazurin solution was prepared by dissolving a 270 mg in 40 mL of sterile distilled water using vortex mixer. The minimum inhibitory concentration (MIC) of the test substance was determined by liquid broth method of two-fold serial dilution technique against *Candida albicans*. Series of assay tubes were prepared by the mixture of equal volume (1 mL) of sterile SD broth and test substance was added. The test substance was serially diluted in two-fold decreasing concentrations. The sixth and seventh test tube left as positive control. The tubes with the test substance, from one to seventh were inoculated with 1 mL of inoculum ($1x10^6$ CFU per mL) and the final concentration of test substance ranged from 1000 to 7.8 µg/mL. Sterility controls were maintained in the experiment and the tubes were incubated at 28^0 C for 48 h. ketoconazole, was tested as standard drug at the concentrations ranging from 1000 to 7.8 µg/ mL and inspected visually to determine the growth of the organism as indicated by turbidity and compared with standard test tube.

Antimycobacterial activity by Luciferase Reporter Phage assay (LRP) (2,23)

Hundred micro litre bacterial suspension equivalent to Macfarland no 2 standard was added to 400 µL of G7H9 with and without the test compound. Two drug-free controls and two drug concentrations (50 and 100 µg/mL) were prepared for each sample and this was incubated for 72 h at 37°C. After incubation, 50 µL of the phage LRP (pH AE129) and 40 µL of 0.1 M $CaCl_2$ were added to all the vials and incubated at 37°C for 4h 100 µL of the mixture was taken from each tube into a luminometer cuvette and equal amount of working D-luciferin solution was added after incubation. The Relative Light Unit (RLU) was measured with 10-S of integration time in the luminometer (Monolight 2010). The percentage reduction in RLU was calculated for each sample and compared with the control. The experiments were repeated when the mean RLU of the control was less than 1000.

RESULTS

The antimicrobial hexapeptide aere designed by SPPS using Fmoc chemistry protocol. The structural integrity and amino acid sequence analysis of the peptides were confirmed by LC-MS-MS method. The Mass spectrum of the compound AIA3 was characterized by the appearance of an M+1 peak in 864.5 and the centroid spectrum. The spectral data of AIA3 is shown in Table 1.

Table 1 *Mass spectrum of hexapeptides related to Apidaecin-IA*

S. No	Compound code	Molecular formula	Molecular weight	Mass peak
01	AIA3	$C_{42}H_{78}N_{12}O_7^{2+}$	863	864.5

The sequence analysis of the acyl hexapeptide (AIA3) was present as the analyte peak named as alanine (A), arginine (R) and histidine (H) at the counts of analyte peak area likes 3400, 34300 & 107000 and the retention time of analytes were found to be 4.72, 13.7 & 13.5 (minutes). The concentration of the analytes was found to be 0.0164 µg/mL (A), 0.0256 µg/mL (F) & 0.0701 µg/mL (L).

The antifungal activity (zone of inhibition) of the compounds were evaluated against *Candida albicans* by well diffusion method. The compound AIA2 data was accepted in bentham science medicinal chemistry. The antifungal activity of the compounds was found to be potent inhibition against *Candida albicans* at both concentrations. The zone of inhibition of the compounds AIA2 and AIA3 were found to be 10 mm & 12 mm at the concentration of 100 µg/mL. The compounds AIA2 and AIA3 were found to be 10 mm & 13 mm at the concentration of 200 µg/mL as compared with the standard is shown in Table 2.

Table 2 *Antifungal activity of hexapeptides related to Apidaecin-IA.*

S. No	**Compound code**	**Con:µg/mL**	**Zone of inhibition in diameter (mm)**
			Candida albicans
01	AIA-2	100	10
		200	10
02	AIA-3	100	12
		200	13
03	Control	100	R
04	Standard	200	22

The antibacterial activity (minimum inhibitory concentration) of the compounds were evaluated against *S. aureus, B. subtilis* (gram positive) and *E. coli, P. aeruginosa* (gram negative) by 96 well plate method. The minimum inhibitory concentration of the compounds AIA2 and AIA3 were found to be extremely potent at 25 µg/mL against *P. aeruginosa* (gram negative). The compounds were found to be complete inhibition at the concentration of 50 µg/mL against *S. aureus* (gram positive) and *Escherichia coli* (gram negative). The compound AIA3 was shown the complete inhibition at the concentration of 50 µg/mL and the compound AIA2 was shown the complete inhibition at the concentration of 100 µg/mL against *Bacillus subtilis* (gram positive) as compared with the standard is shown in Table 3.

Table 3 *Antibacterial activity of hexapeptides related to Apidaecin-IA.*

S.No	**Sample code**	**MIC value (µg/mL)**			
		Organisms			
		Staphylococcus aureus	***Bacillus subtilis***	***Escherichia coli***	***Pseudomonas aeruginosa***
01	AIA 2	50	100	50	25
02	AIA 3	50	50	50	25
03	Standard	25	25	25	25

The antifungal activity (minimum inhibitory concentration) of the compounds were investigated against *C. albicans* by 96 well plate method. The minimum inhibitory concentration of the compounds AIA2 and AIA3 were found to be potent inhibition at 200 µg/mL against C. *albicans* as compared with the standard is shown in **Table 4.**

Table 4 *Antifungal activity of hexapeptides related to Apidaecin-IA.*

S. No	Compound code	Minimal inhibitory concentration (µg/mL)
		Micro organism
		Candida albicans
01	AIA 1	200
02	AIA 2	200
03	Standard	<15.6

Anti-mycobacterial activity was evaluated against *M.tuberculosis* $H_{37}Rv$ by Luciferase Reporter Phage assay (LRP). The percentage of reduction in Relative Light Unit (RLU) of the compounds AIA2 and AIA3 were found to be extremely potent antimycobacterial activity of 89.07% and 60.82% at the concentration of 100 µg/mL and the compounds AIA2 and AIA3 were found potent antimycobacterial activity of 66.90% and 32.07% at the concentration of 50 µg/mL against *M. tuberculosis* $H_{37}Rv$ as compared with control and standard is shown in the **Table 5**.

Table 5 *Antimycobacterial activity of hexapeptides related to Apidaecin IA against* M. tuberculosis *$H_{37}Rv$.*

S. No	Compound code	% of reduction in RLU	
		50 (µg/mL)	100 (µg/mL)
01	AIA 2	66.90	89.07
02	AIA 3	32.07	60.82
03	Rifampicin (2 µg/mL)	99.90	

The compounds AIA3 and AIA2 were found to be extremely potential reduction of 76.15% and 52.43% in the concentration of 100 µg/mL against S, H, R and E resistant as compared with positive control Rifampicin. The compounds AIA3 and AIA2 were found potential reduction of 50.29 % and 48.56% at the concentration of 50 µg/mL against clinical isolate of *M.tuberculosis* S, H, R and E resistant as compared with control and standard is shown in Table 6.

Table 6 *Antimycobacterial activity of hexapeptides related to Apidaecin IA against clinical isolate of M. tuberculosis: S, H, R and E resistant.*

S. No	Compound code	% of reduction in RLU	
		50 (µg/mL)	100 (µg/mL)
01	AIA 2	48.56	52.43
02	AIA 3	50.29	76.15
03	Rifampicin (2 µg/mL)	12.79	

DISCUSSION

Very short peptide (2,23) residues have been identified and reported to be active against bacteria and some of them even show broad spectrum activity (2,24). The possible contributing factor is the high degree of cationic content in a short stretch of peptide that could favor a tight binding of the peptide to the bacterial anionic membrane (2,25). The hexa peptide AIA2 having four hydrophobic amino acid and two cationic amino acids which is used for tight binding of bacterial anionic membrane. The presence of anionic lipids in bacterial membrane enables the selective interaction of cationic AMPs (2,23-24). Antimicrobial peptides can be grouped into haemolytic and non-haemolytic in these the non-haemolytic AMPs are either cyclic short peptides or short linear peptides with moderate hydrophobicity (**26**). Several lipopeptides have been shown to exhibit both antibacterial and antifungal properties (2,23-24). Fatty acylation of linear peptides has, generally, resulted in enhanced membrane association of the peptides. The peptide AIA3 is a short acyl peptide synthesized by the attachment of stearic acid with AIA2 would help to increase the potency of the peptide while retaining its cell selective activity. AIA3 can be expected to behave as a non-lytic antimicrobial lipopeptide. It is worth mentioning here that non-lytic short antimicrobial peptides are better candidates for pharmaceutical applications (27). Compared the antimycobacterial activity of these two peptides, AIA 3 was highly potent than AIA2 due to the acylation of the AIA2 peptide that increases the hydrophobicity of the peptide. Moreover, the molecular docking studies of hexapeptide against glutamine synthetase and RNA polymerase in *Mycobacterium tuberculosis* was performed (28) and *In-silico* result correlated with the *in-vitro* result.

CONCLUSION

The hexa peptide AIA3 related to Apidaecin IA was designed, synthesized, characterized and evaluated against *Candida albicans*, *S. aureus, B. subtilis* (gram positive) and *E. coli, P. aeruginosa* (gram negative) *M. tuberculosis* $H_{37}Rv$ and clinical isolate of *Mycobacterium tuberculosis* S, H, R and E resistant. The compound AIA3 was found to be highly potent antimicrobial and antimycobacterial activity compared with standard. The compound AIA3 is a single drug having highly potent efficacy against *M. tuberculosis* $H_{37}Rv$ and clinical isolate of *Mycobacterium tuberculosis* S, H, R and E resistant than multiple drug regimen. More over single entity can minimize the heavy drugging and drug-drug interaction which are the main advantages over multiple drug regimen.

ACKNOWLEDGEMENTS

Author would like to thank University Grants Commission (UGC) to provide the financial support and Dr Soumya Swaminathan, Director, National Institute for Research in Tuberculosis (NIRT-ICMR), India to carried out my research work in this institute.

CONFLICT OF INTEREST

The author declares that there is no conflict of interest.

References

1. World Health Organisation (WHO), Global Tuberculosis report, 2014.
2. P. Perumal, V. P. Pandey and P. Seenivasan, Antimicrobial Activity of Octapeptides Related to Human Histatin 8, *Eur. J. Pharm. Med. Res.*, 2015, **2**, 526.
3. UNAIDS Report on the Global AIDS Epidemic 2012.
4. H. Hoshimoto, M. Ioue and I. Hayashi, 11th Future Trends in Chemotherapy, Geneva, Switzerland, 1994.
5. P. Harrison, F. Lederberg, *Antimicrobial Resistance,* Institute of Medicine, National Academy Press, Washington, DC, 1998.
6. M. L. Cohen, Changing patterns of infectious disease, *Nature*, 2000, **406**, 762.
7. D. Hultmark, Drosophila immunity: paths and patterns, *Curr. Opin. Immunol.*, 2003, **15**, 2.
8. D. Fearon and R. Locksley, The instructive role of innate immunity in the acquired immune response, *Science*, 1996, **272**, 50.
9. G. Maróti, A. Kereszt, E. Kondorosi and P. Mergaert, Natural roles of antimicrobial peptides in microbes, plants and animals, *Res. Microbiol.*, 2011, **162**, 363.
10. M. Zanetti, R. Gennaro, B. Skerlavaj, L. Tomasinsig and R. Circo, Cathelicidin peptides as candidates for a novel class of antimicrobials, *Curr. Pharm. Des.*, 2002, **8**, 779.
11. A. R. Koczulla and R. Bals, Antimicrobial peptides: current status and therapeutic potential, *Drugs*, 2003, **63**, 389.
12. A. Tossi, Host defense peptides: roles and applications, *Curr. Protein. Pept. Sci.*, 2005, **6**, 1.

13. J. S. Mader and D. W. Hoskin, Cationic antimicrobial peptides as novel cytotoxic agents for cancer treatment, *Expert. Opin. Investig. Drugs.*, 2006, **15**, 933–946.
14. G. Wang, K. M. Watson and R. W. Buckheit, Anti-HIV-1 activity of antimicrobial peptides derived from human and bovine cathelicidins, *Antimicrob. Agents. Chemother.*, 2008, **52**, 3438.
15. P. Perumal, V. P. Pandey and P. Seenivasan, Antimycobacterial Activity of Heterocyclic Dipeptide, *Inven. Imp. Pharma. Biotech. Mic.*, 2015, **5**, 1.
16. P. Casteels, C. Ampe, F. Jacobs, C. Vaeck and P. Tempst, Apidaecins: antibacterial peptides from honeybees, *EMBO J.*, 1989, **8**, 2387.
17. K. Casteels, K. Josson, T. Capaci, P. Casteels and P. Tempst, Apidaecin multipeptide precursor structure: a putative mechanism for amplification of the insect antibacterial response, *EMBO J.*, 1993, **12**, 1569.
18. P. Casteels and P. Tempst, Apidaecin-type peptide antibiotics function through a non-poreforming mechanism involving stereospecificity, *Bioch. Biophys. Res. Commun.*, 1994, **199**, 339.
19. P. Casteels, C. Ampe, F. Jacobs and P. Tempst, Functional and chemical characterization of hymenoptaecin, an antibacterial polypeptide that is infection-inducible in the honeybee (Apis mellifera). *J. Biol. Chem.*, 1993, **268**, 7044.
20. R. E. Hancock, Peptide antibiotics. *Lancet*, 1997, **349**, 418.
21. G. B. Fields, Methods for removing the Fmoc group, in methods in molecular biology, Peptide Synthesis Protocols, *Peptide. Res.*, 1994, **5**, 17.
22. C. G. Fields, D. H. Lloyd, R. L. Macdonald, K. M. Otteson and R. L. Noble, HBTU activation for automated Fmoc solid-phase peptide synthesis. *Peptide. Res.*, 1991, **4**, 95.
23. P. Perumal, V. P. Pandey and P. Seenivasan, Antimicrobial Activity of simple dipeptide, *Inven. Imp. Pharma. Biotech. Mic.*, 2015, **5**, 155.
24. R. A. Houghten, C. Pinilla, S. E. Blondelle, J. R. Appel, C. T. Dooley, and J. H. Cuervo, Generation and use of synthetic peptide combinatorial libraries for basic research and drug discovery, *Nature*, 1991, **354**, 84.
25. S. Thennarasu, T. A. Penumatchu, C. E. Shelburne, D. L. Heyl and A. Ramamoorthy, Antimicrobial and membrane disrupting activities of a peptide derived from the human cathelicidin antimicrobial peptide LL37, *Biophys. J.*, 2010, **98**, 248.
26. A. Majerle, J. Kidric, and P. J. Jerala, Enhancement of antibacterial and lipopolysaccharide binding activities of a human lactoferrin peptide fragment by the addition of acyl chain, *J. Antimicrob. Chemother.*, 2003, **51**, 1159.
27. A. Giuliani, G. Pirri and S. F. Nicoletto, Antimicrobial peptides: An overview of a promising class of therapeutics, *Cent. Eu. J. Biol.*, 2007, **2**, 1.
28. P. Perumal, V. P. Pandey and P. Parasuraman, Docking studies on antimicrobial peptides related to APIDAECIN-IA and Human histatin against glutamine synthetase and RNA polymerase in *Mycobacterium tuberculosis*, *Asian. J. Pharm. Clin. Res.*, 2014, **7**, 195.

BINDING STUDIES OF ANTIPSYCHOTIC DRUG RISPERIDONE WITH β-LACTOGLOBULIN USING FLUORIMETRY AND MOLECULAR MODELLING

Bandoo C. Chatale,[a] Laxmi Ananthanarayan[b] and Mariam S. Degani[a*]

[a]Department of Pharmaceutical Sciences &Technology Institute of Chemical Technology, Mumbai 400019, India; [b]Food Engineering and technology Department, Institute of chemical technology, Mumbai 400019, India
Email: ms.degani@ictmumbai.edu.in

1. INTRODUCTION

Risperidone (RIS) is a highly potent, biopharmaceutic classification system (BCS) class II drug which is often used to treat schizophrenia.[1, 2] The main drawbacks of RIS are that it is practically insoluble in water, has a bitter taste and also shows photoinstability.[2-4] Several studies to improve these drawbacks are as follows: Taste masked orodispersible tablets (ODT) of RIS-cation exchange resin complexes, in ratios 1:1 for RIS: Kyron T-104, 1:2 for RIS-Indion 204 complex, 1:4 for RIS-Amberlite IRP64 resin and complexing RIS with cetyl alcohol in 2:1 ratios were used for masking bitter taste of RIS.[4, 5] Nanocrystal-based drug delivery system of RIS has shown potential for enhancing the apparent solubility of RIS using nanonization technique.[6] Researchers have reported different approaches for enhancement of water solubility of RIS including solid dispersion, self-nanoemulsifying powder and co-solvent.[7-9] Sterically stabilized nanosuspension of RIS has been prepared using high amount of surfactant which enhanced oral bioavailability of RIS by two fold. However, none of them explored the food based protein to modulate the properties of RIS.

β-Lactoglobulin (β-LG) is a major whey protein of the lipocalin family and is present in dairy milk. The native form of β-LG with 162 amino acid residues, composed of eight antiparallel β-strands called calyx or β-barrel is a conical structure with α-helix located at the outer surface of the β-barrel.[11] The presence of two hydrophobic binding sites in the monomer of β-LG has been postulated in literature, one of which is in the internal cavity of β-barrel while the other is located in the outer surface between β-barrel and α-helix.[12-14] Protein has the ability to bind several ligands including retinoids, vitamin D, fatty acids, lipids, folic acid, resveratrol and the antibacterial drug, norfloxacin.[12,15-17] The complexation with the β-LG provides a significant increase in the hydro-solubility of the amphiphilic compound trans-resveratrol thereby increasing its photostability and further β-LG forms water-soluble complexes with retinol, and β-carotene, thus protecting these lipophilic compounds from degradation by heat, oxidation, and irradiation.[18] This has been proven by variety of methods, including equilibrium dialysis, affinity chromatography, electron spin resonance spectroscopy and nuclear magnetic resonance spectroscopy.[12,19] One of the simplest methods for characterizing a protein-ligand complex is perturbation of the intrinsic tryptophanyl (Trp) fluorescence.[20, 21] This effect is particularly noticeable in cases in which the Trp fluorescence decreases on ligand addition.[22] Thus β-LG could be used for complexing with RIS to improve its physicochemical properties. Several studies have shown that there appear to be hydrophobic binding sites on the protein β-LG, but the exact location of the binding has not yet been explored through molecular modelling.[12, 15-17]

The aim of the current research work is to identify the interaction of β-LG with RIS using fluorimetry and molecular modelling. RIS–β-LG binding phenomenon has been studied by fluorimetry in which drug-induced quenching of intrinsic Trp fluorescence of protein (β-LG), has been monitored to elucidate the RIS-β-LG binding strength. The probable binding site, binding stability and binding affinity of RIS with β-LG have been rationalized on the basis of induced fit docking (IFD), molecular dynamic (MD) simulation, and Molecular Mechanics-Generalized Born Surface Area (MMGBSA) methods.

2. MATERIALS AND METHODS

2.1. Materials

RIS was a gift sample from Cipla Ltd., Mumbai, India. β-LG was purchased from Sigma, India.

2.2. Methods

2.2.1. RIS and β-LG interaction assessment using fluorimetry

RIS and β-LG interaction assessment performed using spectrofluorimeter, SYNERGY/H1, microplate reader, (BioTeck, Mumbai, India). Quenching study was carried out in phosphate buffer, pH 6.8 prepared by dissolving sodium chloride, potassium chloride, disodium hydrogen phosphate, potassium dihydrogen phosphate in double distilled water and pH adjusted by dilute HCl.

RIS is practically insoluble in water therefore 2 % ethanol was added to make it soluble for the preparation of 1 mM RIS, 0.5 mM solution of β-LG were prepared in phosphate buffer pH 6.8 and 2 % ethanol was added to this to maintain same components in both the solutions to avoid fluorescence interferences. Fluorescence spectrum scanning was performed using quartz 96 well plate to study the quenching of intrinsic fluorescence of β-LG with RIS. Well plate had maximum capacity of 300 µl per well, while the experimental volume in each well was maintained at 200 µl thereby ensuring that path-length and volume were kept constant during fluorescence quenching experiments. RIS and β-LG, solutions were used as controls, while for complexation trials molar ratios of β-LG to RIS taken were 1: 2, 1:4, 1:6, 1:8, 1:12, 1:14, and 1:18. Quenching of intrinsic fluorescence of β-LG was measured by keeping 50 µM concentration of β-LG constant while RIS (quencher) was added in incremental order of 100, 200, 300, 400, 600, 700, 800, 900 µM concentration. Each quartz well plate was individually recorded for fluorescence from emission range of 350 to 550 nm at an excitation wavelength of 295 nm.

2.2.2. Molecular modelling

All the molecular modelling work was carried out using Schrodinger Suite.[24]

Structure building

The structure of RIS was drawn using 2D sketcher of Schrödinger maestro 2017-1. RIS was optimized for docking using LigPrep module† where the RIS structure was converted to 3D and energy optimized using OPLS2005. Least energy ring conformation was generated for RIS at pH 6.8 using Epik module, in LigPrep with all other default options.

Protein preparation

3D structure of β-LG was retrieved from protein data bank (PDB ID 3BLG)[25] and refined using Protein Preparation Wizard‡ wherein missing hydrogen atoms, amino acid side chains, and missing residues were added and bond order was set. The β-LG was subjected to pre-processing, and optimized to create H-bond network, remove waters with less than 5 Å distance, and minimized to remove any steric clashes using OPLS2015 force field.

Structure optimized using MD simulation for stabilization

The apo structure of β-LG was adequately soaked into simulation box consisting of 21,105 SPC water molecules in orthorhombic boundary conditions in System Builder panel of Desmond module§ MD simulation was then performed for 100 ns using Desmond module.§

Ligand binding site prediction

The fully optimized protein structure was subjected to active site prediction using the Site-Map module** which was used to characterize the binding pockets and functional residues of β-LG. For predicting active probable sites, IFD was performed using the Induced fit docking module†† with ligands cholecalciferol, folic acid, retinol, norfloxacin, and resveratrol as reported in literature[12,15-17] which bind to internal hydrophobic cavity of β-LG.

Binding mode prediction

The fully optimized protein structure was used for the receptor grid generation using the centroid of selected residues of site 1 and site 2 which were predicted by Site-Map module.** Extended sampling protocol of IFD were utilized for docking the ligands on protein, β-LG using the Induced Fit Docking module.††

The Molecular Mechanics-Generalized Born Surface Area (MM-GBSA)

The complex of the RIS with β-LG was subjected to MM-GBSA method using PRIME module.‡‡

---

† LigPrep, Schrödinger, LLC, New York, NY, 2017.
‡ Protein Preparation Wizard, Schrödinger, LLC, New York, NY, 2017.
§ Desmond, Schrödinger, LLC, New York, NY, 2017.
**SiteMap, Schrödinger, LLC, New York, NY, 2017.
†† Induced Fit Docking, Schrödinger, LLC, New York, NY, 2017.
‡‡ PRIME, Schrödinger, LLC, New York, NY, 2017.

Molecular Dynamic (MD) simulation

The induced fit docked pose of RIS with β-LG was adequately soaked into simulation box consisting of 21,105 SPC water molecules in orthorhombic boundary conditions in System Builder panel of Desmond module. § Default values were kept for Van-der Waals, electrostatic interactions, and H-bond. Default relaxation protocol used for simulation. Molecular force field (OPLS3) setup for heating, equilibration used and then MD simulation performed for 100 ns using Desmond module.§ Verlet algorithm was used for integrating the equations of motion in standard Cartesian-space and MD with constant number of molecules (N), pressure (P), and temperature (T) (NPT) ensemble (here p is 1.013 bar, T is 300 K).

3. RESULTS AND DISCUSSION

3.1 Quenching of intrinsic fluorescence of milk protein, β-LG by RIS

The fluorimetry is a highly sensitive technique which serves as a universal tool for the study of protein-ligand interactions. Quenchers are substances capable of reducing the fluorescence emission, whose decline is mathematically expressed using the Stern–Volmer equation, which gives important information about ligand–protein interactions. Ground state, static quenching occurs by addition of quencher because of formation of a nonfluorescent, quencher-flurophore complex which results in shift of the flurophore absorption spectrum thereby confirming complex formation. Another type of static quenching is often observed at high quencher concentrations due to the existence of increasing numbers of quencher-flurophore pairs in which the quencher is close enough to the flurophore to instantaneously quench its excited state. This can be distinguished from quenching due to ground-state complex formation because it does not produce changes in the flurophore absorption spectrum.

β-LG has two tryptophan (Trp) amino acid residues which having excitation at 295 nm and an emission range at longer wavelength of 350 to 550 nm. The intrinsic fluorescence of Trp amino acid of β-LG undergoes progressive quenching with incremental addition of the drug, RIS. The phenomenon of fluorescence has been described in terms of Stern Volmer equation[23] as:

$F_0/F = 1 + Kq\tau(Q) = 1 + Ksv\ (Q)$

Here, F_0 is the intensity of fluorescence without a quencher
F is the intensity of fluorescence with a quencher, (here RIS)
Ksv is the Stern Volmer quenching constant
Kq is the quenching rate constant
τ is the lifetime of the emissive excited state of the protein, β-LG
Q is the molar concentration of quencher (here RIS)

Fluorometric behaviour of RIS-β-LG complex

The interactions of RIS with β-LG were examined by fluorimetry. The maximum excitation of β-LG was located at 295 nm and fluorescence emission was obtained in the range of 350 to 550 nm. The fluorescence emission progressively decreased with incremental addition of RIS, indicating the quenching of fluorescence intensity of β-LG upon binding to RIS. This was attributed to the formation of RIS-β-LG complex (Figure 1).

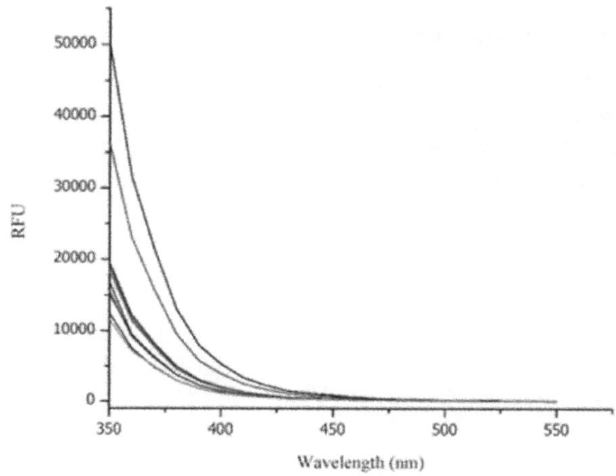

Figure 1 *fluorescence spectra of 50 μM β-LG in incremental addition of RIS, 0, 100, 200, 300, 400, 600, 700, 800, 900 μM in relative fluorescence unit (RFU) versus emission wavelength (nm)*

Experiments were performed for quenching of intrinsic fluorescence of β-LG, by RIS at 25, 30, and 35 ^{0}C temperature. Fluorescence quenching data were analyzed using Stern-Volmer equation to find out the mechanism of quenching by RIS on β-LG.

The slope of the plot decreases with increasing temperature, clearly indicating static quenching (Figure 2).

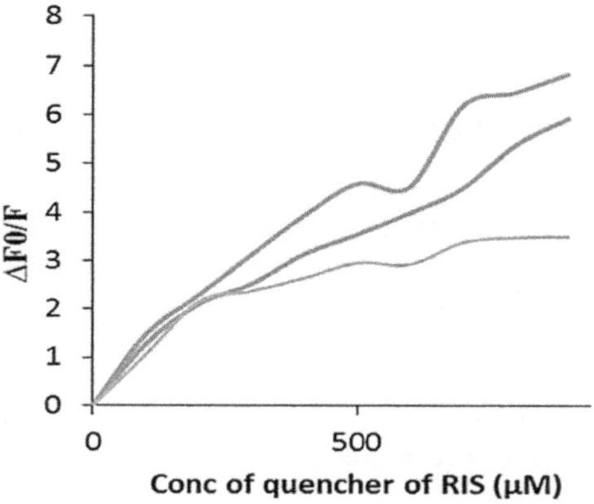

Figure 2 *Change in fluorescence of protein, β-LG versus concentration of quencher, RIS at temp 25, 30 and 35 ^{0}C*

3.2 Molecular modelling

RIS interactions with β-LG were studied using molecular modelling techniques such as IFD, MD simulation and MMGBSA to study binding mode, binding stability, and binding affinity of the RIS with β-LG.

3.2.1. Structure optimized using MD simulation for stabilization

To confirm the stability of apo β-LG, MD simulation was performed for 100 ns during MD simulation, β-LG was took 10 ns for the equibration, once equibrated, RMSD fluctuates minimally within 1.5 Å for β-LG which indicates the conformational stability of the overall β-LG (Figure 3).

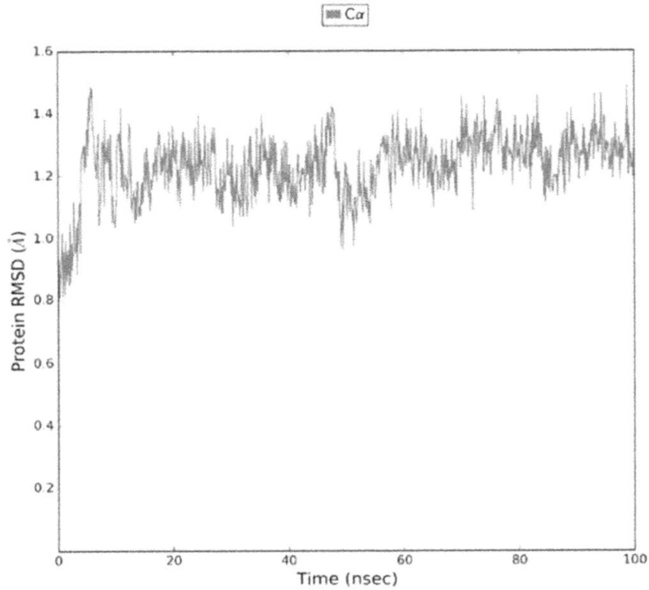

Figure 3 *RMSD of β-LG.*

3.2.2. Binding site prediction and binding affinity analysis

The probable binding sites of the protein, β-LG were mapped by SiteMap[b] (Figure 3) and scored in accordance with various criteria such as cavity size, solvent exposure, protein enclosure, the balance between hydrophilicity and hydrophobicity and the degree to which a ligand might donate or accept hydrogen bonds. Two sites, site 1 and site 2 have been found (Figure 4). Mapping of entire β-LG revealed that yellow color hydrophobic, was found on site 1 (encircled) and blue colour H-bond donor, and red color H-bond acceptor regions were found on site 2 (encircled).

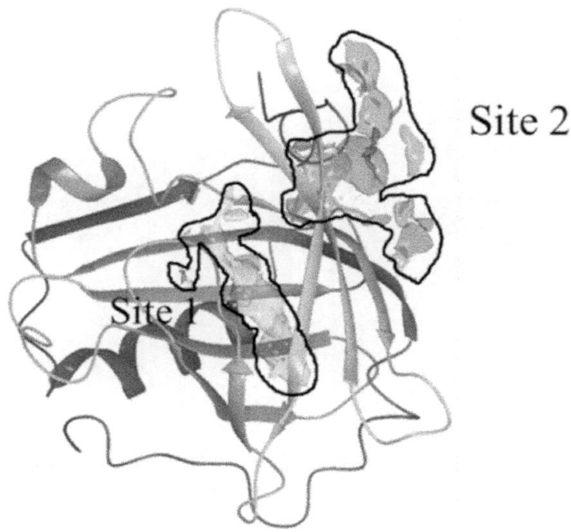

Figure 4 *Site-Map of β-LG.*

Blind docking was used with ligands with known interactions to validate the ligand binding sites on β-LG (Table 1).

Table 1 *Ligand docking scores, and dG Bind with site 1, 2 of β-LG.*

Ligand	Docking score on site 1 of β-LG	dG Bind (kcal/mol) ligand with site 1 of β-LG	Docking score on site 2 of β-LG	dG Bind (kcal/mol) ligand with site 2 of β-LG
Cholecalciferol	-11.380	-83.62	-5.342	-8.02
Folic acid	-9.998	-46.8	-7.785	-39.13
Retinol	-10.191	-77.41	-6.172	-18.36
Norfloxacin	-10.788	-28.95	-6.550	-21.66
Resveratrol	-8.575	-54.25	-6.912	-25.9
Risperidone	-11.893	-118.94	-5.795	-40.02

Site 1 was found to be the most probable active site. This site includes residues of Val92, Leu122, Val94, Ile71, Leu103, Phe105, Met107, Leu54, Ile56, Leu58, Leu10, Ile12, Phe82, Leu39, Ile84, Val15, Ala86, Leu87, Val41, Val43, Lys69, Leu46, Gln120. RIS was first docked using XP protocol at site 1, which gave low docking score -0.457. However on IFD, the score and the orientation improved considerably and docking score for RIS was -11.893 (Table 1). Ligands usually binds to hydrophobic internal cavity of β-barrel of β-LG as reported in literature (Figure 5). Ligands show better binding affinity with site 1 of β-LG as compared with site 2 of β-LG (Table 1).

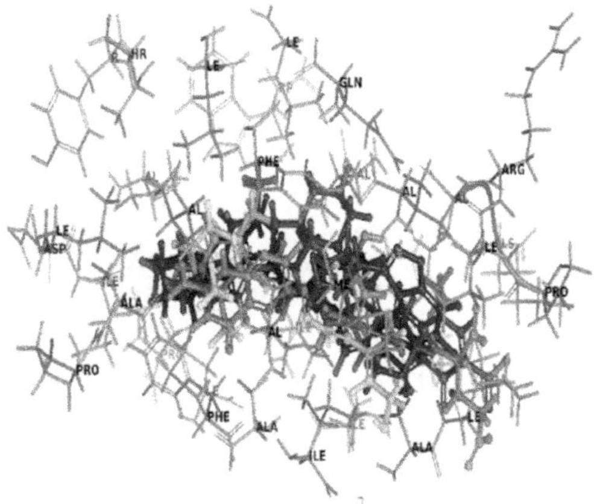

Figure 5 *Ligands docked poses with β-LG (red: cholecalciferol, green: folic acid, yellow: retinol, violet: norfloxacin, orange: resveratrol, and blue: risperidone).*

3.2.3. Binding mode prediction

The knowledge of the binding location of RIS on β-LG is the prerequisite for understanding the molecular level interaction. RIS binds to hydrophobic internal cavity of β-barrel of β-LG. It exhibited hydrophobic interactions resulting in strong binding (Figure 6).

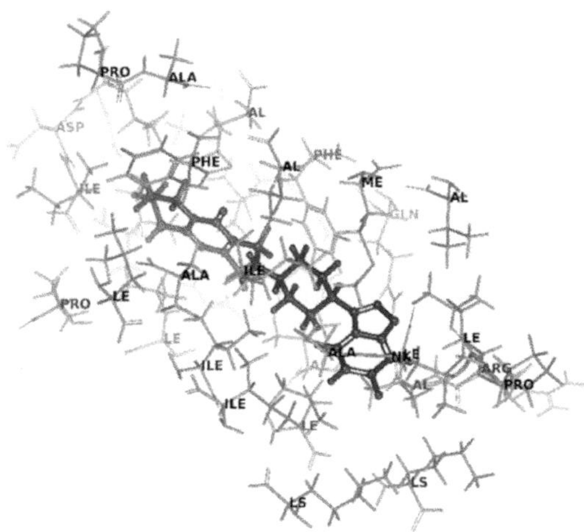

Figure 6 *RIS IFD pose with active site 1 residues of β-LG 9 (blue: risperidone).*

3.2.4. Binding affinity analysis

The Molecular Mechanics-generalized Born Surface area (MM-GBSA) is used to estimate relative binding affinity for the complex of RIS with β-LG. The binding strength of RIS with β-LG determined by the binding free energy, Δ*G* bind, and its shows significant binding Affinity (dG binding energy) with β-LG (Table 2).

Table 2 *MMGBSA bind contributions for RIS-β-LG.*

MMGBSA dG Bind contributions	dG Bind in kcal/mol
MMGBSA dGbind	-118.94
Lipo	-88.85
Coulomb	-14.70
Vdw	-49.14
Hbond	0.53
solv GB	25.70

3.2.5. Molecular dynamics simulation

To confirm the stability of the binding complex during MD simulation, the RIS-β-LG complex was first aligned on the protein backbone of the reference and then the RMSD was calculated with respect to the initial frame. The simulation run was for a total of 100 ns, during which 1000 structures enumerated were saved in the trajectory. RIS-β-LG complex was took 25 ns for the equibration, Once equibrated, RMSD fluctuates minimally within 0.8 Å, 2 Å for β-LG, RIS which indicates the conformational stability of the overall β-LG and RIS and tight binding of RIS with β-LG (Figure 7). The structure of RIS and the nature of its interactions within the active site of β-LG structure itself shows minor variations in the orientations of the active site residues even after 100 ns molecular dynamics perturbations.

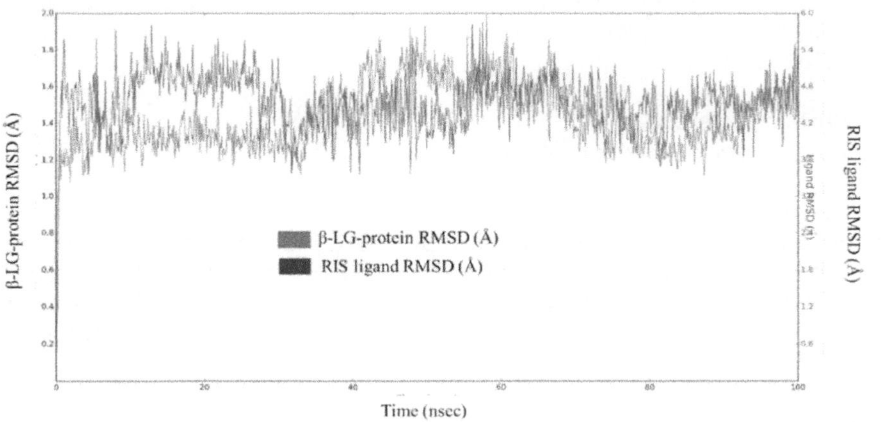

Figure 7 *RMSD plot of the RIS fit β-LG.*

β-LG interactions with RIS were monitored throughout the simulation. These interactions mainly were hydrophobic, and hydrophobic interactions were maintained for more than 30.0% of the simulation time (100 ns) (Figure 8). The hydrophobic residues in β-LG involved in these interactions with RIS are Val-41, Leu-58, Ala- 86, Val-92, Phe-105 (Figure 8).

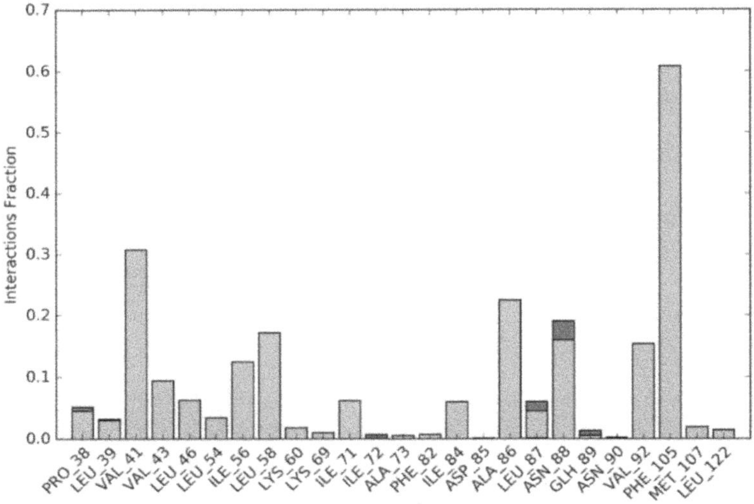

Figure 8 *RIS-β-LG interactions evolved during MD simulation.*

4. CONCLUSIONS

The fluorescence studies as well as molecular modelling indicate good binding of RIS with β-LG. The static quenching indicate that a stable complex of RIS with β-LG is obtained. The extensive molecular modelling suggests the critical role of hydrophobic residues of β-LG in involvement of binding RIS to the central calyx of β-LG. These studies are helpful in the design of suitable formulation of RIS, to improve aqueous solubility, for taste masking, as well as for photostability of RIS.

Conflicts of interest

The authors declare no conflict of interest.

Acknowledgments

Bandoo Chatale would like to thank UGC-BSR, India, for the research grants and Cipla Ltd., Mumbai, India for gift sample of RIS. The authors would also like to thank Mr Pradipkumar R. Wavhule for helping preparation of the extended abstract and thankful to Dr Pritesh Bhat and Dr Prajwal Nandekar from Schrondinger for fruitful discussion.

References

1. S. R. Marder and R. C. Meibach, *Am J Psychiatry*, 1994, **151**, 825–835.
2. N. A. Kasim, M. Whitehouse, C. Ramachandran, M. Bermejo, H. Lennernäs, A. S. Hussain, H. E. Junginger, S. A. Stavchansky, K. K. Midha, V. P. Shah and G. L. Amidon, *Mol. Pharm.*, 2004, **1**, 85–96.

3. Y. Fujisawa, Y. Takahashi, R. Teraoka and S. Kitagawa, *KONA Powder Part. J.*, 2018, **35**, 209–215.
4. S. Reddy, S. Dondeti, R. Manavalan and J. Sreekanth, *Int. J. Pharm. Sci. Res.*, 2010, **1**, 326–346.
5. N. M. Ghuge, A. Bhople, A. R. Thakre, B. V. Bakade, M. A. Channawar, A. V. Chandewar and P. V. Bang, *Indo Am. J. Pharm. Res.*, 2012, **2**.
6. D. Gol, S. Thakkar and M. Misra, *Drug Dev. Ind. Pharm.*, 2018, 1–9.
7. Z. Rahman, A. S. Zidan and M. A. Khan, *Int. J. Pharm.*, 2010, **400**, 49–58.
8. S. Bandi, K. Sanka and V. Bakshi, *J. Microencapsulation*, 2016, **33**, 544–553.
9. F. Shakeel, F. K. Alanazi, I. A. Alsarra and N. Haq, *J. Mol. Liq.*, 2014, **191**, 68–72.
10. H. Salem and R. Kharshoum, *Int. J. Pharm. Pharm. Sci.*, 2016, **2016**, 7.
11. Q. Wang, J. C. Allen and H. E. Swaisgood, *J. Dairy Sci.*, 1997, **80**, 1047–1053.
12. G. Kontopidis, C. Holt and L. Sawyer, *J. Dairy Sci*, 2004, **87**, 785–796.
13. B. Y. Qin, M. C. Bewley, L. K. Creamer, H. M. Baker, E. N. Baker and G. B. Jameson, *Biochemistry*, 1998, **37**, 14014–14023.
14. S. Roufik, S. F. Gauthier, X. Leng and S. L. Turgeon, *Biomacromolecules*, 2006, **7**, 419–426.
15. B. K. Paul, N. Ghosh and S. Mukherjee, *Langmuir*, 2014, **30**, 5921–5929.
16. L. Liang and M. Subirade, *J. Phys. Chem. B,* 2010, **114**, 6707–6712.
17. L. Liang, H. A. Tajmir-Riahi and M. Subirade, *Biomacromolecules*, 2008, **9**, 50–56.
18. M. Hattori, A. Watabe and K. Takahashi, *Bioscience, Biotechnology, and Biochemistry*, 1995, **59**, 2295–2297.
19. L. Ragona, L. Zetta, F. Fogolari, H. Molinari, D. M. Pérez, P. Puyol, K. D. Kruif, F. Löhr and H. Rüterjans, *Protein Sci.*, 2000, **9**, 1347–1356.
20. L. Sawyer, S. Brownlow, I. Polikarpov and S.-Y. Wu, *Int. Dairy J.*, 1998, **8**, 65–72.
21. L. Sawyer, in *Advanced Dairy Chemistry—1 Proteins*, Springer, 2003, pp. 319–386.
22. S. Muresan, A. van der Bent and F. A. de Wolf, *J. Agric. Food Chem.*, 2001, **49**, 2609–2618.
23. J. R. Lakowicz, *Anal. Biochem.*, 2001, **298**, 1–24.
24. Small-Molecule Drug Discovery Suite 2018-2, Schrödinger, LLC, New York, NY, 2018.
25. B. Y. Qin, M. C. Bewley, L. K. Creamer, H. M. Baker, E. N. Baker and G. B. Jameson, *Biochemistry*, 1998, **37**, 14014–14023.

A NOVEL LC-ESI-MS/MS METHOD FOR THE QUANTIFICATION OF FLIBANSERIN IN BULK AND PHARMACEUTICAL FORMULATIONS

G. Kowmudi[a], K. Anbu[a], M. Udhaya Prakash[a], P. Durai Rajan[a], P. Ramalingam[b] and N. Krishnaveni[a*]

[a]Department of Pharmaceutical Analysis, JSS College of Pharmacy Constituent college - JSS Academy of Higher Education & Research, Udhagamandalam, Tamil Nadu – 643001, India; [b]Department of Pharmaceutical Analysis & Quality Assurance, Raghavendra Institute of Pharmaceutical Education & Research, Anantapur, Andhra Pradesh, India

Corresponding Author E-mail: krisath@jssuni.edu.in

1 INTRODUCTION

Flibanserin, a multifunctional serotonin agonist and antagonist (MSAA), which may treat hypoactive sexual desire disorder (HSDD) in women. Women with HSDD have a chronic lack of interest in sex that cannot be explained by another medical condition. Feelings of distress also accompany HSDD. Mechanism of action of flibanserin is different because it works on a women's brain. Flibanserin works on three necessary brain chemicals: dopamine, norepinephrine, & serotonin.[1] It acts as a dual control model of sexual response. Various neurotransmitters, sex steroids, and other hormone effects on the sexual response with significant excitatory and inhibitory effects.

Among neurotransmitters, dopamine and norepinephrine stimulate activity, wh ile serotonin inhibits activity and the balance between before discussed systems are of significance for a healthy sexual response. By modulating serotonin and dopamine activity in some regions of the brain, flibanserin can enhance the equilibrium of sexual response balance between these neurotransmitter systems.[2] The significant side effects of flibanserin are dizziness, sleepiness, nausea, feeling tired & trouble sleeping. Simultaneous consumption of alcohol and flibanserin may result in severely low blood pressure.

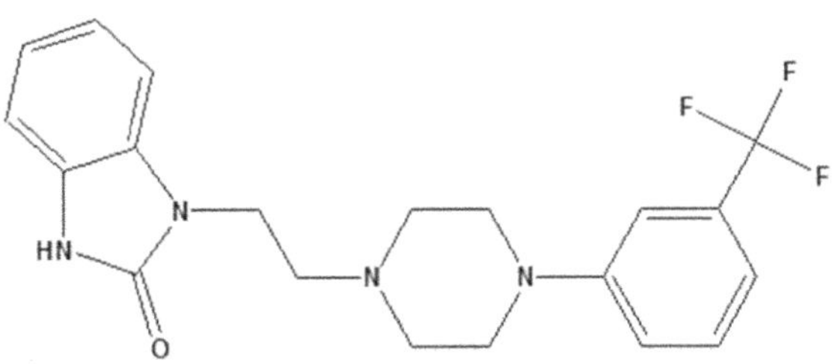

Figure 1 *Structure of Flibanserin*

A thorough review of the literature was carried out to enumerate the reported analytical method for Flibanserin. Muzaffar Iqbal et al.,[3] reported an ultra-performance liquid chromatography (UPLC)–ESI–MS/MS method for the quantitation of Flibanserin in rat plasma samples. Separation has been carried out on C_{18} column using 10mM ammonium formate & acetonitrile (30:70, v/v) as a mobile phase by isocratic elution delivered at a flow rate of 0.3 mL/min. Detection of analyte and internal standard was carried out using a tandem mass spectrometer in multiple reaction monitoring (MRM) mode with positive electrospray ionization. The detection of the ions was carried out at the transition pairs of m/z, i.e., 391.13 → 161.04 and 384.20 → 253.06. Maha A. Sultan and co-workers[4] reported a liquid chromatography – ESI – MS/MS method for the determination of Flibanserin in human plasma. Separation has been carried out on Kinetex C_{18} column (50mm x 2.1 mm, i.d., 2.6µ) and a mixture 20mM ammonium acetate buffer (pH 4.5) & acetonitrile (50:50, v/v) as mobile phase with an isocratic elution mode delivered at a flow rate 0.3 mL/min. Detection method employed was MRM with positive electrospray ionization. The detection of the ions was carried out at the transition pairs of m/z, i.e., 391.1 → 161.1 and 395.20 → 165.1 for Flibanserin and internal standard. Magdalena Poplawska et al.,[5] had reported the LC-MS/MS method for the quantification of Flibanserin and tadalafil in herbal supplements for women sexual desire enhancement. The optimized method utilized three different detectors viz., charged aerosol detector (CAD), diode array detector (DAD), and mass spectrometer (MS). Yuan yang et al.,[6] had reported a method for the detection of avanafil and flibanserin in health food by HPLC-MS/MS. The separation and analysis were performed on an Agilent Eclipse Plus C_{18} column (100 mm x 2.1 mm, i.d., 1.8 µm), with a mobile phase of acetonitrile and 0.1% acetic acid (containing 20 $mmol \cdot L^{-1}$ammonium acetate) (60:40% v/v). Electrospray Ionization (ESI) source was applied and operated in positive mode. Multiple reaction monitoring (MRM) mode was used to quantify avanafil and flibanserin.

As per the literature review, for the drug Flibanserin, few analytical methods were reported for quantification by RP-HPLC and LC-MS/MS in herbal supplements and food products as additive. Due to increasing importance of sensitivity, speed and reliability of analysis in the analytical method, our objective is to develop and report a new approach which is highly sensitive with the cost-effective LC-MS/MS procedure for the quantification of Flibanserin in active pharmaceutical ingredients, pharmaceutical formulation and validate the same as per ICH guidelines.

2 METHODS AND RESULTS

2.1 Chemicals and Reagents Used

Working standards of Flibanserin was provided as a gift sample. LC-MS/MS Grade Solvents Methanol, Acetonitrile was purchased from Merck, (Mumbai, India). Milli-Q-water for LC-MS/MS analysis was obtained using a milli-Q-purification system (Millipore, Bedford, USA).

2.2 Analytical method development and optimization of chromatographic conditions

2.2.1 LC-MS/MS Instrument and Condition

For LC-MS/MS analysis an ultra-fore LC system coupled with a tandem quadrupole mass spectrometer (Shimadzu 8030, Tokyo, Japan) equipped with electrospray ionization (ESI) interface, LC-20AD pump, SPD-M20 PDA detector, CTO-20AC column oven, CBM-20 alite controller and SIL-20AC autosampler was used. LC solutions software was used for data acquisition and processing. The optimized parameters for analysis are as follows: Desolvation Line (DL) temperature 250°C, Block Temperature 200°C, Detector voltage 1.3kv, Nitrogen gas was used as Nebulizer gas (1.5L/min) and drying gas (15 L/min). Argon gas was used for collision-induced dissociation experiments (230kPa).

2.2.2 Chromatographic conditions

A zorbax C_{18} column (50mm x 4.6mm i.d, 5μ particle size) was used as a stationary phase. The mobile phase composition of water: Acetonitrile in the ratio of 15: 85 % (v/v) was pumped at a flow rate of 0.5 mL/min. The injection volume was 10μL, and the samples were injected through an auto-injector. The column temperature was set ambient. The total chromatographic run time was 3.0 min. The eluents were monitored using a triple quadrupole mass analyzer equipped with an electrospray ionization source, operating in positive ionization mode. The detection and quantification of flibanserin were achieved under multiple reaction-monitoring (MRM) mode at a retention time of 1.41 ± 0.1 mins. The precursor to product ion transitions used for quantification and collision energies for flibanserin were as follows: Flibanserin m/z 390.95→161.0; -32.0V.

2.2.3 Solubility of Flibanserin

The solubility of flibanserin was checked in various solvents like water, methanol, ethanol, dimethylformamide, acetonitrile, and acetone.

2.2.4 Preparation of Standard stock solutions

A standard stock solution of flibanserin was prepared by dissolving 10mg of standard in 10mL of methanol to obtain 1mg/ml stock solution. Various working standards ranging from 1-100 μg/mL for flibanserin was prepared in mobile phase [water: Acetonitrile (15:85% v/v)]. From the above working standards, six calibration standards were made in the mobile phase in the concentration range of 1-25 ng/mL.

2.2.5 Assay of marketed formulation

To assess the applicability of the developed LC-MS/MS method on real samples, ten tablets of flibanserin with a dose of 100mg/tablet were taken and crushed into a fine powder. A quantity of powder equivalent to 10 mg of drug was accurately weighed and transferred to a 100 mL volumetric flask, and the analyte was successively extracted three times with 20 mL of HPLC grade methanol. The extracts were filtered, combined, and evaporated to dryness on a water bath at 60°C. The dried residue was reconstituted with 10 mL of LC-MS/MS grade methanol in a 10 mL standard flask, suitably diluted to obtain 10 ng/mL with mobile phase and analyzed by the optimized chromatographic conditions.

2.3 Method Validation

The validation of the developed method was performed based on ICH Q2 guidelines for analytical method validation. The present LC-MS/MS method was validated for specificity, sensitivity, linearity and range, precision, accuracy, ruggedness, and robustness.[7]

2.3.1 Selectivity

The selectivity of the method was performed to prove that the analyte response is not affected by the potential interferences and endogenous compounds coeluting with flibanserin in real samples. It was measured by comparing the chromatograms of the diluent with the corresponding LOQ calibration standard.

2.3.2 Linearity

Linearity for flibanserin was performed using calibration standards ranging from 1.0 – 25.0 ng/ml. The standards were injected in triplicate; the chromatograms were recorded under optimized chromatographic conditions.

2.3.3 Precision and accuracy

Six independent injections of flibanserin were analyzed for Intra-day and inter-day precision at three different concentration levels (LQC, MQC, and HQC) for flibanserin. The precision was recorded as a measure of %CV in which the value does not exceed 2% for nominal QC and 5% for LQC.

The accuracy of the method was expressed as recovery studies. It was performed by the standard addition method, i.e., by adding a known amount of drug to the real sample and analyzed by the optimized chromatographic conditions. The recovery experiments were performed at three different QC levels. The % mean recovery, SD, and % CV were calculated and reported.

2.3.4 Limit of detection (LOD) and limit of quantification (LOQ)

If a method detects deficient concentration, the method is said to be sensitive. According to ICH guidelines, it is based on LOD and LOQ values. LOD and LOQ values were determined based on peak height, with a signal to noise (S/N) ratio of 1:3 and 1:10, respectively.

2.3.5 Robustness and Ruggedness

The robustness of the developed method was determined by altering experimental conditions such as flow rate (±0.1 mL/min), pH (±0.2), column temperature (±2°C), organic phase concentration (±2%) to assess the deviation from the optimized chromatographic conditions.

2.3.6 System Suitability

An essential part of method development and validation is the calculation of system suitability parameters as it ensures the optimal performance of the system. Chromatographic parameters Viz., Retention time (Rt), Theoretical Plates (N), Asymmetry Factor (A) were monitored after six replicate analysis of standard flibanserin at a concentration of 5 ng/mL.

3 DISCUSSIONS

3.1 Method Development and Optimization of Chromatographic Conditions:

The chromatographic conditions, including mobile phase composition and column type, were optimized for better resolution and sensitivity. Several trials were carried out to achieve proper resolution. Water and acetonitrile as the mobile phase improved the detection, ionization, and recovery of the analyte with an ESI interface at positive ionization mode. The versatility, suitability, and robustness of the method were checked with several C_{18} and Cyano columns. The Zorbax C_{18} column (50mm x 4.6mm i.d, 5μ particle size) provided good selectivity, sensitivity, and symmetric peak shape for flibanserin. The detection was carried out in MRM mode by electrospray ionization (ESI) in positive mode, for flibanserin. The retention time of flibanserin was found to be 1.41 ± 0.1 min Figure 2.

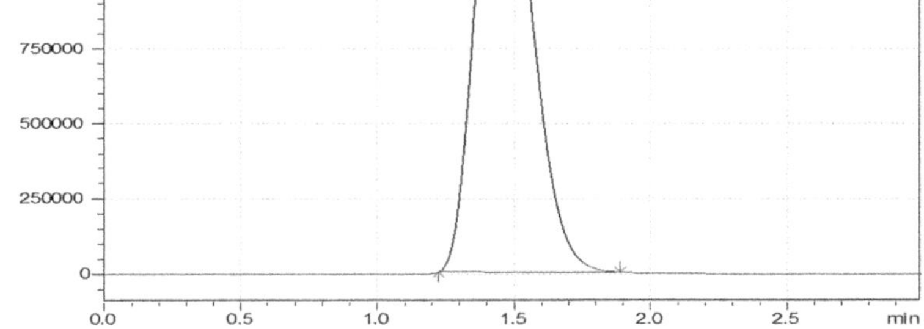

Figure 2 *Typical MRM chromatogram of Flibanserin solution (15 ng/ml)*

The MRM transitions were selected as m/z 390.95→161.0 with a collision energy of -32.0V. The MS scan and product ion scan spectra of standard flibanserin are depicted in Figure 3 and 4.

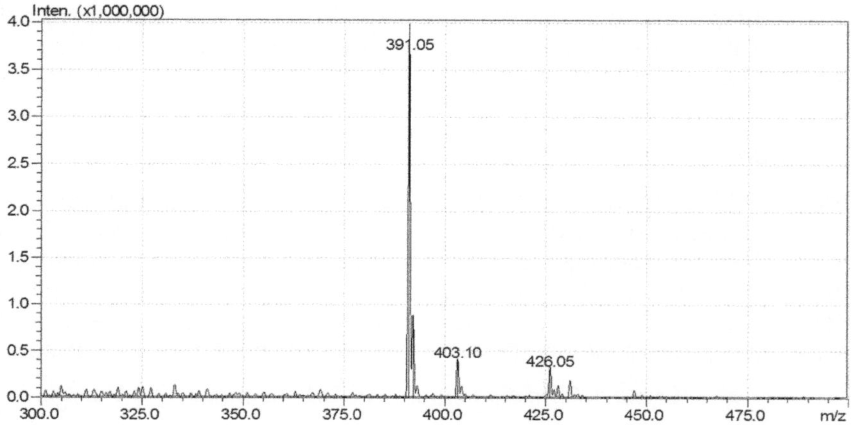

Figure 3 *MS Scan spectra of Flibanserin in positive ionization mode.*

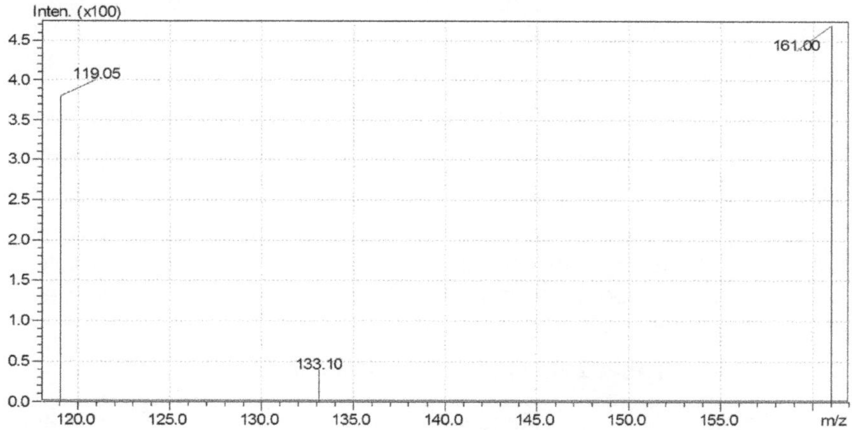

Figure 4 *Product ion scan of Flibanserin.*

3.2 Validation of the developed method:

The real goal of the validation process was to challenge the method and determine the limits of allowed variability for the conditions needed to run the method.

3.2.1 Selectivity

The selectivity of the LC-MS/MS method was assessed by examining peak interferences from other substances present in the sample and components of the mobile phase. This was done by comparing the chromatograms of blank and sample with the pure drug. It was observed that none of the peaks appear at the retention time of the drug. This confirmed the specificity of the proposed method. Besides, the well-shaped peaks also signify the selectivity of the proposed method.

3.2.2 Precision

The precision of the method was evaluated by calculating the intra-day and inter-day precision by analyzing six replicates each at three QC levels {5.0, 15.0 and 25.0 ng/ml}. Each QC was injected on the same day and on the two consecutive days to calculate intraday and interday precision respectively. The results are tabulated in Table 1. The intra and inter-day precision studies showed a %RSD of less than two, proving that the developed analytical method was precise.

Table 1 *Precision studies of Flibanserin.*

Actual concentration(ng/mL)	Intra-day calculated concentration(ng/mL) ±SD; %RSD(n=6)	Inter-day calculated concentration(ng/mL) ±SD; %RSD(n=6)
5	4.98±0.03; 0.63	4.92±0.02; 0.60
15	14.14±0.61; 4.31	14.37±0.26; 1.82
25	24.99±0.83; 3.35	24.69±0.25; 1.02

3.2.3 Accuracy

The accuracy expressed as % mean recovery was performed at three concentration levels by the standard addition method. The samples after extraction from the laboratory mixtures are also analyzed by the developed chromatographic method and reported in table 2.

Table 2 *Accuracy studies of Flibanserin.*

Actual concentration(ng/mL)	Recovered concentration (ng/mL) ± SD; %RSD(n=3)	Percentage recovered (%)
5	4.95±0.04; 0.98	94.21
15	14.57±0.07; 0.52	96.56
25	24.05±0.52; 2.16	97.93

The mean % recovery was observed between 98.5 ± 0.5%, indicating that the developed method was accurate. The obtained result was comparable with a corresponding label claim and are depicted in Figure 5 and Table 3.

Table 3 *Assay results of marketed formulation.*

Sample	Label claim	Amount present (mg/tablet)±SD; %RSD
Formulation –I	100mg	98.87±0.1205; 1.2206

3.2.4 Linearity

The linearity range for flibanserin was found to be within the range of 1.0, 5.0, 10.0, 15.0, 20.0, 25.0 ng /ml. The r^2 value was found to be 0.9936 for flibanserin in the linear curve. The calibration curve was constructed six days over one week to determine the variability of slopes and intercepts. The results indicated that no significant inter-day variability of slope and intercept over the optimized concentration range and are tabulated in Table 4.

Table 4 *Linearity of Flibanserin.*

Concentration of Flibanserin (NG/mL)	Peak Area
1	5155648
5	10729239
10	17039134.5
15	21633156
20	29199646.5
25	37390345.5

3.2.5 Ruggedness & Robustness

The ruggedness & robustness of the method was studied by changing the experimental conditions. No significant changes in chromatographic parameters were observed when changing the experimental conditions (operators, source of reagents & column of similar type) & optimized condition (mobile phase ratio & flow rate). The values are depicted in Table 5.

3.2.6 Limit of Detection

The limit of detection (LOD) was determined by the signal-to-noise ratio of 3:1 for flibanserin which was found to be 0.5 ng/ml. Moreover, the limit of quantification (LOQ) was established by signal- to –noise ratio of 10:1 for flibanserin was found to be 1 ng/ml. The TIC chromatogram of LOQ concentration was shown in Figure 6.

Table 5 *Ruggedness and Robustness.*

Parameters Mobile phase ratio	Retention Time (min)
15.5: 84.5	1.42 ± 0.2
14.5: 85.5	1.42 ± 0.2
16: 84	1.44 ± 0.2
Flow rate	
0.4	1.45 ± 0.2
0.5	1.44 ± 0.2
0.6	1.42 ± 0.2

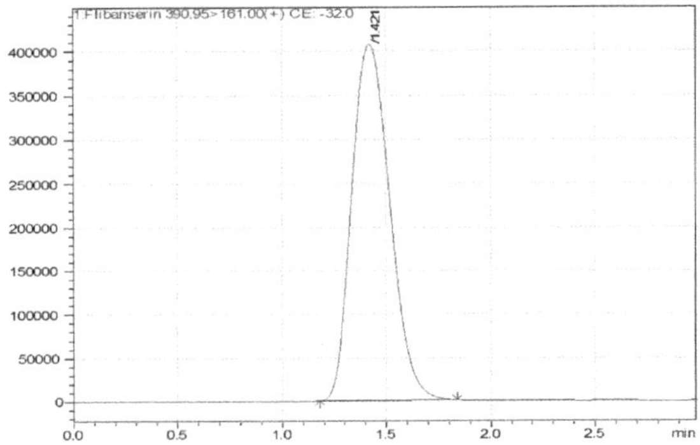

Figure 5 *Typical MRM chromatograms of Flibanserin extracted from marketed formulations.*

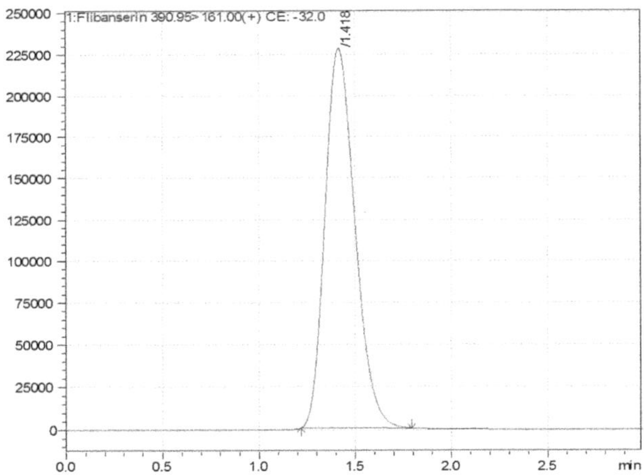

Figure 6 *Limit of quantification (LOQ) for Flibanserin (1 ng/ml).*

Table 6 *System suitability parameters.*

Parameters	Flibanserin
Linearity and Range	1-25 ng/ml
Regression Equation	Y= 13,01,398.26x + 37,06,816.93
Correlation Coefficient	0.9936
Theoretical Plates	3433
Asymmetric factor	1.1
Limit of Detection (LOD)	0.5 ng/ml
Limit of Quantification (LOQ)	1 ng/ml

4. CONCLUSION

A simple, specific, and accurate LC-MS/MS method was developed, validated, and successfully applied for the quantification of flibanserin in pharmaceutical formulation. The validated LC-MS/MS method has an advantage over other methods as the mobile phase is simple consisting of water and acetonitrile was used without the use of pH modifiers like ammonium acetate, ammonium formate, and acetic acid, when compared to the previously reported methods. Acetonitrile constitutes about 85% of the total mobile phase used, which facilitates the maintenance of low pressures during a chromatographic run and a window has been created for extending the flow rates up to 1.5 mL/min without overloading the pressure limits, if needed. A The chromatographic elution of flibanserin was faster by employing this simple, time saving method with simple mobile phase, which is an added advantage in the analytical methods.

Overall, the detection and quantification limits of flibanserin are reported to be as low as 0.5 ng/mL and 1ng/mL respectively, without any observed interference at the retention time of the flibanserin in chromatograms. This method can be claimed to be sensitive when compared to the RP-HPLC. Though there are other chromatographic methods that were reported for the quantification of flibanserin in human biological fluids, herbal formulations, and food products, to our knowledge, this is the first chromatographic method to quantify flibanserin in pharmaceutical formulation. The developed method can be applied successfully in the analysis of quality control samples in manufacturing and BA/BE studies of various formulations of flibanserin.

Acknowledgment

The authors are thankful to JSS Academy of Higher Education & Research, Mysuru for providing the facilities to carry the work.

References

1. J. A. Simon, S. A. Kingsberg, B. Shumel, V. Hanes, M. Garcia Jr and M. Sand, *Menopause*, 2014, **21**, 633.
2. S. M. Stahl, B. Sommer and K. A. Allers, *J. Sex Med.*, 2011, **8**, 15.
3. M. Iqbal, E. Ezzeldin, N. L. Rezk and K. A. Al-Rashood, *Bioanalysis*, 2018, **10**, 1087.
4. M. A. Sultan, R. T. El-Eryan, A. K. Attia and M. J. Eissa, *Biomed. Chromatogr.*, 2019, **33,** e4545.
5. M. Poplawska, A. Blazewicz, P. Zolek and Z. Fijalek, *J. Pharm. Biomed. Anal.*, 2014, **94**, 45.
6. W. U. Yuanyang, W. Ren, M. Zhou and C. Qiaoyuan, *Yiyao Daobao*, 2017, **36**, 783.
7. *Validation of Analytical procedure: Text and Methodology Q2(R1)*, International Conference on Harmonisation of Technical Requirements for Registration of Pharmaceuticals for Human Use, November 2005.

FORMULATION AND CHARACTERIZATION OF KETOCONAZOLE LOADED NANOSPONGES IN HYDROGEL FOR TREATING TOPICAL FUNGAL INFECTIONS

D. Nagasamy Venkatesh,* Sumalatha Devi Balamurugan, M. Manisha and Himangshu Bhowmik

Department of Pharmaceutics, JSS College of Pharmacy, (JSS Academy of Higher Education & Research), Ooty – 643 001. Tamil Nadu, India

1. INTRODUCTION

The background behind the delivery of drugs topically is to render localized action on the skin. Topical drug delivery employing conventional formulations such as ointment, creams suffers with lower uptake of drug due to its hydrophobic nature eventually leads to ineffective treatment. To overcome these issues, topical delivery of β-cyclodextrin-based nanosponges has opted, that facilitates the controlled drug release over a period of time. Topical drug delivery is an intriguing and persuading access to treat skin related infections/diseases owing to their lower risk of reactions a the systemic level duly contrasted with oral or parenteral drug administration and high drug level can be accomplished at the site of ailment. β-Cyclodextrin nanosponges (NS) are composed of permeable nanostructure material comprising of hyper cross-linked β-cyclodextrin system, ensures enhancement in solubility, provides desirable controlled release wider range of drugs by forming inclusion complexes, which diminish the recurrence portion of drug and maintain a distance from systemic absorption[2-3]. Ketoconazole, an antifungal agent classified as BCS II, used to treat fungal infections and candidiasis. The stumbling block in attaining the effective therapeutic application and efficacy of ketoconazole is due to its moderate lipid solubility (Log P 4.35). Ketoconazole, a weak base and very poor permeability whose permeability co-efficient is 1.2×10^{-3} and with specific activity of 4 Ci/mmol. This might result in unsatisfied maintenance of the therapeutic concentration required at the epidermal layers, which in turn results in insufficient quantity to exhibit the satisfied topical delivery of the drug. Though the use of customary topical delivery system might render the cure, but it leads to non-compliance of patient, owing to frequent application[4]. To facilitate topical administration, nanosponges are effectively incorporated into the hydrogel and proven to reduce irritation of the incorporated drug without reducing the therapeutic efficacy. This present work deals with the development ketoconazole nanosponges loaded in hydrogel for treating topical fungal infections. The nanosponges systems have been identified as a potential carrier in prolonging the drug release and found to be applicable for the treatment of various fungal infections. Additionally, they facilitates reduced dose, reduction in the administration frequency and enabling the avoidance of systemic side effects.

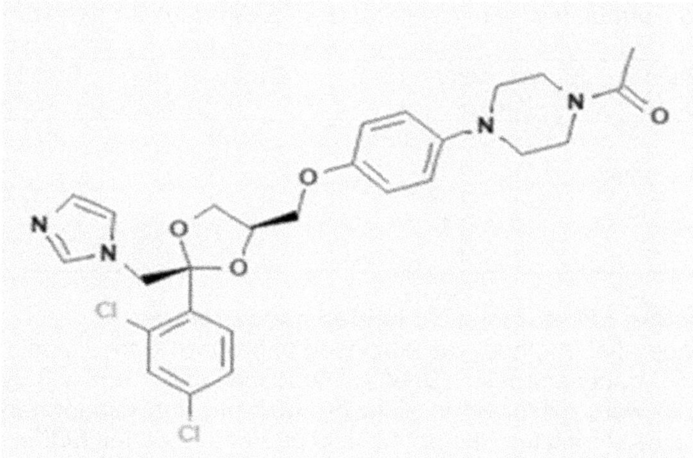

Figure 1 *Structure of ketoconazole.*

2. METHODS

2.1 Materials

Ketoconazole was obtained as a gift from Synergene Active Ingredients Pvt Ltd, Bengaluru, India. β-cyclodextrin and 1, 1-dicarbonyl–imidazole were purchased from Sigma Aldrich, Mumbai. Methanol and Dimethyl formamide was purchased from S.D Fine Chemicals, Mumbai, India. All other chemicals were used in laboratory standards.

2.2 Synthesis of β-cyclodextrin nanosponges

β-cyclodextrin nanosponges were prepared using 1, 1' carbonyl diimidazole as cross linker, briefly, in a 250 ml flask, 15.34 m mol of anhydrous β-cyclodextrin and 61.42 mmol of 1, 1' carbonyl diimidazole (CDI) were dissolved completely using 100 ml of anhydrous N,N dimethyl formamide (DMF) in a round bottom flask. The resulting solution was heated in an oil bath at 100°C for 12 hours until a transparent block of β-cyclodextrin is formed. After the completion of reaction, the mass is powdered in a mortar and washed with deionized water to expel the excess solvent. Finally, the product was rinsed and extracted with ethanol using a soxhlet extractor to remove residual by products or unreacted reagents. The white powder product (nanosponges) so obtained was dried at 60°C in an oven and subsequently ground in mortar for further studies[6,7]. Five different formulations of nanosponges were prepared using fixed quantity of β-cyclodextrin and variable concentrations of N,N dimethyl formamide (cross linker) in the ratio of 1:2,1:4,1:6,1:8 and 1:10 respectively and coded as F1-F5. (Table 1).

Table 1 *Formulation of different ratios of β-cyclodextrin nanosponges.*

Formulation code	β-cyclodextrin (mg)	Cross linker (mg)	Distilled water (ml)
F1	1	2	10
F2	1	4	10
F3	1	6	10
F4	1	8	10
F5	1	10	10

2.3 Preparation of ketoconazole loaded nanosponges

Solvent evaporation method was employed to prepare ketoconazole loaded β-cyclodextrin nanosponges. Briefly, ketoconazole and β-cyclodextrin nanosponges were dissolved in dicholoromethane and evaporated at 50° C with continuous stirring for 1 hour; the solid product was dried further to remove the traces of dichloromethane. The powder was pulverized stored in the desiccator for further analysis.

2.4 Preparation of nanosponges loaded hydrogel

Hydrogel of KET loaded nanosponges was prepared by initially soaking gel forming polymer (Carbopol 971P) in water for 2 hours. To obtain a uniform dispersion, the polymer was agitated continuously with the help of magnetic stirrer (Kemi, India). The stirring was stopped after 15 minutes to remove the entrapped air. To this previously prepared Ketoconazole loaded nanosponges was added and again stirred for 20 minutes, which resulted in the ketoconazole loaded nanosponges incorporated to hydrogel. It was observed that the batch F4 was selected as ideal batch based on size of nanosponges, entrapment efficiency, and zeta potential values.

2.5 Characterization of ketoconazole nanosponges

2.5.1 Determination of particle size and zeta potential

The particle size was assessed by using a Dynamic laser scattering (DLS), to measure the diameter. The zeta potential was measured using zeta sizer (Malvern Instruments, UK) equipped with a 4mW He-Ne laser at a wavelength of 633 nm and a Laser Doppler Anemometry (LDA) was used to determine zeta potential in mille volts (mV) of the drug loaded nanosponges.

2.5.2 Entrapment Efficiency

50 mg of the drug loaded nanosponges was dissolved in methanol and stirred for 30 minutes. Further, solution was centrifuged at 2000 rpm for 10 minutes to remove unbound drug. The supernatant liquid was removed and measured in UV spectrophotometer at 244 nm. The entrapment efficiency was calculated using the formula.

%EE = ((Drug added - Free un-entrapped drug)/Drug added) × 100

2.5.3 Fourier transformer infrared spectroscopy (FTIR)

The compatibility studies between the β-cyclodextrin and drug (ketoconazole) was carried out to find out the possibility of any chemical reaction. The IR spectra of the drug, β-cyclodextrin and drug loaded nanosponges were performed using FTIR. The dry sample was powdered and compressed into a pellet using KBR under hydraulic press at 10,000 psi. The pellets were scanned over a wave number of 4000-400 cm^{-1}. The spectra obtained were compared and interpreted for the appearance and disappearance of major functional peaks.

2.5.4 Scanning electron microscopy (SEM)

Scanning electron microscopy (SEM) was employed to study surface morphology of nanosponges. The aqueous suspension of the samples were spread on a holder and dried under vacuum, subsequently coated with gold (JFC 1200 fine coater, Japan) and examined by a scanning electron microscopy using the refractive index at 30 s run time. The SEM photographs were elaborated in detail using an image processing program and the individual diameters were measured to obtain the mean particle size.

2.6 Evaluation of Ketoconazole loaded nanosponges incorporated Hydrogel

2.6.1 Viscosity and pH determination

The viscosity of hydrogels was measured using Brookfield viscometer (Prime Rheometer DV 1). Viscosity was measured at 25° C at 100 rpm using spindle 61. The pH of different hydrogel formulation was evaluated using calibrated pH meter. One gm of hydrogel containing ketoconazole nanosponges was uniformly dispersed in 100 ml of distilled water and kept for 2 hours at 25°C. Then, pH of the dispersion was measured.

2.6.2 *In vitro* drug release

The release of ketoconazole from the nanosponges in hydrogel was studied using dialysis method, where drug equivalent to 40 mg of gel was suspended in 1 ml phosphate buffer pH 7.4. The bags were kept in 100 ml of dissolution medium and stirred under magnetic stirring at 37 ± 0.5°C. At predetermined time interval, aliquots of the dissolution medium were withdrawn, suitably diluted with phosphate buffer pH 7.4 and the drug content was estimated by spectrophotometrically at 244 nm against blank.

2.6.3 Release kinetics

The data's obtained from the *in vitro* dissolution profile were applied to various kinetic models such as zero order kinetics, first order kinetics, Higuchi and Peppa's equations to understand the mechanism of release of drug from the nanosponges[8].

2.6.4 Antifungal study (zone of inhibition)

The antifungal activity was performed using agar well diffusion method, where Sabouraud's dextrose agar medium was inoculated with 0.2 ml of fungal suspension (*Tinea corporis*) in a sterile glass plates. A care was taken for the uniform thickness of the layer of medium in the plates. After solidification of the medium, the wells were made aseptically with cork bore having a 6 mm diameter. To each plate hydrogel nanosponges of KET and pure KET was introduced, and the plates were incubated for 24 hours. After the incubation, the zone of inhibition was measured [9].

3 RESULTS AND DISCUSSION

3.1 Formulation and optimization of nanosponges

Five batches of nanospongess were prepared by varying the concentration of the cross linker. During the study, it was observed that the batch F4 prepared with 1:8 ratio of β-cyclodextrin and carbonyl diimidazole was selected the best batch based on size of the nanopsponges, entrapment efficiency and zeta potential.

3.2 Evaluation of drug loaded nanosponges

3.2.1 Determination of particle size and zeta potential

The particle size of the nanosponges (F4) was found to be 195.1 nm; the zeta potential was observed to be -0.278mV, Polydispersity index values were found to be 0.326, which indicated that the particles are found to be relatively narrow in distribution (Figure 2).

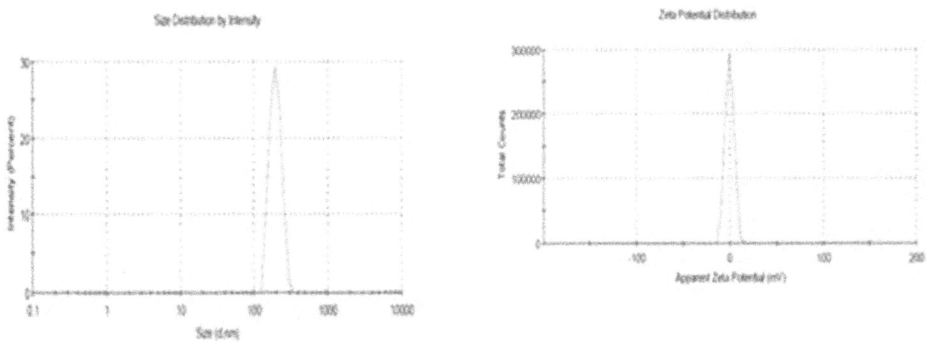

Size and PDI report for Ketoconazole loaded nanosponge

Zeta Potential report for Ketoconazole loaded nanosponge

Figure 2 *Size, PDI and Zeta potential report for ketoconazole loaded Nanosponges.*

3.2.2 Entrapment efficiency (EE)

Percentage EE for different batches are shown in table 2. Among the different formulations, F4 was considered as ideal batch based on the particle size of 195.1 nm with higher entrapment efficiency 72.7%.

Table 2 *Characterization of ketoconazole loaded nanosponges.*

Formulation code	Particle size (nm)	Entrapment efficiency (%)
F1	750	51.8
F2	531	57.3
F3	419	64.5
F4	195.1	72.7
F5	1410	52.2

3.2.3 Fourier transformer infrared spectroscopy

FT-IR study was carried out to check the compatibility between the β-cyclodextrin, cross-linker and drug. The FT-IR study was also helpful to assure the complete physical adsorption of the drug into the β-cyclodextrin without any mutual interaction (Figure 3-5). The spectra obtained from IR studies at wavelength from 4000 cm^{-1} to 400 cm^{-1}. After interpretation of the spectra, it was confirmed that there was no major shifting of peaks and as well as no loss of functional peaks between the spectra of drug and drug loaded nanosponges, indicating that the drug and β-cyclodextrin are compatible with each other.

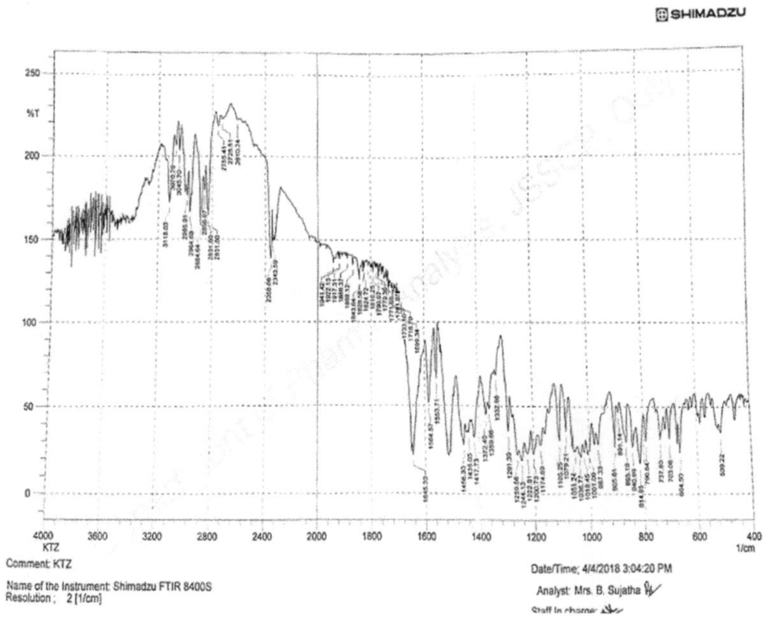

Figure 3 *FT-IR spectrum of ketoconazole.*

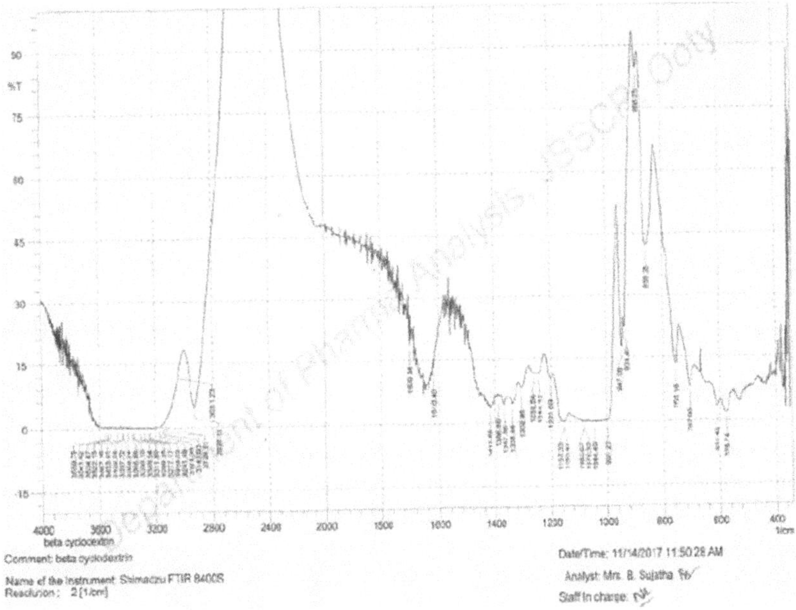

Figure 4 *FT-IR spectrum of β-cyclodextrin.*

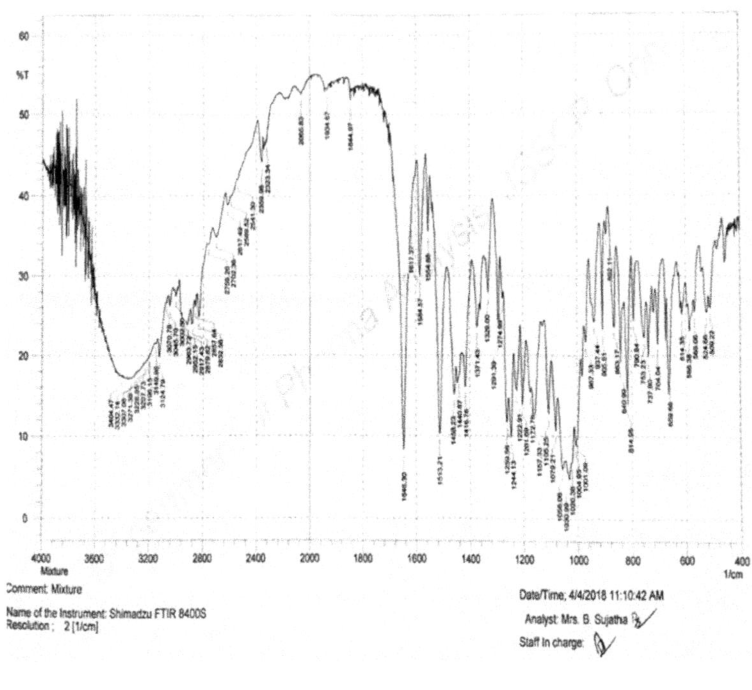

Figure 5 *FT-IR spectrum of ketoconazole loaded nanosponges.*

Table 3 *Functional group analysis of β-Cyclodextrin and Ketoconazole - β-Cyclodextrin nanosponges by FT-IR Technique.*

Functional group details	Wave numbers (in cm^{-1})	
	Ketoconazole	Ketoconazole loaded nanosponges
C-H stretching	2928.53	2919.36
-OH stretching	3370.72	3423.76
-C-C stretching	1157.84	1035.41
-C=C stretching	1705.27	1723.68

3.2.4 Scanning Electron Microscopy (SEM)

SEM was used to study the morphology of the nanosponges. The scanning electron micrographs of ketoconazole loaded nanosponges revealed that the surfaces are found to be open spongy structure with a smooth surface as shown in figure 6.

Figure 6 *Scanning Electron Micrography images of ketoconazole loaded Nanosponges.*

3.3 Evaluation of ketoconazole loaded nanosponges incorporated hydrogel

3.3.1 Viscosity and pH Determination

The viscosity of different formulations in hydrogel was found to be in the range of 12090 to 12300 cps at 25° C. The pH of the different formulations incorporated in to hydrogel was found to be in the range of 6.1 to 6.4.(Table 4)

Table 4 *Viscosity and pH determination of different batches of nanosponges incorporated into hydrogels.*

Formulation code	Viscosity (cps)	pH
F1	12130	6.1
F2	12220	6.4
F3	12208	6.3
F4	12300	6.1
F5	12090	6.2

3.3.2 *In vitro* drug release studies

In vitro drug release study was performed for ideal batch (F4) nanosponges, which was selected based on the higher drug entrapment efficiency and particle size and compared with pure ketoconazole performed in phosphate buffer (pH 7.4) for 24 hours. On comparing the drug release profiles of ketoconazole nanosponges formulation (F4) and pure ketoconazole, the drug release from nanosponges formulations was found to release 85.37±0.05% at the end of 24 hours indicating the sustained release of drug (Figure 7).

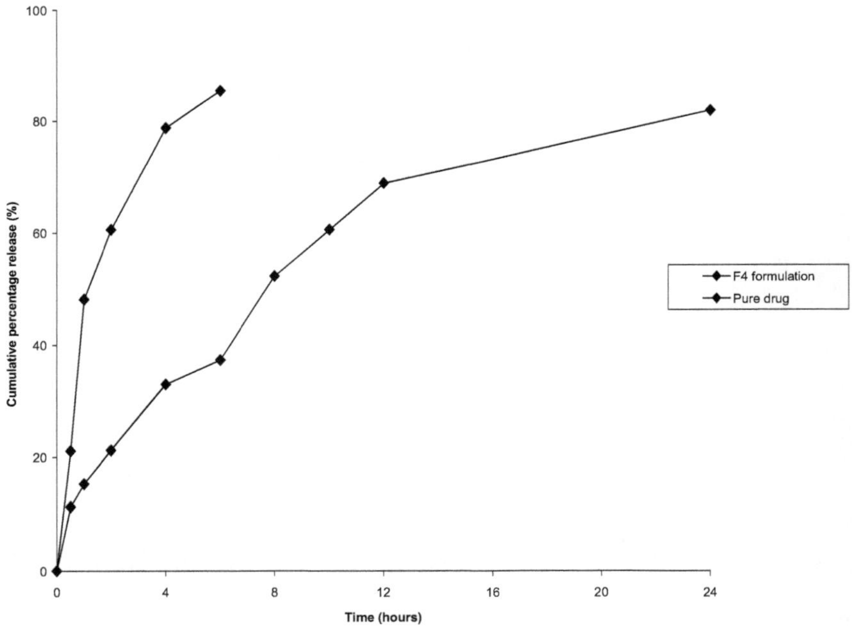

Figure 7 *In vitro Drug Release Profile of Ketoconazole Loaded Nanosponges and Pure drug.*

3.3.3 Release kinetics

To determine the mode and mechanism of drug release from the *in vitro* data, the data was interpreted using various kinetic models (Table 3).The best linearity was observed with first order kinetics and Peppa's model for ketoconazole loaded nanosponges for batch F4 indicating the release of drug from the nanosponges obeying non-fickian diffusion.

Table 5 *Release kinetics of ketoconazole loaded nanosponges.*

Formulation code	Zero order (r^2)	First order (r^2)	Higuchi (r^2)	Peppa's (r^2)	Peppa's n
F1	0.8621	0.9523	0.9011	0.9232	0.551
F2	0.8553	0.9662	0.9487	0.9711	0.562
F3	0.8663	0.9429	0.9348	0.9812	0.544
F4	0.8751	0.9787	0.9781	0.9877	0.584
F5	0.8623	0.9741	0.9671	0.9722	0.522

3.4 Antifungal activity

Antifungal activity of formulated ketoconazole nanosponges incorporated hydrogel and pure drug ketoconazole was observed by inducing the growth of *Tinea corporis* in a sterile medium. The results (Table 4) depicted that the formulated KTZ nanosponges incorporated hydrogel showed superior antifungal activity as compared to pure drug.

Table 6 *Antifungal activity of KTZ nanosponges incorporated hydrogel and pure ketoconazole.*

Fungal species	Zone of inhibition	
	Ketoconazole nanosponges (F4)	Pure ketoconazole
Tinea corporis	23.2 ± 0.5	21.2 ± 0.8

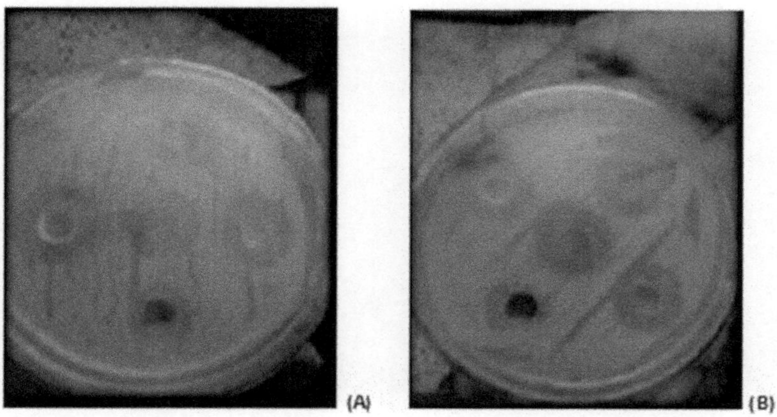

Figure 8 *Anti Fungal Activity of Ketoconazole Loaded Nanosponges (A) and Pure Ketoconazole (B).*

4. CONCLUSION

The nanosponges prepared using β-cyclodextrin and loaded with ketoconazole were sucessfully incorporated into hydrogel for topical administration. Ketoconazole nanosponges were found to be a potential carriers for controlled release of drug and therefore beneficial for application in the topical fungal infections. Further, additional benefits such as dose reduction, reduced frequency of administration and avoidance of systemic side effects can be produced. However, a suitable *in vivo* studies are essential to prove its therapeutic activity.

References

1. A. L. M. Ruela, A. Gravinez Perissinato, M. Esselin de Sousa Lino, P. Silva Mudrik, G. Ribeiro Pereira, *Braz. J. Pharm. Sci.,* 2016, **52**(3), 527.
2. S. P. Atul, R. Bhushan Dravyakar, D. Kadam, M. Jadhav. *Carbohydr. Polym.,* 2017*,* **173**, 37.
3. S.Selva Muthukumar, S.Anandam, K.Krishnamoorthy, R.Manavalan, *J. Pharm. Pharm. Sci.,* 2012, **15**, 103.
4. R. Thiruganesh, S. Umadevi, Khandasami, R. Himabindhu and S. Shanmugam, *Macromol. Res.*, 2012, **20**(7), 682.
5. K. Krishnamoorthy and M. Rajappan, *J. Pharm. Pharm. Sci.*, 2012, **15**(1), 103.
6. K. A. Ansari, P. R. Vavia, F. Trotta, R. Cavalli, *AAPS Pharm. Sci. Tech.*, 2011, **12**(1), 279.
7. M. Bencini, E. Ranucci, P. Ferruti, F. Trotta, M. Donalisio, M. Cornaglia, D. Lembo and R.Cavalli, *J. Control. Rel.*, 2008, **126**, 17.
8. J. Stepmann and N. A. Peppas, *Pharm. Res.*, 2000, **17**(10), 1290.
9. I.Meshram, D.K.Mishra, D.K.Jain, *Asian J. Biomater. Res.*, 2016, **2**(3), 94.

www.ingramcontent.com/pod-product-compliance
Ingram Content Group UK Ltd.
Pitfield, Milton Keynes, MK11 3LW, UK
UKHW020346150926
8624IPUK00003B/28

9 781788 018623